Applied Strength
of Materials

Applied Strength
of Materials

Second Edition

ROBERT L. MOTT, P.E.

Prentice Hall, Englewood Cliffs, New Jersey 07632

Library of Congress Cataloging-in-Publication Data

Mott, Robert L.
   Applied strength of materials / Robert L. Mott.—2nd ed.
     p. cm.
   Includes index.
   ISBN 0-13-043415-9
   1. Strength of materials.   I. Title.
  TA405.M883 1989
  620.1'12—dc19   88-9805
                                  CIP

Editorial/production supervision and
  interior design: **Lillian Glennon**
Manufacturing buyer: **Robert Anderson**

© 1990, 1978 by Prentice-Hall, Inc.
A Division of Simon & Schuster
Englewood Cliffs, New Jersey 07632

Printed in the United States of America

10 9 8 7 6 5 4 3 2 1

ISBN 0-13-043415-9

Prentice-Hall International (UK) Limited, *London*
Prentice-Hall of Australia Pty. Limited, *Sydney*
Prentice-Hall Canada Inc., *Toronto*
Prentice-Hall Hispanoamericana, S.A., *Mexico*
Prentice-Hall of India Private Limited, *New Delhi*
Prentice-Hall of Japan, Inc., *Tokyo*
Simon & Schuster Asia Pte. Ltd., *Singapore*
Editora Prentice-Hall do Brasil, Ltda., *Rio de Janeiro*

To my wife MARGE
To our children LYNNÉ, ROBERT JR., and STEPHEN
And to my MOTHER and FATHER.

Contents

Contents

Contents

Contents

Preface

OBJECTIVES OF THE BOOK

The primary objective of this book is to provide complete coverage of the principles of strength of materials, with an emphasis on the *application* of those principles to practical design and analysis problems. This approach should be attractive to educational programs that emphasize the practical applications in the mechanical, civil, or construction fields, along with practicing designers and engineers. The purpose of writing this second edition is to increase the coverage presented in the first edition and to improve the usefulness of the book for both students and instructors.

Pertinent information, data, and example problems for the civil engineering technology and construction engineering technology fields have been added. But the interests of mechanical engineering technology students have been retained.

STYLE

The style remains the same because the first edition was described by both students and colleagues as being clearly written and easy to use. But some rewriting of sections and additions of material enhance the book. The number of problems for student solution has more than doubled from the first edition and the variety of types of problems has been expanded.

Units used are a mixture of U.S. Customary and SI metric units, in keeping with the dual-unit usage evident in American industry and construction. A change in notation for stresses has been made, using the Greek letter σ (sigma) for normal stresses and τ (tau) for shear stresses.

An increase in the use of calculus is evident, in keeping with the recommendation of accrediting agencies. It is suggested that students complete a basic course in calculus before studying this course. However, the use of the principles presented in the book and all of the problems for student solution can be accomplished with the application of the basic formulas of strength of materials without the use of calculus.

COMPUTER PROGRAMMING ASSIGNMENTS

In keeping with the recommendation of accrediting agencies that students *use* computers in upper-level technical courses after learning computer programming, this book includes suggested computer programming assignments at the end of most chapters. The programs are tied to the presentation of the material in the chapter and are generally in the form of assisting the designer in the analysis of a given case. The benefit to the student is that he or she must clearly understand the principles involved in the problem before a successful program can be written. After completing the program, it can be used to evaluate a wide variety of proposed solutions, which also leads to a better understanding of the principles. Most programs are of such a scope that they can be assigned as normal homework assignments. More enhanced programs could carry more weight.

PREREQUISITES

Statics remains a prerequisite for the material presented in this book, with emphasis on forces, moments, static equilibrium, frames, trusses, centroids, and moments of inertia. There is complete coverage of centroids and moments of inertia in Chapter 7 for those who have not studied those subjects and for those wishing to review. It would be particularly desirable to review this area when dealing with bending and shear stresses in beams. The book presents a significant amount of data on the properties of standard beam sections: wide-flange shapes, American Standard beams, channels, and angles. Composite sections using these shapes are also featured.

ORGANIZATION

The organization of the first part of the book has been changed from that of the first edition so that all of the basic kinds of direct stresses are presented in Chapter 1. This includes direct tension and compression, direct shear, and bearing stress, a new topic. Normal strain, shearing strain, modulus of elasticity, and Poisson's ratio are included.

Chapter 2 presents an expanded treatment of the design properties of materials, including additional topics on the heat treatment of steel, and a new section on advanced composite materials. Then the design of members subjected to direct stresses is presented in Chapter 3. With this approach, the student should first achieve an understanding of the principles of stress analysis and the behavior of materials, then be able to design successful load-carrying members.

In Chapter 5, Torsional Shear Stress and Torsional Deformation, a section has been added on torsion of noncircular members. The development of the torsional shear stress formula and the angle of twist formula has been enhanced.

The coverage of shearing force and bending moment diagrams in Chapter 6 has been expanded to include a wider variety of beam loadings, such as varying distributed loads and concentrated moments on beams. A new section that presents the use of free-body diagrams of *parts* of load-carrying members should improve the student's understanding of internal forces, moments, and torques.

Many additional beam shapes using extrusions and composite sections made from standard structural shapes have been included in Chapter 7, on centroids and moments of inertia.

In Chapters 8 and 9, the developments of the flexure formula and the general shear formula have been strengthened.

Chapter 10, The General Case of Combined Stress and Mohr's Circle, is entirely new. This is followed by several practical, special cases of combined stresses in Chapter 11 which can be covered either after Chapter 10 or entirely separate from it.

The coverage of beam deflections in Chapter 12 has been expanded significantly. Very complete coverage is given for basic formulas for finding beam deflections and for use of the superposition method, the double integration method, and the moment-area method. Development of the fundamental principles controlling the deflection of beams has been added.

In Chapter 13, Statically Indeterminate Beams, a comparison of the performance of beams having different kinds of support is given together with the methods of analyzing statically indeterminate beams. This should help the student or the practicing designer to understand the effect of how the beam is supported: simply supported ends, fixed ends, cantilever, or supported cantilever. Compared are the reaction forces and moments, the variation in shearing forces and bending moments, and the maximum deflection of the beam. Several additional types of beam loading and support are included in appendix tables.

Chapter 15 presents the analysis of pressure vessels, with expanded treatment of thick-walled cylinders and spheres.

Bolted, riveted, and welded connections are presented in Chapter 16.

The Appendix, considered a strong asset to the first edition, has been expanded, updated, and arranged in a more usable form. More material properties are given, more types of beam shapes are listed, more stress concentration factor data are included, and more beam deflection formulas are given.

Robert L. Mott

Basic Concepts in Strength of Materials

1-1 OBJECTIVES OF THIS BOOK

It is essential that any product, machine, or structure be safe and stable under the loads exerted on it during any foreseeable use. The analysis and design of such devices or structures to ensure safety is the primary objective of this book.

Failure of a component of a structure can occur in several ways:

1. The material of the component could fracture completely.
2. The material may deform excessively under load so that the component is no longer suitable for its purpose.
3. The structure could become unstable and buckle, and thus be unable to carry the intended loads.

Design and analysis, using the principles of strength of materials, are required to ensure that a component is safe with regard to strength, rigidity, and stability.

1-2 OBJECTIVES OF THIS CHAPTER

In this chapter we present the basic concepts in strength of materials that will be expanded on in later chapters. After completing this chapter, you should be able to:

1. Use correct units for the quantities encountered in the study of strength of materials in both the SI metric unit system and the U.S. Customary unit system.
2. Use the terms *mass* and *weight* correctly and be able to compute the value of one when the value of the other is given.
3. Define *stress*.
4. Define *direct normal stress* and compute the value of this type of stress for both tension and compression loading.
5. Define *direct shear stress* and compute its value.
6. Identify conditions in which a load-carrying member is in *single shear* or *double shear*.
7. Draw *stress elements* showing the normal and shear stresses acting at a point in a load-carrying member.
8. Define *bearing stress* and compute its value.
9. Define *normal strain* and *shearing strain*.
10. Define *Poisson's ratio* and give its value for typical materials used in mechanical and structural design.
11. Recognize standard structural shapes and standard screw threads and use data concerning them.
12. Define *modulus of elasticity in tension*.
13. Define *modulus of elasticity in shear*.
14. Understand the responsibilities of designers.

1-3 BASIC UNIT SYSTEMS

Computations required in the application of strength of materials involve the manipulation of several sets of units in equations. For numerical accuracy it is of great importance to ensure that consistent units are used in the equations. Throughout this book units will be carried with the applicable numbers.

Because of the present transition in the United States from the traditional U.S. Customary units to metric units, both are used in this book. It is expected that persons entering or continuing an industrial career within the next several years will be faced with having to be familiar with both systems. On the one hand, many new products such as sutomobiles and business machines are being built using metric dimensions. Thus components and manufacturing equipment will be specified in those units. However, the transition is not being made uniformly in all areas. Designers will continue to deal with such items as structural steel, aluminum, and wood whose properties and dimensions are given in English units in standard references. Also, designers, sales and service people, and those in manufacturing must work with equipment which has already been installed and which was made to U.S. Customary system dimensions. Therefore, it seems logical that persons now working in industry should now be capable of working and thinking in both systems.

The formal name for the U.S. Customary unit system is the English Gravitational Unit System (EGU). The metric system, which has been adopted internationally, is called by the French name Systéme International d'Unités, the International System of Units, abbreviated SI in this book.

In most cases, the problems in this book are worked either in the U.S. Customary unit system or the SI system rather than mixing units. In problems for which data are given in both unit systems, it is most desirable to change all data to the same system before completing the problem solution. Appendix tables are included as an aid in performing conversions.

The basic quantities for any unit system are length, time, force, mass, temperature, and angle. Table 1-1 lists the units for these quantities in the SI unit system, and Table 1-2 lists the quantities in the U.S. Customary unit system.

Distinction between force, weight, and mass. Force and mass are separate and distinct quantities. *Mass* refers to the amount of the substance in a body.

TABLE 1-1 BASIC QUANTITIES IN THE SI METRIC UNIT SYSTEM

Quantity	SI unit	Other metric units
Length	meter (m)	millimeter (mm)
Time	second (s)	minute (min), hour (h)
Force	newton (N)	$kg \cdot m/s^2$
Mass	kilogram (kg)	$N \cdot s^2/m$
Temperature	kelvin (K)	degrees Celsius (°C)
Angle	radian	degree

TABLE 1-2 BASIC QUANTITIES IN THE U.S. CUSTOMARY UNIT SYSTEM

Quantity	U.S. Customary unit	Other U.S. units
Length	foot (ft)	inch (in.)
Time	second (s)	minute (min), hour (h)
Force	pound (lb)	kip*
Mass	slug	$lb \cdot s^2/ft$
Temperature	°F	
Angle	degree (deg)	radian (rad)

*1.0 kip = 1000 pounds. The name is derived from the term *kilopound*.

Force is a push or pull effort exerted on a body, either by an external source or by gravity. The force of gravitational pull is called the *weight* of the body. Thus weight has the units of force.

In the SI system, the unit for mass is defined to be the kilogram, while the unit of force is derived using Newton's law:

$$\text{force} = \text{mass} \times \text{acceleration}$$

$$F = m \cdot a = \text{kg} \cdot \text{m/s}^2 = \text{newton}$$

In honor of Sir Isaac Newton, this unit is called the *newton* and its symbol is N. It represents the amount of force required to give a mass of 1.0 kg an acceleration of 1.0 m/s^2.

In the U.S Customary system, the unit of force is defined to be the pound, while the unit of mass is derived from an application of Newton's law:

$$F = m \cdot a$$

where a is acceleration, having the standard units of ft/s^2. Therefore, the derived unit for mass, called the *slug*, is

$$m = \frac{F}{a} = \frac{\text{lb}}{\text{ft/s}^2} = \frac{\text{lb} \cdot \text{s}^2}{\text{ft}} = \text{slug}$$

Because weight is a force, the units are either the pound or the newton. If the quantity (mass) of the substance of a body is given, it is necessary to compute its weight from a special application of Newton's law:

$$W = m \cdot g \qquad (1\text{-}1)$$

where W = weight
  m = mass
  g = acceleration due to gravity

We will use the following values for g:

$$\text{SI units:} \quad g = 9.81 \text{ m/s}^2$$

$$\text{U.S. Customary units:} \quad g = 32.2 \text{ ft/s}^2$$

The conversion of weight and mass is illustrated in the following example problems.

Example Problem 1-1   (SI system)

A hoist lifts 425 kg of concrete. Compute the weight of the concrete that is the force exerted on the hoist by the concrete.

Solution

$$W = m \cdot g$$

$$= 425 \text{ kg} \cdot 9.81 \text{ m/s}^2 = 4170 \text{ kg} \cdot \text{m/s}^2 = 4170 \text{ N}$$

Thus 425 kg of concrete weighs 4170 N.

Example Problem 1-2   (U.S. Customary system)

Determine the mass of a hopper of coal that weighs 8500 lb.

Solution

$$m = \frac{W}{g} = \frac{8500 \text{ lb}}{32.2 \text{ ft/s}^2} = \frac{264 \text{ lb} \cdot \text{s}^2}{\text{ft}} = 264 \text{ slugs}$$

Prefixes for SI units. In the SI system, prefixes should be used to indicate orders of magnitude, thus eliminating insignificant digits and providing a convenient substitute for writing powers of 10, as generally preferred for computation. Prefixes representing steps of 1000 are recommended. Those usually encountered in strength of materials are listed in Table 1-3. Table 1-4 shows how computed results should be converted to the use of the standard prefixes for units.

TABLE 1-3 PREFIXES FOR SI UNITS

Prefix	SI symbol	Factor
giga	G	$10^9 = 1\ 000\ 000\ 000$
mega	M	$10^6 = 1\ 000\ 000$
kilo	k	$10^3 = 1\ 000$
milli	m	$10^{-3} = 0.001$
micro	μ	$10^{-6} = 0.000\ 001$

TABLE 1-4 PROPER METHOD OF REPORTING COMPUTED QUANTITIES

Computed result	Reported result
0.005 48 m	5.48×10^{-3} m, or 5.48 mm
12 750 N	12.75×10^3 N, or 12.75 kN
34 500 kg	34.5×10^3 kg, or 34.5 Mg (megagrams)

1-4 THE CONCEPT OF STRESS

The objective of any strength analysis is to ensure safety. Accomplishing this requires that the *stress* produced in the member being analyzed is below a certain *allowable stress* for the material from which the member is made. Understanding the meaning of stress is of utmost importance in studying strength of materials.

> *Stress is the internal resistance offered by a unit area of the material from which a member is made to an externally applied load.*

We are concerned about what happens *inside* a load-carrying member. The effect of the internal resistance on the material determines whether the member is safe or not. Different materials react differently to loads and thus have different strengths. Therefore, the selection of a material is an important part of the design process. However, it will be shown in later chapters that the *shape* as well as the *size* of a load-carrying member affects its strength. Actually, then, it might be more nearly correct to refer to the study of the "strength of load-carrying members" rather than the "strength of materials."

Stress is a measure of the amount of force exerted on a specific area of the material in a part. In some cases the magnitude of the stress is uniform over a large section of the part. In such cases stress can be computed by simply dividing the total force by the area of the part which resists the force. This concept can be expressed mathematically as

$$\text{stress} = \frac{\text{force}}{\text{area}} \tag{1-2}$$

In the U.S. Customary unit system, the typical unit for force is the pound and the most convenient unit of area is the square inch. Thus stress is indicated in lb/in^2, abbreviated psi. Typically encountered stress levels in machine design and structural analysis are several thousand psi. For this reason, the unit of $kips/in^2$ is often used, abbreviated ksi. For example, if a stress is computed to be 26 500 psi, it might be reported as

$$\text{stress} = \frac{26\ 500\ lb}{in^2} \cdot \frac{1\ kip}{1000\ lb} = \frac{26.5\ kips}{in^2} = 26.5\ ksi$$

In the SI unit system, the standard unit for force is the newton and area is in square meters. Thus the standard unit for stress is the N/m^2, given the name *pascal* and abbreviated Pa. Typical levels of stress are several million pascals, so the most convenient unit for stress is the megapascal or MPa. This is convenient for another reason. In calculating the cross-sectional area of load-carrying members, measurements of dimensions in mm are usually used. Then the stress would be in N/mm^2 and it can be shown that this is numerically equal to the unit of MPa. For example, assume that a force of 15 000 N is exerted over a square area 50 mm on a side. The resisting area would be 2 500 mm^2 and the resulting stress would be

$$\text{stress} = \frac{\text{force}}{\text{area}} = \frac{15\ 000\ N}{2500\ mm^2} = \frac{6.0\ N}{mm^2}$$

Converting this to pascals would produce

$$\text{stress} = \frac{6.0\ N}{mm^2} \cdot \frac{(1000)^2\ mm^2}{m^2} = 6.0 \times 10^6\ N/m^2 = 6.0\ MPA$$

This demonstrates that the unit of N/mm^2 is identical to the MPa, an observation that will be used throughout this book.

1-5 DIRECT NORMAL STRESS

One of the most fundamental types of stress that exists is the normal stress, indicated by the lowercase Greek letter σ (sigma), in which the stress acts perpendicular, or normal, to the cross section of the load-carrying member. If the stress is also uniform across the resisting area, the stress is called a direct normal stress.

Basic Concepts in Strength of Materials   Chap. 1

Normal stresses can be either compressive or tensile. A compressive stress is one that tends to crush the material of the load-carrying member and to shorten the member itself. A tensile stress is one that tends to stretch the member and pull the material apart.

An example of a member subjected to compressive stress is shown in Figure 1-1. The support stand is designed to be placed under heavy equipment during assembly and the weight of the equipment tends to crush the square shaft of the support, placing it in compression.

Example Problem 1-3

Consider the support stand shown in Figure 1-1. If it is to carry 27 500 pounds, compute the stress in the square shaft at the top of the stand.

Solution Figure 1-2 shows an arbitrarily selected part of the shaft with the applied load on the top and an internal resisting force equally distributed over the cut section. Each unit area of the cut section would support the same load and thus the stress would be

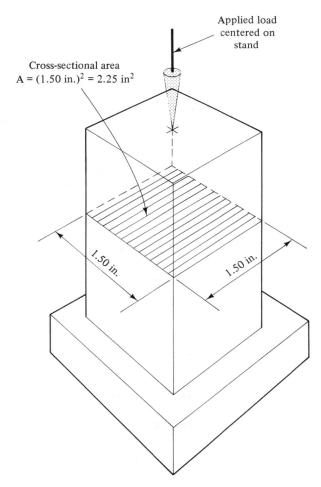

Figure 1-1 Example of direct compressive stress.

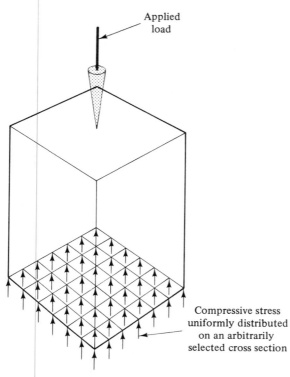

Applied
load

Compressive stress
uniformly distributed
on an arbitrarily
selected cross section

Figure 1-2 Compressive stress on an arbitrary cross section of support stand shaft.

uniform over the cross section. The load applied to the stand acts downward on the top surface and is centered on the square pad. The stress produced in the shaft is a compressive stress, and the magnitude of the stress can be computed from Equation (1-2). The area resisting the force is the cross sectional area of the square shaft.

$$\text{Area} = (1.50 \text{ in})^2 = 2.25 \text{ in}^2$$

$$\text{Stress} = \sigma = \frac{\text{force}}{\text{area}} = \frac{27\ 500 \text{ lb}}{2.25 \text{ in}^2} = 12\ 222 \text{ lb/in}^2 = 12\ 222 \text{ psi}$$

This level of stress would be present on any cross section of the shaft between the ends.

An example of a member subjected to tensile stress is shown in Figure 1-3. The rods suspend the heavy casting from a conveyor and the weight of the casting tends to stretch the rods and pull them apart.

Example Problem 1-4

Figure 1-3 shows two circular rods carrying a casting weighing 11.2 kN. If each rod is 12 mm in diameter, compute the stress in the rods.

Solution   Each rod carries one-half of the weight of the casting, 5.6 kN or 5600 N. The cross sectional area of one rod is

$$A = \pi D^2/4 = \pi(12 \text{ mm})^2/4 = 113 \text{ mm}^2$$

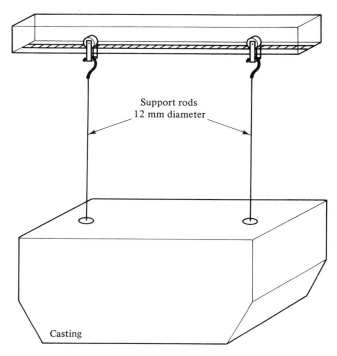

Support rods
12 mm diameter

Casting

Figure 1-3 Example of direct tensile stress.

The stress in the rod is tensile with a magnitude of

$$\text{stress} = \sigma = \frac{\text{force}}{\text{area}} = \frac{5600 \text{ N}}{113 \text{ mm}^2} = 49.5 \text{ N/mm}^2 = 49.5 \text{ MPa}$$

Figure 1-4 shows an arbitrarily selected part of the rod with the applied load on the bottom and the internal tensile stress distributed uniformly over the cut section.

1-6 DIRECT SHEAR STRESS

Shear refers to a cutting-like action, and such an action occurs frequently in load-carrying members. Figure 1-5 shows two examples.

Figure 1-5(a) shows a simple pin joint in which the link rotates about the pin while carrying a load. The force on the link is transferred through the pin to the support structure, and the force tends to cut the pin across the shear plane. Because only one cross section of the pin resists the force, it is said to be in *single shear*.

The clevis joint in Figure 1-5(b) is similar to the simple pin joint except that the force is transferred to two ears rather than directly to the support structure. It acts like a hinge. If the link is a close fit between the ears, a force on the link tends to cut or shear the pin at the two cross sections between the sides of the link and the ears. Because two cross sections share the load, the pin is in *double shear*.

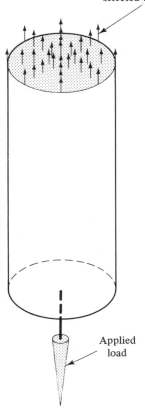

Tensile stress
uniformly distributed
over an arbitrarily
selected cross section

Applied
load

Figure 1-4 Tensile stress on an
arbitrary cross section of a circular rod.

Figure 1-6 shows a punching operation in which a cylindrical slug is to be sheared out from a plate by a punch. In this case the area of metal that is sheared is the surface of the sides of the cylinder, shown shaded in the figure.

The symbol for shear stress is the lowercase Greek letter tau (τ). In computing the direct shear stress on parts such as those shown in Figures 1-5 and 1-6, it is assumed that the shear force on a section is uniformly distributed across the entire area resisting the shear force. Thus the direct shear stress can be found from

$$\text{shear stress} = \tau = \frac{\text{applied force}}{\text{shear area}} = \frac{F}{A_s} \tag{1-3}$$

Example Problem 1-5

The force on the link in the simple pin joint shown in Figure 1-5(a) is 3550 N. If the pin has a diameter of 10.0 mm, compute the shear stress in the pin.

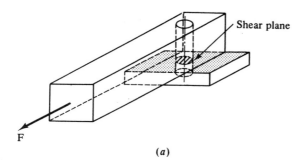

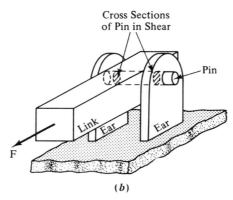

(a)

Cross Sections
of Pin in Shear

Pin

Link

Ear

Ear

F

(b)

Figure 1-5   Direct shear stress
examples. (a) Pin joint in a single shear.
(b) Clevis joint with pin in double shear.

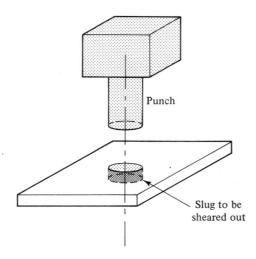

Punch

Slug to be
sheared out

Figure 1-6   Direct shear stress in a
punching operation.

Solution   The pin is in direct shear with one cross section of the pin resisting the applied force. The shear area is shown in Figure 1-7 and is computed from

$$A_s = \frac{\pi D^2}{4} = \frac{\pi (10.0 \text{ mm})^2}{4} = 78.5 \text{ mm}^2$$

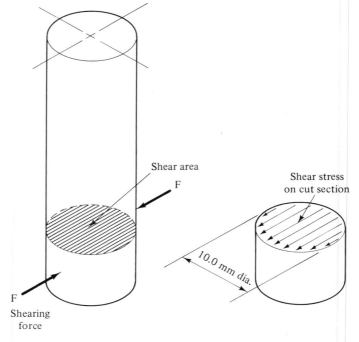

Figure 1-7   Direct shear stress in a pin.

Then the shear stress is

$$\tau = \frac{F}{A_s} = \frac{3550 \text{ N}}{78.5 \text{ mm}^2} = 45.2 \text{ N/mm}^2 = 45.2 \text{ MPa}$$

This stress is shown in Figure 1-7 on a cut section of the pin at the shear plane.

Example Problem 1-6

The link in the clevis joint shown in Figure 1-5(b) also carries 3550 N and the pin has a diameter of 10.0 mm. Compute the shear stress in the pin.

Solution   The pin is in double shear because two cross sections of the pin resist the applied load. Then the shear area is

$$A_s = 2\left(\frac{\pi D^2}{4}\right) = 2\left[\frac{\pi (10.0 \text{ mm})^2}{4}\right] = 157 \text{ mm}^2$$

The shear stress in the pin is

$$\tau = \frac{F}{A_s} = \frac{3550 \text{ N}}{157 \text{ mm}^2} = 22.6 \text{ N/mm}^2 = 22.6 \text{ MPa}$$

Example Problem 1-7

The punching operation in Figure 1-6 requires 850 lb of force to punch the 0.50-in.-diameter slug from an aluminum sheet having a thickness of 0.040 in. Compute the shear stress in the aluminum as it was being sheared.

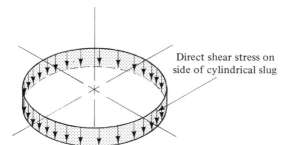

Direct shear stress on
side of cylindrical slug

Figure 1-8 Shear stress on slug for
Example Problem 1-7.

Solution The sides of the cylindrical slug are in direct shear. Figure 1-8 shows the shear area, calculated from

$$A_s = \pi Dt = \pi(0.50 \text{ in.})(0.040 \text{ in.}) = 0.0628 \text{ in}^2$$

Then the shear stress is

$$\tau = \frac{F}{A_s} = \frac{850 \text{ lb}}{0.0628 \text{ in}^2} = 13\ 535 \text{ psi}$$

1-7 STRESS ELEMENTS

The illustrations of stresses in Figures 1-2, 1-4, and 1-7 are useful for visualizing the nature of the internal resistance to the externally applied force, particularly for these cases in which the stresses are uniform across the entire cross section. In other cases it is more convenient to visualize the stress condition on a small (infinitesimal) element. Consider a small cube of material anywhere inside the square shaft of the support stand shown in Figure 1-1. There must be a net compressive force acting on the top and bottom faces of the cube as shown in Figure 1-9. If the faces are considered to be unit areas, these forces can be considered to be the stresses acting on the faces of the cube. Such a cube is called a *stress element*.

Because the element is taken from a body in equilibrium, the element itself is also in equilibrium and the stresses on the top and bottom faces are the same. A simple stress element like this one is often shown as a two-dimensional square element rather than the three-dimensional cube, as shown in Figure 1-10.

Similarly, the tensile stress on any element of the rod in Figure 1-3 can be shown as in Figure 1-11, with the stress vector acting outward from the element. Note that either compressive or tensile stresses are shown acting perpendicular (normal) to the surface of the element.

An element taken from the shear plane of the pin in Figure 1-5(a) would look as shown in Figure 1-12. The shear stresses act parallel to the surfaces of the cube with those on one side acting in the opposite direction to those on the other side of the cube. This is the cutting action characteristic of shear. The shear stresses on the top and bottom of the element cannot exist alone because the element would tend to rotate

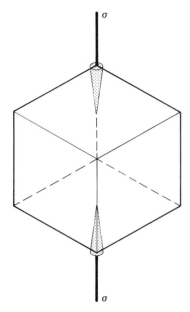

Figure 1-9 Stress element with compressive stress on top and bottom faces.

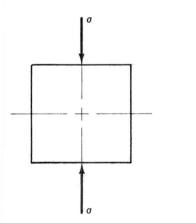

Figure 1-10 Two-dimensional stress element showing compressive stress.

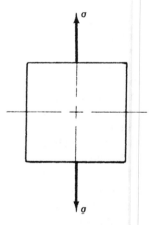

Figure 1-11 Two dimensional stress element showing tensile stress.

under the influence of the couple formed by the two shear forces acting in opposite directions. The shear stresses acting on the sides of the element are created simultaneously with those on the top and bottom, and set up an opposing couple to keep the element in equilibrium.

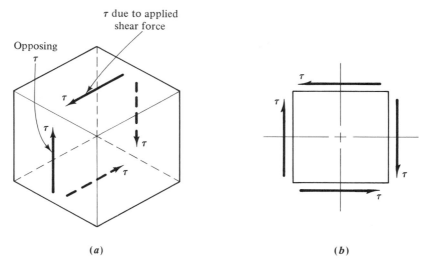

Figure 1-12 Stress element showing shear stress. (a) Three dimensional stress element. (b) Two dimensional stress element.

1-8 BEARING STRESS

When one solid body rests on another and transfers a load to it, the form of stress called *bearing stress* is developed at the surfaces in contact. Similar to direct compressive stress, the bearing stress is a measure of the tendency for the applied force to crush the supporting member.

Bearing stress is computed in a manner similar to direct normal stresses:

$$\sigma_b = \frac{\text{applied load}}{\text{bearing area}} = \frac{P}{A_b} \tag{1-4}$$

For flat surfaces in contact, the bearing area is simply the area over which the load is transferred from one member to the other. If the two parts have different areas, the smaller area is used. Another requirement is that the materials transmitting the loads must remain nearly rigid and flat in order to maintain their ability to carry the loads. Excessive deflection will reduce the effective bearing area.

Figure 1-13 shows an example from building construction in which bearing stress is important. A hollow steel column 4.00 inches square, rests on a thick steel plate 6.00 inches square. The plate rests on a concrete pier, which in turn rests on a base of gravel. These successively larger areas are necessary to limit bearing stresses to reasonable levels for the materials involved.

Example Problem 1-8

Refer to Figure 1-13. The square steel tube carries 30 000 lb of axial compressive force. Compute the compressive stress in the tube and the bearing stress between each mating surface. Consider the weight of the concrete pier to be 338 lb.

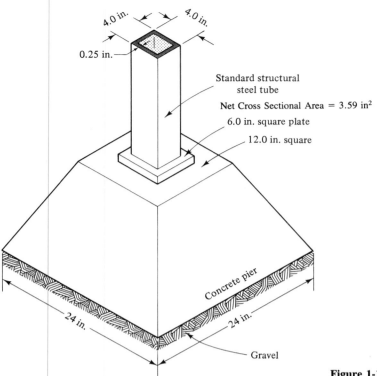

4.0 in.

4.0 in.

0.25 in.

Standard structural
steel tube

Net Cross Sectional Area = 3.59 in²

6.0 in. square plate

12.0 in. square

Concrete pier

24 in.

24 in.

Gravel

Figure 1-13 Bearing stress example.

Solution Using the direct compressive stress formula, the stress in the wall of the tube is

$$\sigma = \frac{P}{A} = \frac{30\ 000\ \text{lb}}{3.59\ \text{in}^2} = 8356\ \text{psi}$$

The bearing stress between the tube and the plate has this same value because the cross-sectional area of the tube is the smallest contact area.

The bearing stress between the plate and the top of the concrete pier is

$$\sigma_b = \frac{P}{A_b} = \frac{30\ 000\ \text{lb}}{(6.00\ \text{in.})^2} = 833\ \text{psi}$$

Considering the weight of the concrete pier, the total load on the gravel at the base of the pier is 30 338 lb. For soil and gravel, the bearing stress is often called *bearing pressure* and is measured in pounds per square foot rather than psi. Then the bearing stress (pressure) is

$$\sigma_b = \frac{P}{A_b} = \frac{30\ 338\ \text{lb}}{(2.00\ \text{ft})^2} = 7585\ \text{lb/ft}^2$$

Bearing stresses in pin joints. Frequently in mechanical and structural design, cylindrical pins are used to connect components together. One design for such

Basic Concepts in Strength of Materials   Chap. 1

a connection is shown in Figure 1-5(b). When transferring a load across the pin, the bearing stress between the pin and each mating member should be computed.

The effective bearing area for the cylindrical pin requires that the *projected area* be used, computed as the product of the diameter of the pin and the length of the surface in contact.

Example Problem 1-9

In Figure 1-5(b), compute the bearing stress between the pin and the link and between the pin and each ear of the clevis joint. The pin has a diameter of 10.0 mm; the link is 40 mm wide; and each ear is 12 mm thick. The load P applied to the link is 3550 N.

Solution First consider the bearing stress between the link and the pin. The bearing area is

$$A_b = (\text{pin diameter} \times \text{link width}) = (10.0 \times 40.0) = 400 \text{ mm}^2$$

Then the bearing stress is

$$\sigma_b = \frac{3550 \text{ N}}{400 \text{ mm}^2} = 8.88 \text{ N/mm}^2 = 8.88 \text{ MPa}$$

For the bearing stress between the pin and one ear of the clevis, note that only one-half of the load, 1775 N, acts on each ear. The bearing area is

$$A_b = (\text{pin diameter} \times \text{ear thickness}) = (10.0 \times 12.0) = 120 \text{ mm}^2$$

Then the bearing stress is

$$\sigma_b = \frac{1775 \text{ N}}{120 \text{ mm}^2} = 14.8 \text{ N/mm}^2 = 14.8 \text{ MPa}$$

1-9 THE CONCEPT OF STRAIN

Any load-carrying member deforms under the influence of the load applied. The square shaft of the support stand in Figure 1-1 gets shorter as the heavy machine is placed on the stand. The rods supporting the casting in Figure 1-3 get longer as the casting is hung onto them.

The total deformation of a load-carrying member can, of course, be measured. It will also be shown later how the deformation can be calculated.

Figure 1-14 shows an aluminum bar, 0.75 in. in diameter, subjected to an axial tensile force of 10 000 lb. Before the load was applied, the bar was 10.000 in. long.

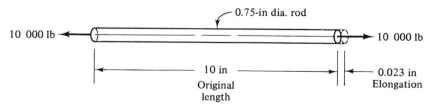

Figure 1-14 Elongation of a bar in tension.

After the load is applied, the length is 10.023 in. Thus the total deformation is 0.023 in.

Strain, also called *unit deformation*, is found by dividing the total deformation by the original length of the bar. The lowercase Greek letter epsilon (ϵ) is used to denote strain:

$$\text{strain} = \epsilon = \frac{\text{total deformation}}{\text{original length}} \tag{1-5}$$

For the case shown in Figure 1-14,

$$\epsilon = \frac{0.023 \text{ in.}}{10.000 \text{ in.}} = 0.0023 \text{ in./in.}$$

Strain could be said to be dimensionless because the units in the numerator and denominator could be canceled. However, it is better to report the units as in./in. or mm/mm to maintain the definition of deformation per unit length of the member. More will be said about strain and deformation in later chapters.

1-10 POISSON'S RATIO

A more complete understanding of the deformation of a member subjected to normal stress can be had by referring to Figure 1-15. The element shown is taken from the bar in Figure 1-14. The tensile force on the bar causes an elongation in the direction of the applied force, as would be expected. But at the same time, the width of the bar is being shortened. Thus a simultaneous elongation and contraction occurs in the stress element. From the elongation, the axial strain can be determined, and from the contraction, the lateral strain can be determined.

The ratio of the lateral strain on the element to the axial strain is called *Poisson's ratio* and is a property of the material from which the load-carrying member is made. In this book the lowercase Greek letter nu (ν) is used to denote Poisson's ratio. Note that some references use mu (μ).

Most commonly used metallic materials have a Poisson's ratio value between 0.25 and 0.35. For concrete, ν varies widely depending on the grade and on the applied stress, but it usually falls between 0.1 and 0.25. Elastomers and rubber may have Poisson's ratio as high as 0.50. Approximate values for Poisson's ratio are listed in Table 1-5.

1-11 SHEARING STRAIN

Earlier discussions of strain described normal strain because it is caused by the normal tensile or compressive stress developed in a load-carrying member. Under the influence of a shear stress, shearing strain would be produced.

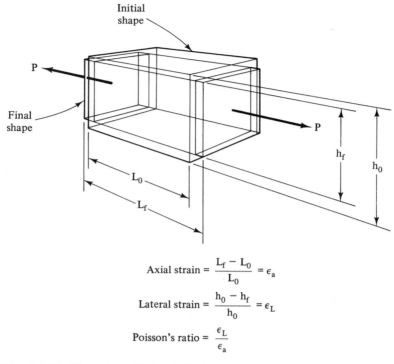

Initial shape

Final shape

P

P

h_f

h_0

L_0

L_f

Axial strain = $\dfrac{L_f - L_0}{L_0}$ = ϵ_a

Lateral strain = $\dfrac{h_0 - h_f}{h_0}$ = ϵ_L

Poisson's ratio = $\dfrac{\epsilon_L}{\epsilon_a}$

Figure 1-15 Illustration of Poisson's Ratio for an element in tension.

TABLE 1-5 APPROXIMATE VALUES OF POISSON'S RATIO

Material	Poisson's ratio, ν
Aluminum (most alloys)	0.33
Brass	0.33
Cast iron	0.27
Concrete	0.10–0.25
Copper	0.33
Phosphor bronze	0.35
Carbon and alloy steel	0.29
Stainless steel (18-8)	0.30
Titanium	0.30

Figure 1-16 shows a stress element subjected to shear. The shearing action on parallel faces of the element tends to deform it angularly as shown to an exaggerated degree. The angle γ (gamma), measured in radians, is the *shearing strain*. Only very small values of shearing strain are encountered in practical problems, and thus the dimensions of the element are virtually unchanged.

Sec. 1-11   Shearing Strain

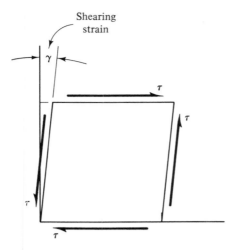

Shearing
strain

γ

τ

τ

τ

τ

Figure 1-16 Shearing strain shown on a stress element.

1-12 MODULUS OF ELASTICITY

A measure of the stiffness of a material is found by computing the ratio of the normal stress on an element to the corresponding strain on the element. This ratio is called the *modulus of elasticity* and is denoted by E. That is,

$$\text{modulus of elasticity} = \frac{\text{normal stress}}{\text{normal strain}}$$

$$E = \frac{\sigma}{\epsilon} \tag{1-6}$$

A material with a higher value of E will deform less for a given stress than one with a lower value of E. A more complete term for E would be modulus of elasticity in tension or compression because it is defined in terms of normal stress. However, the term "modulus of elasticity" without a modifier is usually assumed to be the tensile modulus. More will be said about modulus of elasticity in Chapter 2 and typical values are identified there.

1-13 MODULUS OF ELASTICITY IN SHEAR

The ratio of the shearing stress to the shearing strain is called the *modulus of elasticity in shear* or the *modulus of rigidity*, and is denoted by G. That is,

$$G = \frac{\text{shearing stress}}{\text{shearing strain}} = \frac{\tau}{\gamma} \tag{1-7}$$

G is a property of the material and is related to the tensile modulus and Poisson's ratio by

$$G = \frac{E}{2(1 + \nu)} \tag{1-8}$$

1-14 PREFERRED SIZES AND STANDARD SHAPES

In the preceding section it was stated that one of the responsibilities of a designer is to specify a convenient or standard dimension for the component being designed. Several guides are available to aid in this decision.

Preferred basic sizes. When the component being designed will be made to the designer's specifications, it is recommended that final dimensions be specified from a set of *preferred basic sizes*. Appendix A-2 lists such data for fractional inch dimensions, decimal inch dimensions, and metric dimensions.

American Standard screw threads. Threaded fasteners and machine elements having threaded connections are manufactured according to standard dimensions to ensure interchangeability of parts and to permit convenient manufacture with standard machines and tooling. Appendix A-3 gives the dimensions of American Standard Unified threads. Sizes smaller than $\frac{1}{4}$ in. are given numbers from 0 to 12, while fractional-inch sizes are specified for $\frac{1}{4}$ in. and larger sizes. Two series are listed; UNC is the designation for coarse threads and UNF designates fine threads. Standard designations are illustrated below.

6-32 UNC   (number size 6, 32 threads per inch, coarse thread)
12-28 UNF   (number size 12, 28 threads per inch, fine thread)
$\frac{1}{2}$-13 UNC   (fractional size $\frac{1}{2}$ in., 13 threads per inch, coarse thread)
$1\frac{1}{2}$-12 UNF   (fractional size $1\frac{1}{2}$ in., 12 threads per inch, fine thread)

Given in the tables are the basic major diameter (D), number of threads per inch (n), and the tensile stress area found from

$$A_t = 0.7854 \left(D - \frac{0.9743}{n} \right)^2 \tag{1-9}$$

When a threaded member is subjected to direct tension, the tensile stress area is used to compute the average tensile stress. It corresponds to the smallest area that would be produced by a transverse cut across the threaded rod. For convenience, some standards use the root area or gross area and adjust the value of the allowable stress.

Metric screw threads. Appendix A-3 gives similar dimensions for metric threads. Standard metric thread designations are of the form

$$M10 \times 1.5$$

where M stands for metric, the following number is the basic major diameter in mm, and the last number is the pitch between adjacent threads in mm. Thus the designation above would denote a metric thread with a basic major diameter of 10.0 mm and a pitch of 1.5 mm. Note that pitch = $1/n$.

Commercially available structural shapes. Steel and aluminum manufacturers provide a large variety of standard shapes for use as beams. Some are also used as columns, as will be shown in a later chapter. It is important to know the standard terminology for these shapes and how to use the tables of data which report their geometrical properties.

For steel shapes, the standard symbols used are listed in Table 1-6, along with a sketch of the shape and an example of how it is designated. In the designation, the first letter or letters give the basic shape symbol. The first numbers give the nominal size, and the last set of numbers gives the weight per foot of length of a beam having this shape and size. For example, the W12×16.5 shape would be a wide flange shape with a nominal depth (vertical height) of 12 in., and a 1-ft length would weigh 16.5

TABLE 1-6 DESIGNATIONS FOR STEEL SHAPES

Name of Shape	Shape	Symbol	Example Designation
Wide flange beam		W	W12 × 16.5
American Standard beam		S	S10 × 35
Channel		C	C15 × 50
Angle		L	L4 × 3 × $\frac{1}{2}$
Tees, cut from W shape		WT	WT5 × 8.5
Tees, cut from S shape		ST	ST6 × 15.9
Zees		Z	Z5 × 16.4
Structural tubing		TS	TS4 × 0.188

lb. This format is typical for all sections listed in Table 1-6 except the angles and the structural tubing. For angles, the designation L4×3×$\frac{1}{2}$ would be for an unequal leg angle whose legs are 4 in. and 3 in. and whose thickness is $\frac{1}{2}$ in. For the round tubing, the designation TS4×0.188 would be for a tube with a nominal size of 4 in. and a wall thickness of 0.188 in. It must be noted that the actual dimensions of standard shapes are not always the same as the nominal sizes. Reference must be made to standard tables of dimensions for actual sizes.

The properties of several standard shapes are listed in the Appendix. The data include such items as area of the section, depth, flange width and thickness, web thickness, and moment of inertia with respect to major axes. Other properties such as section modulus and radius of gyration, used in later chapters, are also listed. For channels and angles, the dimensions to the centroidal axes are also given.

The properties for Aluminum Association standard structrural I-beams and channels are also listed in the Appendix. These are extruded shapes having flanges with uniform thickness. In contrast, rolled shapes have tapered flanges, making joining somewhat more difficult. The proportions for flange width and thickness are also somewhat different for the extruded sections. The designations for aluminum shapes are similar to those for steel shapes. For example, I8×6.181 would be a standard I-beam with a depth of 8.00 in., and a 1-ft length would weigh 6.181 lb. A C4×1.738 would be a standard channel with a depth of 4.00 in., and a 1-ft length would weigh 1.738 lb. For these aluminum shapes, the actual depth is the same as the nominal depth.

In addition to the Aluminum Association structural shapes described above, structural shapes similar in size and shape to the rolled steel shapes are commercially available in aluminum. The Aluminum Association references should be consulted for a complete listing.

REFERENCES

1. Aluminum Association, *Aluminum Standards and Data,* Washington, D.C., 1984.

2. Aluminum Association, *Engineering Data for Aluminum Structures,* Washington, D.C., 1981.

3. Aluminum Association, *Specifications for Aluminum Structures,* Washington, D.C., 1982.

4. American Institute of Steel Construction, *Manual of Steel Construction,* Chicago, 1980.

5. Oberg, E., F. D. Jones, and H. L. Horton, *Machinery's Handbook,* 23rd ed., Industrial Press, New York, 1988.

6. United States Steel Corporation, *Structural Steel Shapes,* Pittsburgh, Pa., 1982.

PROBLEMS

1-1. Define *mass* and state the units for mass in both the U.S. Customary unit system and the SI metric unit system.

1-2. Define *weight* and state its units in both systems.

1-3. Define *stress* and state its units in both systems.

1-4. Define *direct normal stress*.

1-5. Explain the difference between compressive stress and tensile stress.

1-6. Define *direct shear stress*.

1-7. Explain the difference between single shear and double shear.

1-8. Draw a stress element subjected to direct tensile stress.

1-9. Draw a stress element subjected to direct compressive stress.

1-10. Draw a stress element subjected to direct shear stress.

1-11. Define *normal strain* and state its units in both systems.

1-12. Define *shearing strain* and state its units in both systems.

1-13. Define *Poisson's ratio* and state its units in both systems.

1-14. Define *modulus of elasticity in tension* and state its units in both systems.

1-15. Define *modulus of elasticity in shear* and state its units in both systems.

1-16. A truck carries 1800 kg of gravel. What is the weight of the gravel in newtons?

1-17. A four-wheeled truck having a total mass of 4000 kg is sitting on a bridge. If 60% of the weight is on the rear wheels and 40% is on the front wheels, compute the force exerted on the bridge at each wheel.

1-18. A total of 6800 kg of a bulk fertilizer is stored in a flat-bottomed bin having side dimensions 5.0 m × 3.5 m. Compute the loading on the floor in newtons per square meter, or pascals.

1-19. A mass of 25 kg is suspended by a spring that has a spring scale of 4500 N/m. How much will the spring be stretched?

1-20. Measure the length, width, and thickness of this book in millimeters.

1-21. Determine your own weight in newtons and your mass in kilograms.

1-22. Express the weight found in Problem 1-16 in pounds.

1-23. Express the forces found in Problem 1-17 in pounds.

1-24. Express the loading in Problem 1-18 in pounds per square foot.

1-25. For the data in Problem 1-19, compute the weight of the mass in pounds, the spring scale in pounds per inch, and the stretch of the spring in inches.

1-26. A cast iron base for a machine weighs 2750 lb. Compute its mass in slugs.

1-27. A roll of steel hanging on a scale causes a reading of 12 800 lb. Compute its mass in slugs.

1-28. Determine your own weight in pounds and your mass in slugs.

1-29. A pressure vessel contains a gas at 1200 psi. Express the pressure in pascals.

1-30. A structural steel has an allowable stress of 21 600 psi. Express this in pascals.

1-31. The stress at which a material will break under a direct tensile load is called the *ultimate strength*. The range of ultimate strengths for aluminum alloys ranges from about 14 000 to 76 000 psi. Express this range in pascals.

1-32. An electric motor shaft rotates at 1750 rpm. Express the rotational speed in radians per second.

1-33. Express an area of 14.1 in^2 in the units of square millimeters.

1-34. An allowable deformation of a certain beam is 0.080 in. Express the deformation in millimeters.

1-35. A base for a building column measures 18.0 in. by 18.0 in. on a side and 12.0 in. high.

Compute the cross-sectional area in both square inches and square millimeters. Compute the volume in cubic inches, cubic feet, cubic millimeters, and cubic meters.

1-36. Compute the area of a rod having a diameter of 0.505 in. in square inches. Then convert the result to square millimeters.

Direct Tensile and Compressive Stresses

1-37. Compute the stress in a round bar subjected to a direct tensile force of 3200 N if the diameter of the bar is 10 mm.

1-38. Compute the stress in a rectangular bar having cross-sectional dimensions of 10 mm by 30 mm if a direct tensile force of 20 kN is applied.

1-39. A link in a mechanism for an automated packaging machine is subjected to a tensile force of 860 lb. If the link is square, 0.40 in. on a side, compute the stress in the link.

1-40. A circular rod, $\frac{3}{8}$ in. in diameter supports a heater assembly weighing 1850 lb. Compute the stress in the rod.

1-41. A shelf is being designed to hold crates having a total mass of 1840 kg. Two support rods like those shown in Figure 1-17 hold the shelf. Each rod has a diameter of 12.0 mm. Assume that the center of gravity of the crates is at the middle of the shelf. Compute the stress in the middle portion of the rods.

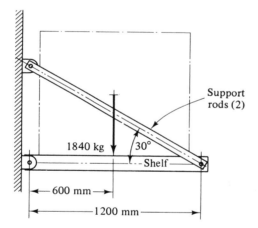

Figure 1-17 Shelf support for Problem 1-41.

1-42. A concrete column base is circular, 8 in. in diameter, and carries a direct compressive load of 70 000 lb. Compute the compressive stress in the concrete.

1-43. Three short, square, wood blocks, $3\frac{1}{2}$ in. on a side, support a machine weighing 29 500 lb. Compute the compressive stress in the blocks.

1-44. A short link in a mechanism carries an axial compressive load of 3500 N. If it has a square cross section, 8.0 mm on a side, compute the stress in the link.

1-45. A machine having a mass of 4200 kg is supported by three solid steel rods arranged as shown in Figure 1-18. Each rod has a diameter of 20 mm. Compute the stress in each rod.

1-46. A centrifuge is used to separate liquids according to their densities using centrifugal force. Figure 1-19 illustrates one arm of a centrifuge having a bucket at its end to hold the liquid. In operation, the bucket and the liquid have a mass of 0.40 kg. The centrifugal

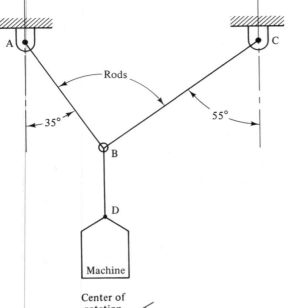

Figure 1-18 Support rods for Problem 1-45.

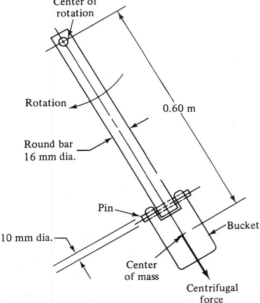

Figure 1-19 Centrifuge for Problem 1-46.

force has the magnitude in newtons of

$$F = 0.010\ 97 \cdot m \cdot R \cdot n^2$$

where m = rotating mass of bucket and liquid (kilograms)

R = radius to center of mass (meters)

n = rotational speed (revolutions per minute) = 3000 rpm

Compute the stress in the round bar. Consider only the force due to the container.

1-47. A bar carries a series of loads as shown in Figure 1-20. Compute the stress in each segment of the bar. All loads act along the central axis of the bar.

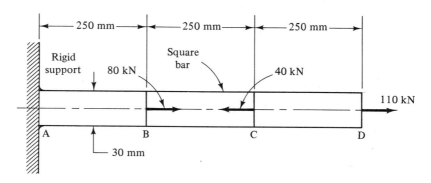

Figure 1-20   Bar carrying axial loads for Problem 1-47.

1-48. Repeat Problem 1-47 for the bar in Figure 1-21.

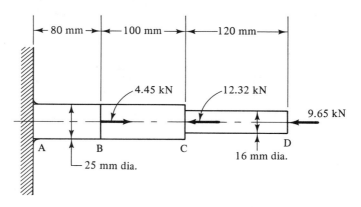

Figure 1-21   Bar carrying axial loads for Problem 1-48.

1-49. Repeat problem 1-47 for the pipe in Figure 1-22. The pipe is a $1\frac{1}{2}$ in. schedule 40 steel pipe.

1-50. Compute the stress in member *BD* shown in Figure 1-23 if the applied force *F* is 2800 lb.

For Problems 51 and 52 using the trusses shown, compute the forces in all members and the stresses in the midsection, away from any joint. Refer to the Appendix for the cross-sectional area of the members indicated in the figures. Consider all joints to be pinned.

1-51. Use Figure 1-24.

1-52. Use Figure 1-25.

1-53. Find the tensile stress in member *AB* shown in Figure 1-26.

1-54. Figure 1-27 shows the shape of a test specimen used to measure the tensile properties of metals (as described in Chapter 2). An axial tensile force is applied through the threaded ends and the test section is the reduced-diameter part near the middle. Compute the stress in the middle portion when the load is 12 600 lb.

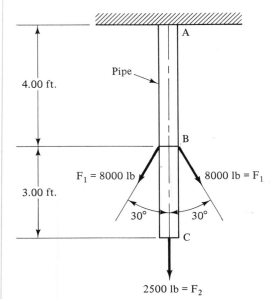

4.00 ft.

Pipe

A

B

$F_1 = 8000$ lb   8000 lb $= F_1$

3.00 ft.

30°   30°

C

2500 lb $= F_2$

Figure 1-22   Pipe for Problem 1-49.

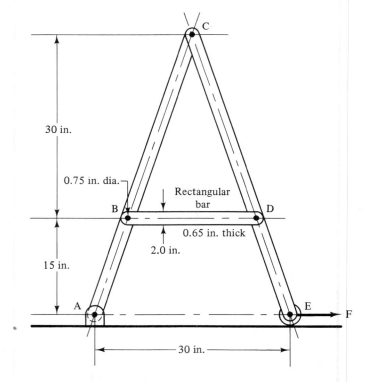

C

30 in.

0.75 in. dia.

Rectangular
bar

B   D

0.65 in. thick

2.0 in.

15 in.

A   E   F

30 in.

Figure 1-23   Frame for Problem 1-50.

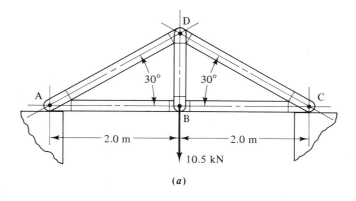

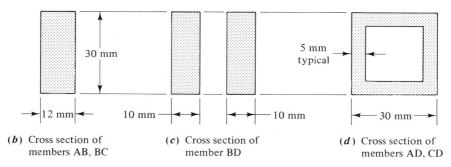

(*b*) Cross section of   (*c*) Cross section of   (*d*) Cross section of
members AB, BC   member BD   members AD, CD

Figure 1-24   Truss for Problem 1-51.

1-55. A short compression member has the cross section shown in Figure 1-28. Compute the stress in the member if a compressive force of 52 000 lb is applied in line with its centroidal axis.

1-56. A short compression member has the cross section shown in Figure 1-29. Compute the stress in the member if a compressive force of 640 kN is applied in line with its centroidal axis.

Direct Shearing Stresses

1-57. A clevis joint like that shown in Figure 1-5(b) is subjected to a force of 16.5 kN. Determine the shear stress in the 12.0-mm-diameter pin.

1-58. In a pair of pliers, the hinge pin is subjected to direct shear, as indicated in Figure 1-30. If the pin has a diameter of 3.0 mm and the force exerted at the handle, F_h, is 55 N, compute the stress in the pin.

1-59. For the centrifuge shown in Figure 1-19 and the data from Problem 1-46, compute the shear stress in the pin between the rod and the bucket.

1-60. A notch is made in a piece of wood, as shown in Figure 1-31, in order to support an external load F of 1800 lb. Compute the shear stress in the wood.

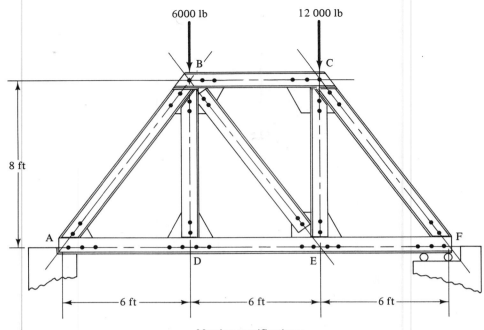

Member specifications:

AD, DE, EF   L2 × 2 × $\frac{1}{8}$ – doubled ——— ⌐L

BD, CE, BE   L2 × 2 × $\frac{1}{8}$ – single ——— ⌐

AB, BC, CF   C3 × 4.1 – doubled ——— ⌐⌐

Figure 1-25   Truss for Problem 1-52.

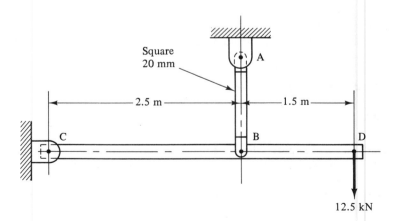

Figure 1-26   Support for Problem 1-53.

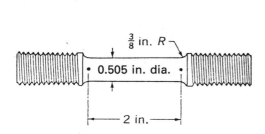

Figure 1-27 Tensile test specimen for Problem 1-54.

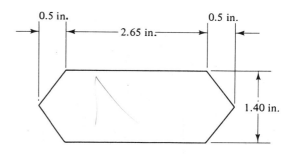

Figure 1-28 Short compression member for Problem 1-55.

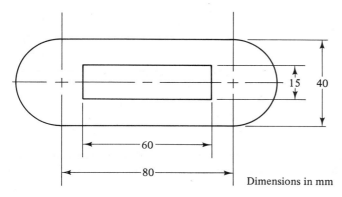

Dimensions in mm

Figure 1-29 Short compression member for Problem 1-56.

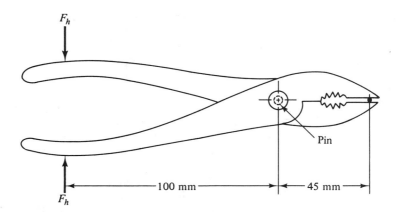

Figure 1-30 Pliers for Problem 1-58.

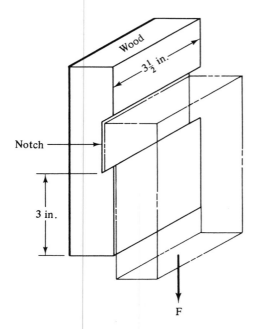

Figure 1-31 Wood notch loaded in shear for Problem 1-60.

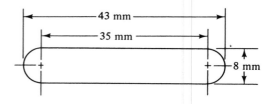

Figure 1-32 Shape of a slug for Problem 1-61.

1-61. Figure 1-32 shows the shape of a slug to be punched from a sheet of aluminum 5.0 mm thick. Compute the shear stress in the aluminum if a punching force of 38.6 kN is applied.

1-62. Figure 1-33 shows the shape of a slug to be punched from a sheet of steel 0.194 in. thick. Compute the shear stress in the steel if a punching force of 45 000 lb is applied.

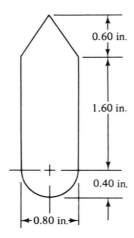

Figure 1-33 Shape of a slug for Problem 1-62.

1-63. The key in Figure 1-34 has the dimensions $b = 10$ mm, $h = 8$ mm, and $L = 22$ mm. Determine the shear stress in the key when 95 N·m of torque is transferred from the 35-mm-diameter shaft to the hub.

1-64. A key is used to connect a hub of a gear to a shaft, as shown in Figure 1-34. It has a rectangular cross section with $b = \frac{1}{2}$ in. and $h = \frac{3}{8}$ in. The length is 2.25 in. Compute the shear stress in the key when it transmits 8000 lb·in. of torque from the 2.0-in.-diameter shaft to the hub.

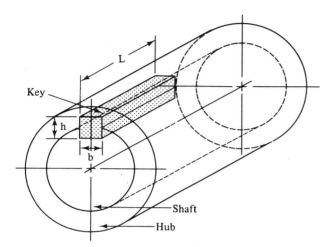

Figure 1-34 Key, shaft and hub for Problems 1-63 and 1-64.

1-65. A set of two tubes is connected in the manner shown in Figure 1-35. Under a compressive load of 20 000 lb, the load is transferred from the upper tube through the pin to the connector, then through the collar to the lower tube. Compute the shear stress in the pin and in the collar.

1-66. A small, hydraulic crane like that shown in Figure 1-36 carries an 800-lb load. Determine the shear stress that occurs in the pin at B which is in double shear. The pin diameter is $\frac{3}{8}$ in.

1-67. A ratchet device on a jack stand for a truck has a tooth configuration as shown in Figure 1-37. For a load of 88 kN compute the shear stress at the base of the tooth.

1-68. Figure 1-38 shows an assembly in which the upper block is brazed to the lower block. Compute the shear stress in the brazing material if the force is 88.2 kN.

1-69. Figure 1-39 shows a bolt subjected to a tensile load. One failure mode would be if the circular shank of the bolt pulled out from the head, a shearing action. Compute the shear stress in the head for this mode of failure if a force of 22.3 kN is applied.

1-70. Figure 1-40 shows a riveted lap joint connecting two steel plates. Compute the shear stress in the rivets due to a force of 10.2 kN applied to the plates.

1-71. Figure 1-41 shows a riveted butt joint with cover plates connecting two steel plates. Compute the shear stress in the rivets due to a force of 10.2 kN applied to the plates.

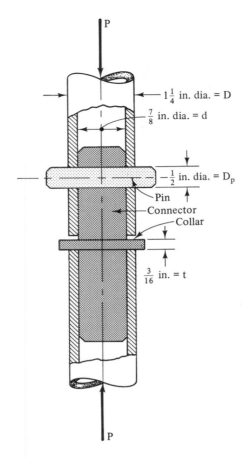

$1\frac{1}{4}$ in. dia. = D

$\frac{7}{8}$ in. dia. = d

$-\frac{1}{2}$ in. dia. = D$_p$

Pin

Connector

Collar

$\frac{3}{16}$ in. = t

P

Figure 1-35 Connector for Problem 1-65.

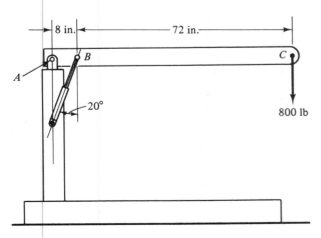

8 in.

72 in.

A

B

C

20°

800 lb

Figure 1-36 Hydraulic crane for Problem 1-66.

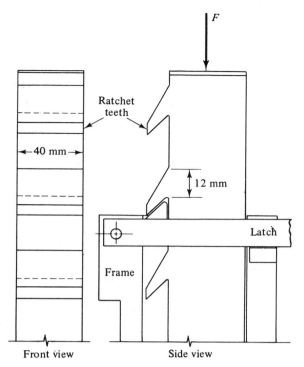

Front view   Side view

Figure 1-37 Ratchet for a jack stand for Problem 1-67.

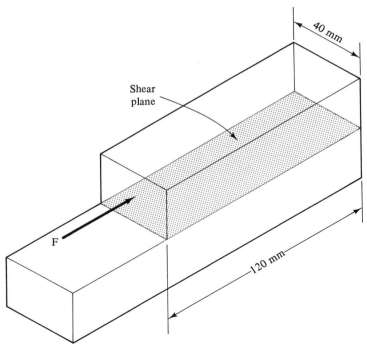

Figure 1-38 Brazed components for Problem 1-68.

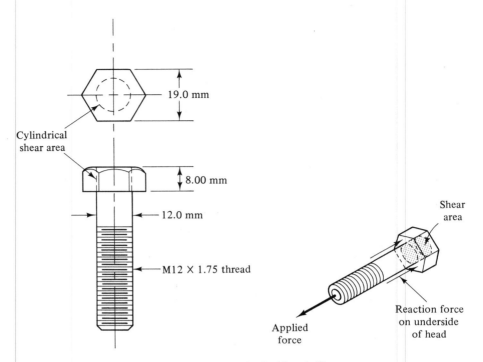

Figure 1-39 Bolt for Problem 1-69.

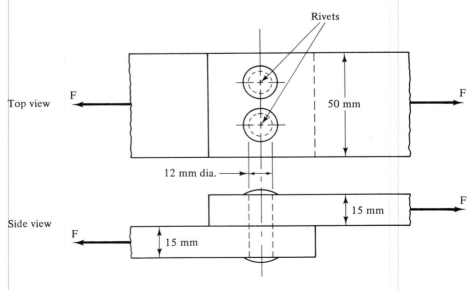

Figure 1-40 Riveted lap joint for Problem 1-70.

Basic Concepts in Strength of Materials   Chap. 1

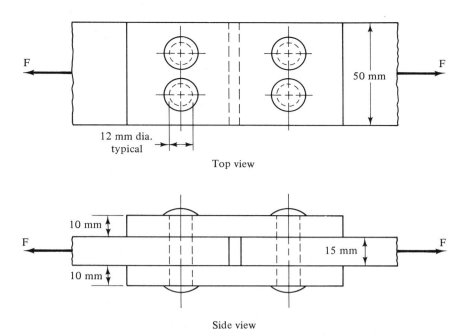

Figure 1-41 Riveted butt joint for Problem 1-71.

Bearing Stress

1-72. Compute the bearing stresses at the mating surfaces A, B, C, and D, in Figure 1-42.

1-73. A 2-in. schedule 40 steel pipe is used as a leg for a machine. The load carried by the leg is 2350 lb.

   (a) Compute the bearing stress on the floor if the pipe is left open at its end.

   (b) Compute the bearing stress on the floor if a flat plate is welded to the bottom of the pipe having a diameter equal to the outside diameter of the pipe.

1-74. A bolt and washer are used to fasten a wooden board to a concrete foundation as shown in Figure 1-43. A tensile force of 385 lb is created in the bolt as it is tightened. Compute the bearing stress (a) between the bolt head and the steel washer, and (b) between the washer and the wood.

1-75. For the data of Problem 1-64, compute the bearing stress on the side of the key.

1-76. For the data of Problem 1-65, compute the bearing stress on the tube at the interfaces with the pin and the collar.

1-77. For the data of Problem 1-70, compute the bearing stress on the rivets.

1-78. For the data of Problem 1-71, compute the bearing stress on the rivets.

1-79. The heel of a woman's shoe has the shape shown in Figure 1-44. If the force on the heel is 535 N, compute the bearing stress on the floor.

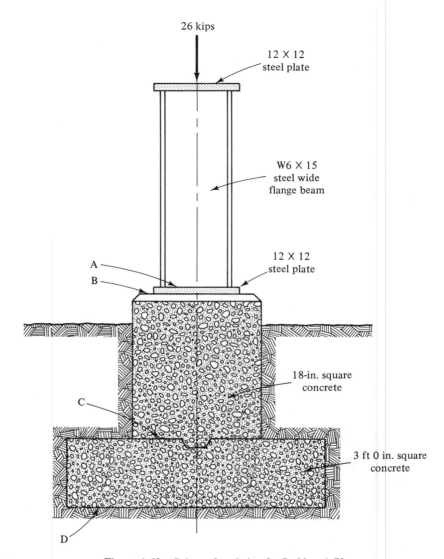

Figure 1-42 Column foundation for Problem 1-72.

Basic Concepts in Strength of Materials   Chap. 1

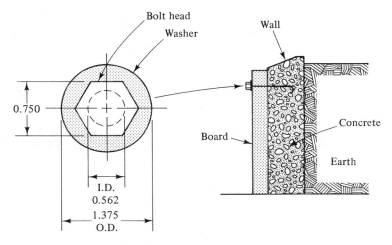

Washer dimensions (in.)

Figure 1-43 Bolt and washer for Problem 1-74.

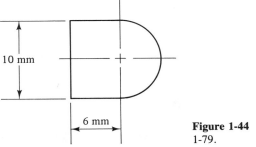

Figure 1-44 Shoe heel for Problem 1-79.

2

Design Properties of Materials

2-1 OBJECTIVES OF THIS CHAPTER

The study of strength of materials requires a knowledge of how external forces and moments affect the stresses and deformations developed in the material of a load-carrying member. In order to put this knowledge to practical use, however, a designer needs to know how such stresses and deformations can be withstood safely by the material. Thus, material properties as they relate to design must be understood along with the analysis required to determine the magnitude of stresses and deformations.

In this chapter we present information concerning the materials most frequently used to make components for structures and mechanical devices, emphasizing the design properties of the materials rather than their metallurgical structure or chemical composition. Although it is true that a thorough knowledge of the structure of materials is an aid to a designer, it is most important to know how the materials behave when carrying loads. This is the behavior on which we concentrate in this chapter.

First, we discuss metals, the most widely used materials in engineering design. The important properties of metals are described, along with the special characteristics of several different metals.

Nonmetals presented include wood, concrete, plastics, and composites. The manner in which the behavior of these materials differs from that of metals is discussed, along with some of their special properties.

40

After completing this chapter, you should be able to:

1. List typical uses for engineering materials.
2. Define *ultimate tensile strength*.
3. Define *yield point*.
4. Define *yield strength*.
5. Define *elastic limit*.
6. Define *proportional limit*.
7. Define *modulus of elasticity* and describe its relationship to the stiffness of materials
8. Define *Hooke's law*.
9. Describe ductile and brittle behavior of materials.
10. Define *percent elongation* and describe its relationship to the ductility of materials.
11. Describe the Unified Numbering System (UNS) for metals and alloys.
12. Describe the four-digit designation system for steels.
13. Describe the important properties of carbon steels, alloy steels, stainless steels, and structural steels.
14. Describe the four-digit designation system for wrought and cast aluminum alloys.
15. Describe the aluminum temper designations.
16. Describe the design properties of copper, brass, bronze, zinc, magnesium, and titanium.
17. Describe the design properties of cast irons, including gray iron, ductile iron, austempered ductile iron, malleable iron, and white iron.
18. Describe the design properties of wood, concrete, plastics, and composites.

2-2 METALS IN MECHANICAL DESIGN

Metals are most widely used for load-carrying members in buildings, bridges, machines, and a wide variety of consumer products. Beams and columns in commercial buildings are made of structural steel or aluminum. In automobiles, a large number of steels are used, including carbon steel sheet for body panels, free-cutting alloys for machined parts, and high strength alloys for gears and heavily loaded parts. Cast iron is used in engine blocks, brake drums, and cylinder heads. Tools, springs, and other parts requiring high hardness and wear resistance are made from steel alloys containing a large amount of carbon. Stainless steels are used in transportation equipment, chemical plant products, and kitchen equipment where resistance to corrosion is required.

Aluminum sees many of the same applications as steel. Aluminum is used in many architectural products and frames for mobile equipment. Its corrosion resistance allows its use in chemical storage tanks, cooking utensils, marine equipment, and products such as highway signposts. Automotive pistons, trim, and die-cast housings for pumps and alternators are made of aluminum. Aircraft structures, engine parts, and sheet-metal skins use aluminum because of its high strength-to-weight ratio.

Copper and its alloys such as brass and bronze are used in electric conductors, heat exchangers, springs, bushings, marine hardware, and switch parts. Magnesium is often cast into truck parts, wheels, and appliance parts. Zinc sees similar service and may also be forged into machinery components and industrial hardware. Titanium has a high strength-to-weight ratio and good corrosion resistance, and thus is used in aircraft parts, pressure vessels, and chemical equipment.

Material selection requires consideration of many factors. Generally, strength, stiffness, ductility, weight, corrosion resistance, machinability, workability, weldability, weight, appearance, cost, and availability must all be evaluated. Relative to the study of strenght of materials, the first three of these factors are most important: strength, stiffness, and ductility.

Strength. Reference data listing the mechanical properties of metals will almost always include the *ultimate tensile strength* and *yield strength* of the metal. Comparison of the actual stresses in a part with the ultimate or yield strength of the material from which the part is made is the usual method of evaluating the suitability of the material to carry the applied loads safely. More is said about the details of stress analysis in Chapter 3 and subsequent chapters.

The ultimate tensile strength and yield strength are determined by testing a sample of the material in a tensile-testing machine such as that shown in Figure 2-1.

Figure 2-1 Universal testing machine for obtaining stress-strain data for materials. (Source: Tinius Olsen Testing Machine Co., Inc., Willow Grove, Pa.)

Figure 2-2 Tensile test specimen mounted in a holder. (Source: Tinius Olsen Testing Machine Co., Inc., Willow Grove, Pa.)

A round bar or flat strip is placed in the upper and lower jaws. Figure 2-2 shows a photograph of a typical tensile test specimen. A pulling force is applied slowly and steadily to the sample, stretching it until it breaks. During the test, a graph is made which shows the relationship between the stress in the sample and the strain or unit deformation. A typical stress–strain diagram for a low-carbon steel is shown in Figure 2-3. It can be seen that during the first phase of loading, the plot of stress versus strain is a straight line, indicating that stress is directly proportional to strain. After point A on the diagram, the curve is no longer a straight line. This point is called the *proportional limit*. As the load on the sample is continually increased, a point called the *elastic limit* is reached, marked B in Figure 2-3. At stresses below this point, the material will return to its original size and shape if the load is removed. At higher stresses, the material is permanently deformed. The *yield point* is the stress at which a noticeable elongation of the sample occurs with no apparent increase in load. The yield point is at C in Figure 2-3, about 36 000 psi (248 MPa). Applying still higher loads after the yield point has been reached causes the curve to rise again. After reaching a peak, the curve drops somewhat until finally the sample breaks, terminating the plot. The highest apparent stress taken from the stress–strain diagram is called the *ultimate strength*. In Figure 2-3 the ultimate strength would be about 53 000 psi (365 MPa).

The fact that the stress–strain curves in Figures 2-3 and 2-4 drop off after reaching a peak tends to indicate that the stress level decreases. Actually, it does not; the *true stress* continues to rise until ultimate failure of the material. The reason for the apparent decrease in stress is that the plot taken from a typical tensile test machine is actually *load versus elongation* rather than *stress versus strain*. The vertical axis is converted to stress by dividing the load (force) on the specimen by the *original* cross-sectional area of the specimen. When the specimen nears its breaking load, there is a reduction in diameter and consequently a reduction in the cross-sectional area. The reduced area requires a lower force to continue stretching the specimen, even though the actual stress in the material is increasing. This results in the dropping curve shown in Figures 2-3 and 2-4. Because it is very difficult to monitor the decreasing diameter,

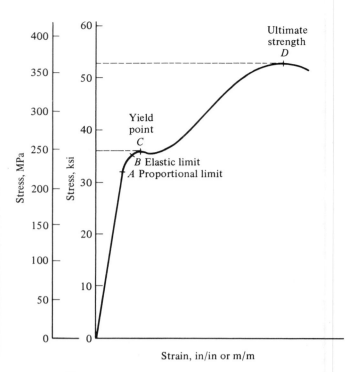

Figure 2-3 Typical stress-strain curve for steel.

and because experiments have shown that there is little difference between the true maximum stress and that found from the peak of the *apparent stress* versus strain curve, the peak is accepted as the ultimate tensile strength of the material.

Many metals do not exhibit a well-defined yield point like that in Figure 2-3. Some examples are high-strength alloy steels, aluminum, and titanium. However, these materials do in fact yield in the sense of deforming a sizable amount before fracture actually occurs. For these materials, a typical stress–strain diagram would look as shown in Figure 2-4. The curve is smooth with no pronounced yield point. For such materials, the yield strength is defined by a line like *M-N* drawn parallel to the straight-line portion of the test curve. Point *M* is usually determined by finding that point on the strain axis representing a strain of 0.002 in./in. This point is also called the point of 0.2% offset. The point *N*, where the offset line intersects the curve, defines the yield strength of the material, about 55 000 psi in Figure 2-4. The ultimate strength is at the peak of the curve, as was described before. *Yield strength* is used in place of yield point for these materials.

In most wrought metals, the behavior of the materials in compression is similar to that in tension and so separate compression tests are not usually performed. However, for cast materials and nonhomogeneous materials such as wood and concrete, there are large differences between the tensile and compressive properties and compressive testing should be done.

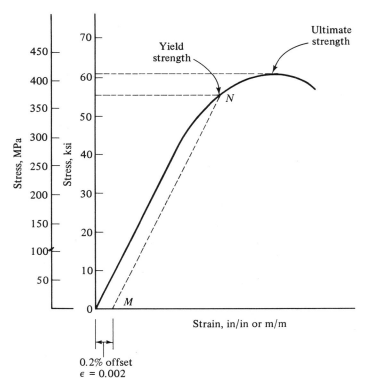

Figure 2-4 Typical stress-strain curve for aluminum.

Stiffness. It is frequently necessary to determine how much a part will deform under load in order to ensure that excessive deformation does not destroy the usefulness of the part. This can occur at stresses well below the yield strength of the material, especially in very long members or in high-precision devices. Stiffness of a material is a function of its *modulus of elasticity*, sometimes called *Young's modulus*. The symbol E is used for modulus of elasticity, which is measured by determining the slope of the stress–strain curve during the first part, where it is a straight line. This can be stated mathematically as

$$E = \frac{\text{stress}}{\text{strain}} = \frac{\sigma}{\epsilon} \tag{2-1}$$

Therefore, a material having a steeper slope on its stress–strain curve will be stiffer and will deform less under load than a material having a less steep slope. Figure 2-5 illustrates this concept by showing the straight-line portions of the stress–strain curves for steel, titanium, aluminum, and magnesium. It can be seen that if two otherwise identical parts were made of steel and aluminum, respectively, the aluminum part would deform about three times as much when subjected to the same load.

Expressing Equation (2-1) as $\sigma = E\epsilon$, we can say that stress is proportional to strain as long as the stress is below the proportional limit of the material. This is called

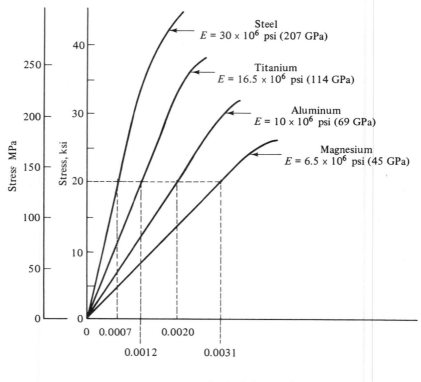

Figure 2-5 Modulus of elasticity for different metals.

Hooke's law. The assumption that stress is proportional to strain is used in developing many of the formulas in strength of materials.

Ductility. When metals break, their fracture can be classified as either ductile or brittle. A ductile material will stretch and yield prior to actual fracture, causing a noticeable decrease in the cross-sectional area at the fractured section. Conversely, a brittle material will fracture suddenly with little or no change in the area at the fractured section. Ductile materials are preferred for parts that carry repeated loads or are subjected to impact loading because they are usually more resistant to fatigue failure and because they are better at absorbing impact energy.

Ductility in metals is usually measured during the tensile test by noting how much the material has elongated permanently after fracture. At the start of the test, a set of gage marks is scribed on the test sample as shown in Figure 2-6. Most tests use 2.000 in. or 50.0 mm for the gage length as shown in the figure. Very ductile structural steels sometimes use 8.000 in. or 200.0 mm for the gage length. After the sample has been pulled to failure, the broken parts are fitted back together and the distance between the marks is again measured. From these data, the *percent elongation* is

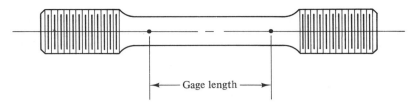

Gage length

Typically 2.00 in. or 50.0 mm

(*a*)

(*b*)

Figure 2-6 Gage length for tensile test specimen. (a) Gage length marked on a test specimen (b) Test specimen in a fixture used to mark gage length (Source: Tinius Olsen Testing Machine Co., Inc., Willow Grove, Pa.)

computed from

$$\text{percent elongation} = \frac{\text{final length - gage length}}{\text{gage length}} \times 100\% \qquad (2\text{-}2)$$

A metal is considered to be *ductile* if its percent elongation is greater than about 5.0%. A material with a percent elongation under 5.0% is considered to be *brittle* and does not exhibit the phenomenon of yielding. Failure of such materials is suddden, without noticeable deformation prior to ultimate fracture. In most structural and mechanical design applications, ductile behavior is desirable and the percent elongation of the material should be significantly greater than 5.0%. A high-percentage elongation indicates a highly ductile material.

Virtually all wrought forms of steel and aluminum alloys are ductile. But the higher-strength forms tend to have lower ductility and the designer is often forced to compromise strength and ductility in the specification of a material. Gray cast iron, many forms of cast aluminum, and some high-strength forms of wrought or cast steel are brittle.

Failure modes. Obviously, in most designs, a part is considered to have failed if it breaks or if it deforms excessively and permanently by yielding. Therefore, the ultimate tensile strength and yield strength are the two most important properties of the material.

Deformation of the material before yielding occurs is dependent on the stiffness of the material, indicated by the modulus of elasticity. Methods for computing the total deformation of load-carrying members are presented in later chapters. There are no absolute standards for the level of deformation that would constitute failure. Rather, the designer must make a judgment based on the use of the structure or machine. Reference 13 presents some guidelines.

Classifiction of metals and alloys. Various industry associations take responsibility for setting standards for the classification of metals and alloys. Each has its own numbering system, convenient to the particular metal covered by the standard. But this leads to confusion at times when there is overlap between two of more standards and when widely different schemes are used to denote the metals. Order has been brought to the classification of metals by the use of the Unified Numbering Systems (UNS) as defined in the Standard E 527-74 (Reapproved 1981), **Standard Practice for Numbering Metals and Alloys (UNS)**, by the American Society for Testing and Materials (ASTM). Besides listing materials under the control of ASTM itself, the UNS coordinates designations of:

The Aluminum Association (AA)
American Iron and Steel Institute (AISI)
Copper Development Association (CDA)
Society of Automotive Engineers (SAE)

The primary series of numbers within UNS are listed in Table 2-1 along with the organization having responsibility for assigning numbers within each series.

Many alloys within UNS retain the familiar numbers from the systems used for many years by the individual association. For example, the following section describes the four-digit designation system of the AISI for carbon and alloy steels. Figure 2-7 shows two examples; AISI 1020, a carbon steel, and AISI 4140, an alloy steel. These steels would carry the UNS designations, G10200 and G41400 respectively.

TABLE 2-1 UNIFIED NUMBERING SYSTEM (UNS)

Number Series	Types of Metals and Alloys	Responsible Organization
Nonferrous Metals and Alloys		
A00001-A99999	Aluminum and aluminum alloys	AA
C00001-C99999	Copper and copper alloys	CDA
E00001-E99999	Rare earth metals and alloys	ASTM
L00001-L99999	Low melting metals and alloys	ASTM
M00001-M99999	Misc. nonferrous metals and alloys	ASTM
N00001-N99999	Nickel and nickel alloys	SAE
P00001-P99999	Precious metals and alloys	ASTM
R00001-R99999	Reactive and refractory metals and alloys	SAE
Z00001-Z99999	Zinc and zinc alloys	ASTM
Ferrous Metals and Alloys		
D00001-D99999	Steels, mechanical properties specified	SAE
F00001-F99999	Cast irons and cast steels	ASTM
G00001-G99999	Carbon and alloy steels (Includes former SAE carbon and alloy steels)	AISI
H00001-H99999	H-steels; specified hardenability	AISI
J00001-J99999	Cast steels (except tool steels)	ASTM
K00001-K99999	Misc. steels and ferrous alloys	ASTM
S 00001-S99999	Heat and corrosion resistant (stainless) steels	ASTM
T00001-T99999	Tool steels	AISI

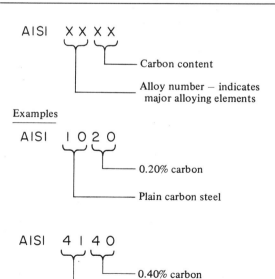

Examples

Figure 2-7 Steel Designation.

2-3 STEEL

The term *steel* refers to alloys of iron and carbon and, in many cases, other elements. Because of the large number of steels available, they will be classified in this section as carbon steels, alloy steels, stainless steels, and structural steels.

For *carbon steels* and *alloy steels*, a four-digit designation code is used to define each alloy. Figure 2-7 shows the significance of each digit. The four digits would be the same for steels classified by the American Iron and Steel Institute (AISI) and the Society of Automotive Engineers (SAE). Classification by the American Society for Testing and Materials (ASTM) will be discussed later.

Usually, the first two digits in a four-digit designation for steel will denote the major alloying elements, other than carbon, in the steel. The last two digits denote the average percent (or points) of carbon in the steel. For example, if the last two digits are 40, the steel would have about 0.4% carbon content. Carbon is given such a prominent place in the alloy designation because, in general, as carbon content increases, the strength and hardness of the steel also increases. Carbon content usually ranges from a low of 0.1% to about 1.0%. It should be noted that while strength increases with increasing carbon content, the steel also becomes more brittle.

Table 2-2 shows the major alloying elements, which correspond to the first two digits of the steel designation. Table 2-3 lists come common alloys along with the principal uses for each.

Conditions for steels. The mechanical properties of carbon and alloy steels are very sensitive to the manner in which they are formed and to heat-treating processes. Appendix A-13 lists the ultimate strength, yield strength, and percent elongation for a variety of steels in a variety of conditions. Note that these are typical or example properties and may not be relied on for design. Material properties are dependent on many factors, including section size, temperature, actual composition, variables in processing, and fabrication techniques. It is the responsibility of the designer to investigate the possible range of properties for a material and to design load-carrying members to be safe regardless of the combination of factors present in a given situation.

TABLE 2-2 MAJOR ALLOYING ELEMENTS IN STEEL ALLOYS

Steel AISI No.	Alloying elements	Steel AISI No.	Alloying elements
10xx	Plain carbon	46xx	Molybdenum-nickel
11xx	Sulfur (free-cutting)	47xx	Molybdenum-nickel-chromium
13xx	Manganese	48xx	Molybdenum-nickel
14xx	Boron	5xxx	Chromium
2xxx	Nickel	6xxx	Chromium-vanadium
3xxx	Nickel-chromium	8xxx	Nickel-chromium-molybdenum
4xxx	Molybdenum	9xxx	Nickel-chromium-molybdenum (except 92xx)
41xx	Molybdenum-chromium	92xx	Silicon-manganese
43xx	Molybdenum-chromium-nickel		

TABLE 2-3 COMMON STEEL ALLOYS AND TYPICAL USES

Steel AISI No.	Typical uses
1020	Structural steel, bars, plate
1040	Machinery parts, shafts
1050	Machinery parts
1095	Tools, springs
1137	Shafts, screw machine parts (free-cutting alloy)
1141	Shafts, machined parts
4130	General-purpose, high-strength steel; shafts, gears, pins
4140	Same as 4130
4150	Same as 4130
5160	High-strength gears, bolts
8760	Tools, springs, chisels

Generally, the more severely a steel is worked, the stronger it will be. Some forms of steel, such as sheet, bar, and structural shapes, are produced by *hot rolling* while still at an elevated temperature. This produces a relatively soft, low-strength steel which has a very high ductility and is easy to form. Rolling the steel to final form while at or near room temperature is called *cold rolling* and produces a higher strength and somewhat lower ductility. Still higher strength can be achieved by *cold drawing*, drawing the material through dies while it is at or near room temperature. Thus, for these three popular methods of producing steel shapes, the cold-drawn (CD) form results in the highest strength, followed by the cold-rolled (CR) and hot-rolled (HR) forms. This can be seen in Appendix A-13 by comparing the strength of the same steel, say AISI 1040, in the hot-rolled and cold-drawn conditions.

Alloy steels are usually heat treated to develop specified properties. Heat treatment involves raising the temperature of the steel to above about 1450 to 1650°F (depending on the alloy) and then cooling it rapidly by quenching in either water or oil. After quenching, the steel has a high strength and hardness, but it may also be brittle. For this reason a subsequent treatment called *tempering* (or *drawing*) is usually performed. The steel is reheated to a temperature in the range 400 to 1300°F and then cooled. The effect of tempering an alloy steel can be seen by referring to Figure 2-8. Thus the properties of a heat-treated steel can be controlled by specifying a tempering temperature. In Appendix A-13, the condition of heat-treated alloys is described in a manner like OQT 400. This means that the steel was heat treated by quenching in oil and then tempered at 400°F. Similarly, WQT 1300 means water quenched and tempered at 1300°F.

The properties of heat-treated steels at tempering temperatures of 400 and 1300°F shows the total range of properties that the heat-treated steel can have. However, typical practice would specify tempering temperatures no lower than 700°F because steels tend to be too brittle at the lower tempering temperatures. The properties of several alloys are listed in Appendix A-13 at tempering temperatures of 700°F, 900°F, 1100°F, and 1300°F to give you a feeling for the range of strengths available. These alloys are good choices for selecting materials in problems in later chapters. Strengths at intermediate temperatures can be found by interpolation.

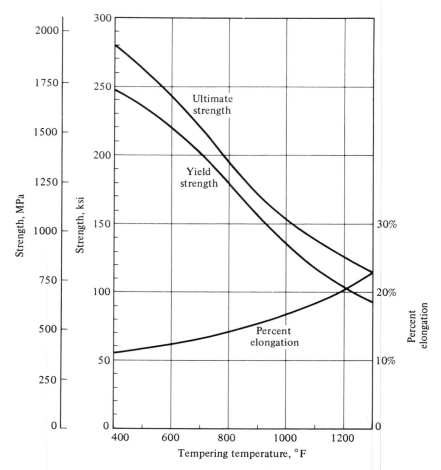

Figure 2-8 Effect of tempering temperature on the strength and ductility of an alloy steel.

Annealing and *normalizing* are thermal treatments designed to soften steel, give it more uniform properties, make it easier to form, or relieve stresses developed in the steel during such processes as welding, machining, or forming. Two of the types of annealing processes used are full anneal and stress-relief anneal. Figure 2-9 illustrates these thermal treatment processes, along with quenching and tempering.

Normalizing of steel starts by heating it to approximately the same temperature (called the *upper critical temperature*) as would be required for through-hardening by quenching as described before. But rather than quenching, the steel is cooled in still air to room temperature. This results in a uniform, fine-grained structure, improved ductility, better impact resistance, and improved machinability.

Full annealing involves the heating to above the upper critical temperature followed by very slow cooling to the lower critical temperature and then cooling in

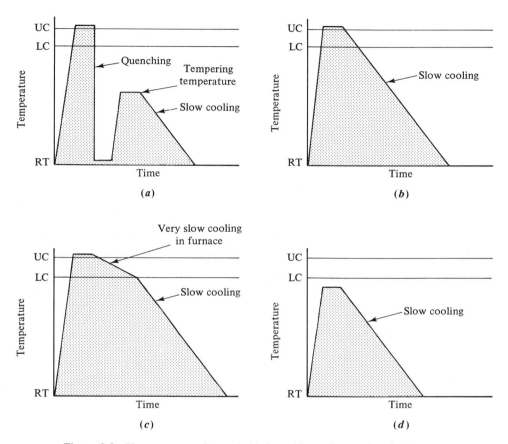

Note:
RT = room temperature
LC = lower critical temperature
UC = upper critical temperature

Figure 2-9 Heat treatments for steel. (a) Quenching and tempering (b) Normalizing (c) Full Annealing (d) Stress relief annealing

still air to room temperature. This is one of the softest forms of the steel and makes it more workable for shearing, forming, and machining.

Stress-relief annealing consists of heating to below the lower critical temperature, holding to achieve uniform temperature throughout the part, and then cooling to room temperature. This relieves residual stresses and prevents subsequent distortion.

Stainless steels get their name because of their corrosion resistance. The primary alloying element in stainless steels is chromium, being present at about 17% in most alloys. A minimum of 10% chromium is used, and it may range as high as 27%.

Although over 40 grades of stainless steel are available from steel producers, they are usually categorized into three series containing alloys with similar properties. The 200 and 300 series steels have high strength and good corrosion resistance. They can be used at temperatures up to about 1200°F with good retention of properties. Because of their structure, these steels are essentially nonmagnetic. Good ductility and toughness and good weldability make them useful in chemical processing equipment, architectural products, and food-related products. They are not hardenable by heat treatment, but they can be strengthened by cold working.

The AISI 400 series steels are used for automotive trim and for chemical processing equipment such as acid tanks. Certain alloys can be heat-treated so they can be used as knife blades, springs, ball bearings, and surgical instruments. These steels are magnetic. The properties of some stainless steels are listed in Appendix A-14.

Precipitation hardening steels, such as 17-4PH and PH13-8Mo, are hardened by holding at an elevated temperature, about 900 to 1100°F (480 to 600°C). Such steels are generally classed as *high-strength stainless steels* having yield strengths about 180 000 psi (1240 MPa).

Structural steels are produced in the forms of sheet, plate, bars, tubing, and structural shapes such as I-beams, wide-flange beams, channels, and angles. The American Society for Testing and Materials (ASTM) assigns a numbr designation to these steels which is the number of the standard that defines the required minimum properties. Appendix A-15 lists four frequently used grades of structural steels and their properties.

A very popular steel for structural applications is ASTM A36, a carbon steel used for many commercially available shapes, plates, and bars. It has a minimum yield point of 36 kips per square inch (ksi) (248 MPa), is weldable, and is used in bridges, buildings, and for general structural purposes.

Steels called ASTM A242, A440, and A441 are members of a class called high-strength, low-alloy steels. Produced as shapes, plates, and bars, they may be specified instead of A36 steel to allow the use of a smaller, lighter member. In sizes up to $\frac{3}{4}$ in. thick, they have a minimum yield point of 50 ksi (345 MPa). From $\frac{3}{4}$ to $1\frac{1}{2}$ in. thick, the minimum yield point of 46 ksi (317 MPa) is specified. Alloy A242 is for general structural purposes and is called a *weathering steel* since its corrosion resistance is about four times that of plain carbon steel. Alloy A441 is used primarily for welded construction, and alloy A440 is used in riveted and bolted construction. Cost, of course, must be considered before specifying these alloys.

Alloy A514 is a high-strength alloy, heat treated by quenching and tempering to a yield point of 100 ksi (690 MPa) minimum. Produced as plates, it is used in welded bridges and similar structures.

Structural tubing is either round, square, or rectangular and is frequently made from ASTM A501 steel (hot formed) or ASTM A500 (cold formed) (see Appendixes A-9 and A-15).

Another general-purpose structural steel is ASTM A572, available as shapes, plates, and bars, and in grades 42 to 65. The grade number refers to the minimum yield point of the grade in ksi, and can be 42, 45, 50, 55, 60, and 65.

In summary, steels come in many forms and have a wide range of strengths and other properties. Selection of a suitable steel is indeed an art, supported by a knowledge of the significant features of each alloy.

2-4 CAST IRON

The attractive properties of cast iron include low cost, good wear resistance, good machinability, and its ability to be cast into complex shapes. Five varieties will be discussed here: gray iron, ductile iron, austempered ductile iron, malleable iron, and white iron.

Gray iron is used in automotive engine blocks, machinery bases, brake drums, and large gears. It is usually specified by giving a grade number corresponding to the minimum ultimate tensile strength. For example, grade 20 gray cast iron has a minimum ultimate strength of 20 000 psi (138 MPa); grade 60 has $s_u = 60\ 000$ psi (414 MPa), and so on. The usual grades available are from 20 to 60. Gray iron is somewhat brittle, so that yield strength is not usually reported as a property. An outstanding feature of gray iron is that its compressive strength is very high, about three to five times as high as the tensile strength. This should be taken into account in design, especially when a part is subjected to bending stresses, as discussed in Chapter 8.

Because of variations in the rate of cooling after the molten cast iron is poured into a mold, the actual strength of a particular section of a casting is dependent on the thickness of the section. Figure 2-10 illustrates this for grade 40 gray iron. The range

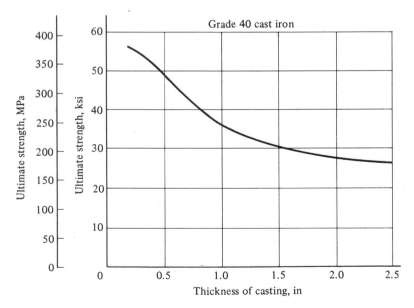

Figure 2-10 Strength versus thickness for grade 40 gray cast iron.

of in-place strength may range from as high as 52 000 psi (359 MPa) to as low as 27 000 psi (186 MPa).

Ductile iron differs from gray iron in that it does exhibit yielding, has a greater percent elongation, and has generally higher tensile strength. Grades of ductile iron are designated by a three-number system such as 80-55-6. The first number indicates the minimum ultimate tensile strength in ksi; the second is the yield strength in ksi; and the third is the percent elongation. Thus grade 80-55-6 has an ultimate strength of 80 000 psi, a yield strength of 55 000 psi, and a percent elongation of 6%. Uses for ductile iron include crankshafts and heavily loaded gears.

The strength of ductile iron can be incresed by nearly a factor of 2 by a process called *austempering*. First the castings are heated to between 1500 and 1700°F and held to achieve a uniform structure. Then they are quenched rapidly to a lower temperature, 450 to 750°F, and held again. After several hours of isothermal soaking, the castings are allowed to cool to room temperture.

Austempered ductile iron (ADI) has higher strength and better ductility than standard ductile irons, as can be seen in Appendix A-16. This allows parts to be smaller and lighter and makes ADI desirable for uses such as automotive gears, crankshafts, and structural members for construction and transporation equipment, replacing wrought or cast steels.

White iron is produced by rapidly chilling a casting of either gray iron or ductile iron during the solidification process. The chilling is typically applied to selected areas which become very hard and have a high wear resistance. The chilling does not allow the carbon in the iron to precipitate out during solidification, resulting in the white appearance. Areas away from the chilling medium solidify more slowly and acquire the normal properties of the base iron. One disadvantage of the chilling process is that the white iron is very brittle.

Malleable iron is used in automotive and truck parts, construction machinery, and electrical equipment. It does exhibit yielding, has tensile strengths comparable to ductile iron, and has ultimate compressive strengths which are somewhat higher than ductile iron. Generally, a five-digit number is used to designate malleable iron grades. For example, grade 40010 has a yield strength of 40 000 psi (276 MPa) and a percent elongation of 10%.

Appendix A-16 lists the mechanical properties of several grades of gray iron, ductile iron, ADI, and malleable iron.

2-5 ALUMINUM

Alloys of aluminum are designed to achieve optimum properties for specific uses. Some are produced primarily as sheet, plate, bars, or wire. Standard structural shapes and special sections are often extruded. Several alloys are used for forging, while others are special casting alloys. Appendix A-17 lists the properties of selected aluminum alloys.

Aluminum in wrought form uses a four-digit designation to define the several alloys available. The first digit indicates the alloy group according to the principal alloying element. The second digit denotes a modification of the basic alloy. The last two digits identify a specific alloy within the group. A brief description of the seven major series of aluminum alloys follows.

1000 series, 99.0% aluminum or greater. Used in chemical and electrical fields. Excellent corrosion resistance, workability, and thermal and electrical conductivity. Low mechanical properties.

2000 series, copper alloying element. Heat–treatable with high mechanical properties. Lower corrosion resistance than most other alloys. Used in aircraft skins and structures.

3000 series, manganese alloying element. Non–heat–treatable, but moderate strength can be obtained by cold working. Good corrosion resistance and workability. Used in chemical equipment, cooking utensils, residential siding, and storage tanks.

4000 series, silicon alloying element. Non–heat–treatable with a low melting point. Used as welding wire and brazing alloy. Alloy 4032 used as pistons.

5000 series, magnesium alloying element. Non–heat–treatable, but moderate strength can be obtained by cold working. Good corrosion resistance and weldability. Used in marine service, pressure vessels, auto trim, builder's hardware, welded structures, TV towers, and drilling rigs.

6000 series, silicon and magnesium alloying elements. Heat–treatable to moderate strength. Good corrosion resistance, formability, and weldability. Used as heavy-duty structures, truck and railroad equipment, pipe, furniture, architectural extrusions, machined parts, and forgings. Alloy 6061 is one of the most versatile available.

7000 series, zinc alloying element. Heat–treatable to very high strength. Relatively poor corrosion resistance and weldability. Used mainly for aircraft structural members. Alloy 7075 is among the highest strength alloys available. It is produced in most rolled, drawn, and extruded forms and is also used in forgings.

Aluminum Temper Desigantions.   Since the mechanical properties of virtually all aluminum alloys are very sensitive to cold working or heat treatment, suffixes are applied to the four-digit alloy designations to describe the temper. The most frequently used temper designations are described as follows:

O temper. Fully annealed to obtain the lowest strength. Annealing makes most alloys easier to form by bending or drawing. Parts formed in the annealed condition are frequently heat–treated later to improve properties.

H temper, strain-hardened. Used to improve the properties of non–heat–treatable alloys such as those in the 1000, 3000, and 5000 series. The H is always followed by a two- or three-digit number to designate a specific degree of strain hardening or special processing. The second digit following the H ranges from 0 to 8 and indicates a successively greater degree of strain hardening, resulting in higher strength. Appendix A-17 lists the properties of several aluminum alloys. Referring to alloy 3003 in that table shows that the yield strength is increased from 18 000 psi (124 MPa) to 27 000 psi (186 MPa) as the temper is changed from H12 to H18.

T temper, heat-treated. Used to improve strength and achieve a stable condition. The T is always followed by one or more digits indicating a particular heat treatment. For wrought products such as sheet, plate, extrusions, bars, and drawn tubes, the most frequently used designations are T4 and T6. The T6 treatment produces higher strength but generally reduces workability. In Appendix A-17 several heat-treatable alloys are listed in the O, T4, and T6 tempers to illustrate the change in properties.

Cast aluminum alloys are designated by a modified four-digit system of the form, XXX.X, in which the first digit indicates the main alloy group according to the major alloying elements. Table 2-4 shows the groups. The second two digits indicate the specific alloy within the group or indicate the aluminum purity. The last digit, after the decimal point, indicates the product form; 0 for castings, and 1 or 2 for ingots.

Aluminum is also sensitive to the manner in which it is produced. The size of the section, and temperature. Appendix A-17 lists typical properties and cannot be relied on for design. References 1 and 2 give extensive data on minimum strengths.

TABLE 2-4 CAST ALUMINUM ALLOY GROUPS

Group	Major alloying elements
1XX.X	99% or greater aluminum
2XX.X	Copper
3XX.X	Silicon, copper, magnesium
4XX.X	Silicon
5XX.X	Magnesium
6XX.X	(Unused series)
7XX.X	Zinc
8XX.X	Tin
9XX.X	Other elements

2-6 COPPER, BRASS, AND BRONZE

The name *copper* is properly used to denote virtually the pure metal having 99% or more copper. Its uses are primarily as electric conductors, switch parts, and motor parts which carry electric current. Copper and its alloys have good corrosion resistance, are readily fabricated, and have an attractive appearance. The main alloys of copper are the brasses, bronzes, and beryllium copper. Each has its special properties and applications.

Beryllium copper has very high strength and good electrical conductivity. Its uses include switch parts, fuse clips, electric connectors, bellows, Bourdon tubing for pressure gauges, and springs.

Brasses are alloys of copper and zinc. They have good corrosion resistance, workability, and a pleasing appearance, which leads to applications in automobile radiators, lamp bases, heat-exchanger tubes, marine hardware, ammunition cases, and home furnishings. Adding lead to brass improves its machinability, which makes it attractive to use for screw machine parts.

The major families of bronzes include phosphor bronze, aluminum bronze, and silicon bronze. Their high strength and corrosion resistance make them useful in marine applications, screws, bolts, gears, pressure vessels, springs, bushings, and bearings.

The strength of copper and its alloys is dependent on the hardness which is achieved by cold working. Successively higher strengths would result from the tempers designated annealed, quarter hard, half hard, three-quarters hard, hard, extra hard, spring, and extra spring tempers. The strengths of four copper alloys in the hard temper are listed in Appendix A-14.

2-7 ZINC, MAGNESIUM, AND TITANIUM

Zinc has moderate strength and toughness and excellent corrosion resistance. It is used in wrought forms such as rolled sheet and foil and drawn rod or wire. Dry-cell battery cans, builder's hardware, and plates for photoengraving are some of the major applications.

Many zinc parts are made by die casting because the melting point is less than 800°F (427°C), much lower than other die-casting metals. The as-cast finish is suitable for many applications, such as business machine parts, pump bodies, motor housings, and frames for light-duty machines. Where a decorative appearance is required, electroplating with nickel and chromium is easily done. Such familiar parts as radio grilles, lamp housings, horn rings, and body moldings are made in this manner.

Appendix A-14 lists the properties of particular rolled and cast zinc alloys.

Magnesium is the lightest metal commonly used in load-carrying parts. Its density of only 0.066 lb/in.^{3} (1830 kg/m^3) is only about one-fourth that of steel and zinc, one-fifth that of copper, and two-thirds that of aluminum. It has moderate strength and lends itself well to applications where the final fabricated weight of the part or structure should be light. Ladders, hand trucks, conveyor parts, portable power tools, and lawn-mower housings use magnesium. In the automtotive industry, body parts, blower wheels, pump bodies, and brackets are often made of magnesium. In aircraft, its lightness makes this metal attractive for floors, frames, fuselage skins, and wheels. The stiffness (modulus of elasticity) of magnesium is low, which is an advantage in parts where impact energy must be absorbed. Also, its lightness results in low-weight designs when compared with other metals on an equivalent rigidity basis.

Titanium has very high strength, and its density is only about half that of steel. Although aluminum has a lower density, titanium is superior to both aluminum and most steels on a strength-to-weight basis. It retains a high percentage of its strength at elevated temperatures and can be used up to about 1000°F (538°C). Most applications of titanium are in the aerospace industry in engine parts, fuselage parts and skins, ducts, spacecraft structures, and pressure vessels. Because of its corrosion resistance and high-temperature strength, the chemical industries use titanium in heat exchangers and as a lining for processing equipment. High cost is a major factor to be considered.

2-8 NONMETALS IN ENGINEERING DESIGN

Wood and concrete are widely used in construction. Plastics and composites are found in virtually all fields of design, including consumer products, industrial equipment, automobiles, aircraft, and architectural products. To the designer, the properties of strength and stiffness are of primary importance with nonmetals, as they are with metals. Because of the structural differences in the nonmetals, their behavior is quite different from the metals.

Wood, concrete, composites, and many plastics have structures that are *anisotropic*. This means that the mechanical properties of the material are different depending on the direction of the loading. Also, because of natural chemical changes, the properties vary with time and often with climatic conditions. The designer must be aware of these factors.

2-9 WOOD

Since wood is a natural material, its structure is dependent on the way it grows and not on manipulation by human beings, as is the case in metals. The long, slender, cylindrical shape of trees results in an internal structure composed of longitudinal cells. As the tree grows, successive rings are added outside the older wood. Thus the inner core, called heartwood, has different properties than the sapwood, near the outer surface.

The species of the wood also affects its properties, as different kinds of trees produce harder or softer, stronger or weaker wood. Even in the same species variability occurs because of different growing conditions, such as differences in soil and amount of sun and rain.

The cellular structure of the wood gives it the grain which is so evident when sawn into boards and timber. The strength of the wood is dependent on whether it is loaded perpendicular to or parallel to the grain. Also, going across the grain, the strength is different in a radial direction than in a tangential direction with respect to the original cylindrical tree stem from which it was cut.

Another important variable affecting the strength of wood is moisture content. Changes in relative humidity can vary the amount of water absorbed by the cells of the wood.

Most construction lumber is stress-graded by standard rules adopted by the U.S. Forest Products Laboratory. Appendix A-18 lists allowable stresses for several species and grades of lumber. These allowable stresses account for variability due to natural imperfections.

2-10 CONCRETE

The components of concrete are cement and an aggregate. The addition of water and the thorough mixing of the components tend to produce a uniform structure with cement coating all the aggregate particles. After curing, the mass is securely bonded together. Some of the variables involved in determining the final strength of the concrete are the type of cement used, the type and size of aggregate, and the amount of water added.

A higher quantity of cement in concrete yields a higher strength. Decreasing the quantity of water relative to the amount of cement increases the strength of the concrete. Of course, sufficient water must be added to cause the cement to coat the aggregate and to allow the concrete to be poured and worked before excessive curing takes place. The density of the concrete, affected by the aggregate, is also a factor. A mixture of sand, gravel, and broken stone is usually used for construction grade concrete.

Concrete is graded according to its compressive strength, which varies from about 2000 psi (14 MPa) to 7000 psi (48 MPa). The tensile strength of concrete is extremely low, and it is common practice to assume that it is zero. Of course, reinforcing concrete with steel bars allows its use in beams and wide slabs since the steel resists the tensile loads.

Concrete must be cured to develop its rated strength. It should be kept moist for at least 7 days, at which time is has about 75% of its rated compressive strength. Although its strength continues to increase for years, the strength at 28 days is often used to determine its rated strength.

The allowable working stresses in concrete are typically 25% of the rated 28-day strength. For example, concrete rated at 2000 psi (14 MPa) would have an allowable stress of 500 psi (3.4 MPa) (Reference 6).

2-11 PLASTICS

Plastics are being used in an increasing array of products. Although some applications are decorative, the use of plastics in sophisticated load-carrying parts accounts for a large part of the production. Plastics are composed of long chain-like molecules called polymers. They are synthetic organic materials which can be formulated and processed in literally thousands of ways.

One classification which can be made is between *thermoplastic* materials and *thermosetting* materials. Thermoplastics can be softened repeatedly by heating with no change in properties or chemical composition. Conversely, after initial curing of thermosetting plastics, they cannot be resoftened. A chemical change occurs during curing with heat and pressure.

Some examples of thermoplastics include ABS, acetals, acrylics, cellulose acetate, TFE fluorocarbons, nylon, polyethylene, polypropylene, polystyrene, and vinyls. Thermosetting plastics include phenolics, epoxies, polyesters, silicones, urethanes, alkyds, allylics, and aminos.

TABLE 2-5 APPLICATIONS OF PLASTIC MATERIALS

Application	Desired properties	Suitable plastics
Housings, containers, ducts	High impact strength, stiffness, low cost, formability, environmental resistance, dimensional stability	ABS, polystyrene, polypropylene, polyethylene, cellulose acetate, acrylics
Low friction—bearings, slides	Low coefficient of friction; resistance to abrasion, heat, corrosion	TFE fluorocarbons, nylon, acetals
High-strength components, gears, cams, rollers	High tensile and impact strength, stability at high temperatures, machinable	Nylon, phenolics, TFE-filled acetals
Chemical and thermal equipment	Chemical and thermal resistance, good strength, low moisture absorption	Fluorocarbons, polypropylene, polyethylene, epoxies, polyesters, phenolics
Electrostructural parts	Electrical resistance, heat resistance, high impact strength, dimensional stability, stiffness	Allylics, alkyds, aminos, epoxies, phenolics, polyesters, silicones
Light-transmission componenets	Good light transmission in transparent and translucent colors, formability, shatter resistance	Acrylics, polystyrene, cellulose acetate, vinyls

A particular plastic is often selected for a combination of properties such as light weight, flexibility, color, strength, chemical resistance, low friction, or transparency. Since the available products are so numerous, only a brief table of properties of plastics is included as Appendix A-19, Table 2-5 lists the primary plastic materials used for six different types of applications. An extensive comparative study of the design properties of plastics can be found in References 10 and 11.

2-12 COMPOSITES

Composites are materials having two or more constituents blended together in a way that results in mechanical or adhesive bonding between the materials. This differs from other materials in which metallurgical or chemical bonding at the elemental or molecular level determines the final properties.

To form a composite, a filler material, usually in the form of a high-strength fiber, is distributed in a matrix, usually a plastic resin. The resulting composite typically has high strength, high stiffness, and low density compared to most metals. In fact, it is the high strength-to-weight and stiffness-to-weight ratios that make composites attractive in an increasing number of transportation and aerospace applications.

An unlimited number of varieties of composites can be made by combining different resins with different fillers in different quantities and with different orientations of the fibers in the matrix. This is one of the desirable characteristics of composites because the properties can be tailored to meet specific needs for load-carrying ability and stiffness of the structure. In general terms, the filled plastics that have been used for many years are composites. Recent developments have included metal-matrix composites in which reinforcing fibers are placed in light metals such as aluminum, titanium, and magnesium.

Plastic resins used for the matrix include epoxies, polyesters, phenolics, silicones, nylons, polypropylenes, polystyrenes, and ABS. Filler materials include glass, carbon, graphite, boron, aramid, silicon carbide, and alumina. The fillers have very high strength and high modulus of elasticity, while the matrix materials have low densities and serve to support and maintain the orientation of the fibers in advantageous directions. The composite is then an optimized combination of the two materials.

Uses for composites include gears and other mechanical parts, housings, fans, exterior panels for land vehicles, boat hulls, luggage, shipping containers, sports equipment, and structural components and skins for aircraft and aerospace vehicles.

The general-purpose filled plastics use fibers which are very small, permitting them to be randomly dispersed in the resin. The material can then usually be molded in a manner similar to the unreinforced plastic. Other glass/plastic composites use a woven fabric or mat that is applied in layers to form complex curved shapes and to benefit from the orientation of the longer fibers. The matrix, usually polyester, is placed on the fabric between layers. The resulting composite is cured, either at room temperature or with moderate heating. These materials are generically called fiberglass.

The term *advanced composites* is usually used to describe the epoxy/carbon composite family and uses are found primarily in the aerospace industry. Some high-quality sports equipment such as tennis rackets, golf clubs, and skis use advanced composites. Applications are also being researched for automotive structures and engines.

The carbon fibers have tensile strengths in the range 300 to 400 ksi (2.1 to 2.8 GPa) and the modulus of elasticity ranges from 33×10^6 to 75×10^6 psi (230 to 520 GPa). The fibers can be milled to short lengths, chopped to approximately $\frac{1}{4}$ to 2 in. long (6 to 50 mm), or used as continuous filiments. The long fibers produce the highest strengths and stiffnesses but are more costly to fabricate.

While wide ranges of strengths and stiffnesses for composites are to be expected depending on materials and fabrication, advanced composites can achieve very high strength-to-weight and stiffness-to-weight ratios compared to those of typical alloy steels, aluminum, or titanium. This is the basis for their attractiveness in aerospace applications. Figure 2-11 illustrates this attractiveness by plotting specific tensile strength versus specific tensile modulus for several metals, typical fiberglass, advanced fiber/polymer composites, and metal matrix composites. *Specific strength* and *specific modulus* are defined as

$$\text{specific strength} = \frac{\sigma}{\rho} \qquad (2\text{-}3)$$

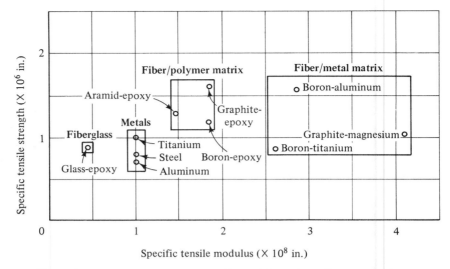

Figure 2-11 A comparison of the specific modulus and specific strength of several composite materials and metals.

$$\text{specific modulus} = \frac{E}{\rho} \qquad (2\text{-}4)$$

where σ = tensile strength (psi)
  E = tensile modulus of elasticity (psi)
  ρ = density (lb/in^3)

The resulting unit for either the specific strength or specific modulus is inches.

Disadvantages of composites must be considered. The technology is still relatively new and it is difficult to obtain reliable design data. The composites are inherently anisotropic, having radically different properties in different directions. Failure can occur by delamination or separation of the fibers from the matrix.

REFERENCES

1. Aluminum Association, *Aluminum Standards and Data*, New York, 1985.
2. Aluminum Association, *Standards for Aluminum Sand and Permanent Mold Castings*, New York, 1984.
3. American Institute of Chemical Engineers, *New Composite Materials and Technology*, AIChE Symposium Series No. 217, Vol. 78, New York, 1982.
4. Aronson, R. B., "Metal-Matrix Composites—Materials of the Future," *Machine Design*, August 8, 1985, pp. 68–73.
5. Askeland, Donald R., *The Science and Engineering of Materials*, Brooks/Cole, Monterey, Calif., 1984.
6. Baumeister, T., Editor-in-Chief, *Marks' Standard Handbook for Mechanical Engineers*, 8th ed., McGraw-Hill, New York, 1978.

7. Bennett, B. A., "Carbon-Fiber Composites: A Light-Weight Alternative," *Mechanical Engineering*, September 1985.

8. Bethlehem Steel Corporation, *Modern Steels and Their Properties*, Bethlehem, Pa., 1980.

9. Forrest, R. D., "Austempered Ductile Iron for Both Strength and Toughness," *Machine Design*, September 26, 1985, pp. 95–99.

10. *Materials Reference Issue, Machine Design*, April 18, 1985.

11. *Materials Selector, 1986, Materials Engineering*, December 1985.

12. American Society for Metals, *Metals Handbook*, 8th ed., Metals Park, Ohio.

13. Mott, R. L., *Machine Elements in Mechanical Design*, Charles E. Merrill, Columbus, Ohio, 1985.

PROBLEMS

2-1. Name four kinds of metals commonly used for load-carrying members.

2-2. Name 11 factors that should be considered when selecting a material for a product.

2-3. Define *ultimate tensile strength*.

2-4. Define *yield point*.

2-5. Define *yield strength*.

2-6. When is yield strength used in place of yield point?

2-7. Define *stiffness*.

2-8. What material property is a measure of its stiffness?

2-9. State Hooke's law.

2-10. What material property is a measure of its ductility?

2-11. How is a material classified as to whether it is ductile or brittle?

2-12. Name four types of steels.

2-13. What does the designation AISI 4130 for a steel mean?

2-14. What are the ultimate strength, yield strength, and percent elongation of AISI 1040 hot-rolled steel? Is it a ductile or a brittle material?

2-15. Which has a greater ductility: AISI 1040 hot-rolled steel or AISI 1020 hot-rolled steel?

2-16. What does the designation AISI 1141 OQT 700 mean?

2-17. If the required yield strength of a steel is 150 ksi, could AISI 1141 be used? Why?

2-18. What is the modulus of elasticity for AISI 1141 steel? For AISI 5160 steel?

2-19. A rectangular bar of steel is 1.0 in. by 4.0 in. by 14.5 in. How much does it weigh in pounds?

2-20. A circular bar is 50 mm in diameter and 250 mm long. How much does it weigh in newtons?

2-21. If a force of 400 N is applied to a bar of titanium and an identical bar of magnesium, which would stretch more?

2-22. Name four types of structural steels and list the yield point for each.

2-23. What does the aluminum alloy designation 6061-T6 mean?

2-24. List the ultimate strength, yield strength, modulus of elasticity, and density for 6061-O, 6061-T4, and 6061-T6 aluminum.

2-25. List five uses for bronze.

2-26. List three desirable characteristics of titanium as compared with aluminum or steel.

2-27. Name five varieties of cast iron.

2-28. Which type of cast iron is usually considered to be brittle?

2-29. What are the ultimate strengths in tension and in compression for ASTM A48 grade 40 cast iron?

2-30. How does a ductile iron differ from gray iron?

2-31. List the allowable stresses in bending, tension, compression, and shear for No. 2 grade Douglas fir.

2-32. What is the normal range of compressive strengths for concrete?

2-33. Describe the difference between thermoplastic and thermosetting materials.

2-34. Name three suitable plastics for use as gears or cams in mechanical devices.

2-35. Describe the term *composite*.

2-36. Name three filler materials used for composites.

2-37. Name three matrix materials used for composites.

2-38. The general term *fiberglass* refers to what type of composite material?

2-39. Describe an advanced composite.

2-40. What are the characteristics of composites that make them desirable compared to metals.

2-41. Discuss the variables related to the form of the fillers for composites that affect the resulting properties of the composites.

2-42. Describe a metal-matrix composite.

2-43. Discuss some of the disadvantages of composites.

3

Design of Members
under Direct Stresses

3-1 OBJECTIVES OF THIS CHAPTER

In Chapter 1 the concept of direct stress was presented together with examples of the calculation of direct tensile stress, direct compressive stress, direct shear stress, and bearing stress. The emphasis was on the understanding of the basic phenomena, units, terminology, and the magnitude of stresses encountered in typical structural and mechanical applications. Nothing was said about the acceptability of the stress levels which were computed or about the design of members to carry a given load.

In this chapter the primary emphasis is on *design* in which you, as the designer, must make decisions about whether or not a proposed design is satisfactory; what the shape and size of the cross section of a load-carrying member should be; and what material the member should be made from.

After completing this chapter, you should be able to:

1. Describe the conditions that must be met for satisfactory application of the direct stress formulas.
2. Define *design stress* and tell how to determine an acceptable value for it.
3. Define *design factor* and select appropriate values for it depending on the conditions present in a particular design.

4. Discuss the relationship among the terms *design stress, allowable stress,* and *working stress.*
5. Discuss the relationship among the terms *design factor, factor of safety,* and *margin of safety.*
6. Describe 11 factors that affect the specification of the design factor.
7. Describe various types of loads experienced by structures or machine members, including static load, repeated load, impact, and shock.
8. Design members subjected to direct tensile stress, direct compressive stress, direct shear stress, and bearing stress.
9. Determine when stress concentrations exist and specify suitable values for stress concentration factors.
10. Use stress concentration factors in design.

3-2 DESIGN OF MEMBERS UNDER DIRECT TENSION OR COMPRESSION

In Chapter 1 the direct stress formula was developed and stated as follows:

$$\sigma = \frac{P}{A} \tag{3-1}$$

where σ = direct normal stress; tension or compression
P = direct axial load
A = cross-sectional area of member subjected to P

For Equation (3-1) to be valid, the following conditions must be met.

1. The loaded member must be straight.
2. The loaded member must have a uniform cross section over the length under consideration.
3. The material from which the member is made must be homogeneous.
4. The load must be applied along the centroidal axis of the member so there is no tendency to bend it.
5. Compression members must be short so that there is no tendency to buckle. (See Chapter 14 for the special analysis required for long, slender members under compressive stress and for the method to decide when a member is to be considered long or short.)

It is important to recognize that the concept of stress refers to the internal resistance provided by a *unit area,* that is, an infinitely small area. Stress is considered to act at a point and may, in general, vary from point to point in a particular body. Equation (3-1) indicates that for a member subjected to direct axial tension or compression, the stress is uniform across the entire area if the five conditions are met. In many practical applications the minor variations that could occur in the local stress levels are accounted for by carefully selecting the allowable stress, as discussed later.

3-3 DESIGN NORMAL STRESSES

Failure occurs in a load-carrying member when it breaks or deforms excessively, rendering it unacceptable for the intended purpose. Therefore, it is essential that the level of applied stress never exceed the ultimate tensile strength or the yield strength of the material. Consideration of excessive deformation without yielding is discussed in later chapters.

Design stress is that level of stress which may be developed in a material while ensuring that the loaded member is safe. To compute design stress, two factors must be specified: the *design factor N* and the *property of the material on which the design will be based*. Usually, for metals, the design stress is based on either the yield strength s_y or the ultimate strength s_u of the material.

The *design factor N* is a number by which the reported strength of a material is divided to obtain the *design stress* σ_d. The following equations can be used to compute the design stress for a certain value of N.

$$\sigma_d = \frac{s_y}{N} \text{ based on yield strength} \tag{3-2}$$

or

$$\sigma_d = \frac{s_u}{N} \text{ based on ultimate strength} \tag{3-3}$$

The value of the design factor is normally determined by the designer, using judgment and experience. In some cases, codes, standards, or company policy may specify design factors or design stresses to be used. When the designer must determine the design factor, his or her judgment must be based on an understanding of how parts may fail and the factors that affect the design factor. Sections 3-4 and 3-5 give additional information about the design factor and about the choice of methods for computing design stresses.

Other references may use the term *factor of safety* in place of *design factor*. Also, *allowable stress* or *working stress* may be used in place of *design stress*. The choice of terms for use in this book is made to emphasize the role of the designer in specifying the design stress.

Theoretically, a material could be subjected to a stress up to s_y before yield would occur. This condition corresponds to a value of the design factor of $N = 1$ in Equation (3-2). Similarly, with a design factor of $N = 1$ in Equation (3-3), the material would be on the brink of ultimate fracture. Thus $N = 1$ is the lowest value we can consider.

A different approach to evaluating the acceptability of a given design, used primarily in the aerospace industry, is the *margin of safety*, defined as follows:

$$\text{margin of safety} = \frac{\text{yield strength}}{\text{maximum stress}} - 1.0 \tag{3-4}$$

when the design is based on yielding of the material. When based on the ultimate strength, the margin of safety is

$$\text{margin of safety} = \frac{\text{ultimate strength}}{\text{maximum stress}} - 1.0 \tag{3-5}$$

Then the lowest feasible margin of safety is 0.0.

In this book we use the concept of design stresses and design factors as opposed to the margin of safety.

3-4 DESIGN FACTOR

Many different aspects of the design problem are involved in the specification of the design factor. In some cases the precise conditions of service are not known. The designer must then make conservative estimates of the conditions, that is, estimates which would cause the resulting design to be on the safe side when all possible variations are considered. The final choice of a design factor depends on the following 11 conditions.

Codes and standards.   If the member being designed falls under the jurisdiction of an existing code or standard, obviously the design factor or design stress must be chosen to satisfy the code or standard. Examples of standard-setting bodies are:

American Institute of Steel Construction (AISC): buildings, bridges, and similar structures using steel

Aluminum Association (AA): buildings, bridges, and similar structures using aluminum

American Society of Mechanical Engineers (ASME): boilers, pressure vessels, and shafting

State building codes: buildings, bridges, and similar structures affecting the public safety

Department of Defense—Military Standards: aerospace vehicle structures and other military products

American National Standards Institute (ANSI): a wide variety of products

American Gear Manufacturers Association (AGMA): gears and gear systems

It is the designer's responsibility to determine which, if any, standards or codes apply to the member being designed and to ensure that the design meets those standards.

Material strength basis.   Most designs using metals are based on either yield strength or ultimate strength or both, as stated previously. This is because most theories of metal failure show a strong relationship between the stress at failure and these material properties. Also, these properties will almost always be reported for materials used in engineering design. The value of the design factor will be different depending on which material strength is used as the basis for design, as will be shown later.

Type of material.   A primary consideration with regard to the type of material is its ductility. The failure modes for brittle materials are quite different than those for ductile materials. Since brittle materials such as cast iron do not exhibit yielding, designs are always based on ultimate strength. Generally a metal is considered to be brittle if its percent elongation in a 2-in. gage length is less than 5%. Except for highly hardened alloys, virtually all steels are ductile. Except for castings, aluminum is ductile. Other material factors which can affect the strength of a part are its uniformity and the confidence in the stated properties.

Manner of loading.   Three main types of loading can be identified. A *static load* is one which is applied to a part slowly and gradually and which remains applied, or at least is applied and removed only infrequently during the design life of the part. *Repeated loads* are those which are applied and removed several thousand times during the design life of the part. Under repeated loading a part fails by the mechanism of fatigue at a stress level much lower than that which would cause failure under a dead load. This calls for the use of a higher design factor for repeated loads than for static loads. Parts subject to *impact or shock* require the use of a large design factor for two reasons. First, a suddenly applied load causes stresses in the part which are several times higher than those which would be computed by standard formulas. Second, under impact loading the material in the part is usually required to absorb energy from the impacting body. The certainty with which the designer knows the magnitude of the expected loads also must be considered when specifying the design factor.

Possible misuse of the part.   In most cases the designer has no control over actual conditions of use of the product he or she designs. Legally, it is the responsibility of the designer to consider any reasonably foreseeable use or *misuse* of the product and to ensure the safety of the product. The possibility of an accidental overload on any part of a product must be considered.

Complexity of stress analysis.   As the manner of loading or the geometry of a structure or a part becomes more complex, the designer is less able to perform a precise analysis of the stress condition. Thus the confidence in the results of stress analysis computations has an effect on the choice of a design factor.

Environment.   Materials behave differently in different environmental conditions. Consideration should be given to the effects of temperature, humidity, radiation, weather, sunlight, and corrosive atmospheres on the material during the design life of the part.

Size effect, sometimes called *mass effect*.   Most metals exhibit different strengths as the cross-sectional area of a part varies. Most material property data were obtained using standard specimens about $\frac{1}{2}$ in. in diameter. Parts with larger sections

TABLE 3-1 SIZE EFFECT FOR AISI 4140 OQT 1100 STEEL

Specimen size (in.)	Tensile strength ksi	MPa	Yield strength ksi	MPa	Percent elongation (% in 2 in.)
1/2	158	1089	149	1027	18
1	140	965	135	931	20
2	128	883	103	710	22
4	117	807	87	600	22

usually have lower strengths. Parts of smaller size, for example drawn wire, have significantly higher strengths. An example of the size effect is shown in Table 3-1.

Quality control. The more careful and comprehensive a quality control program is, the better a designer knows how the product will actually appear in service. With poor quality control, a larger design factor should be used.

Hazard presented by a failure. The designer must consider the consequences of a failure to a particular part. Would a catastrophic collapse occur? Would people be placed in danger? Would other equipment be damaged? Such considerations may justify the use of a higher than normal design factor.

Cost. Compromises must usually be made in design in the interest of limiting cost to a reasonable value under market conditions. Of course, where danger to life or property exists, compromises should not be made which would seriously affect the ultimate safety of the product or structure.

Experience in design and knowledge about the conditions above must be applied to determine a design factor. Table 3-2 includes guidelines which will be used in this book for selecting design factors. These should be considered to be average values. Special conditions or uncertainty about conditions may justify the use of other values.

The design factor is used to determine design stress as shown in Equations (3-2) and (3-3).

If the stress in a part is already known and one wishes to choose a suitable material for a particular application, the computed stress is considered to be the design

TABLE 3-2 DESIGN FACTOR GUIDELINES

Manner of loading	Design factor, N		
	Ductile metals		Brittle metals
	Yield strength basis	Ultimate strength basis	Ultimate strength basis
Static load	2	—	6
Repeated load	—	8	10
Impact or shock	—	12	15

stress. The required yield or ultimate strength is then found from

$$s_y = N \cdot \sigma_d \qquad (N \text{ based on yield strength})$$

or

$$s_u = N \cdot \sigma_d \qquad (N \text{ based on ultimate strength})$$

3-5 METHODS OF COMPUTING DESIGN STRESS

As mentioned in Section 3-3, an important factor to be considered when computing the design stress is the manner in which a part may fail when subjected to loads. In this section we discuss failure modes relevant to parts subjected to tensile and compressive loads. Other kinds of loading are discussed later.

The failure modes and the consequent methods of computing design stresses can be classified according to the type of material and the manner of loading. Ductile materials, having more than 5% elongation, exhibit somewhat different modes of failure than do brittle materials. Static loads, repeated loads, and shock loads produce different modes of failure.

Ductile materials under static loads. Ductile materials will undergo large plastic deformations when the stress reaches the yield strength of the material. Under most conditions of use, this would render the part unfit for its intended use. Therefore, for ductile materials subjected to static loads, the design stress is usually based on yield strength. That is,

$$\sigma_d = \frac{s_y}{N}$$

As indicated in Table 3-2, a design factor of $N = 2$ would be a reasonable choice under average conditions.

Ductile materials under repeated loads. Under repeated loads, ductile materials fail by a mechanism called *fatigue*. The level of stress at which fatigue occurs is lower than the yield strength. By testing materials under repeated loads, the stress at which failure will occur can be measured. The terms *fatigue strength* or *endurance strength* are used to denote this stress level. However, fatigue-strength values are often not available. Also, factors such as surface finish, the exact pattern of loading, and the size of a part have a marked effect on the actual fatigue strength. To overcome these difficulties, it is often convenient to use a high value for the design factor when computing the design stress for a part subjected to repeated loads. It is also recommended that the ultimate strength be used as the basis for the design stress because tests show that there is a good correlation between fatigue strength and the ultimate strength. Therefore, for ductile materials subjected to repeated loads, the design stress can be computed from

$$\sigma_d = \frac{s_u}{N}$$

A design factor of $N = 8$ would be reasonable under average conditions. Also, stress concentrations, which are discussed in Section 3-8, must be accounted for since fatigue failures often originate at points of stress concentrations.

Where data are available for the endurance strength of the material, the design stress can be computed from

$$\sigma_d = \frac{s_n}{N} \tag{3-6}$$

where s_n is the symbol for endurance strength.

Ductile materials under impact or shock loading. The failure modes for parts subjected to impact or shock loading are quite complex. They depend on the ability of the material to absorb energy and on the flexibility of the part. Because of the general inability of designers to perform precise analysis of stresses under shock loading, large design factors are recommended. In this book we will use

$$\sigma_d = \frac{s_u}{N}$$

with $N = 12$ for ductile materials subjected to impact or shock loads.

Brittle materials. Since brittle materials do not exhibit yielding, the design stress must be based on ultimate strength. That is,

$$\sigma_d = \frac{s_u}{N}$$

with $N = 6$ for static loads, $N = 10$ for repeated loads, and $N = 15$ for impact or shock loads.

Structural steel. According to the AISC specifications, the allowable stress in tension for structural steel is computed from

$$\sigma_d = 0.60 s_y$$

or

$$\sigma_d = 0.50 s_u$$

whichever is lower. In bending (discussed in Chapter 8),

$$\sigma_d = 0.66 s_y \qquad \text{(for compact beam cross sections)}$$

These design stresses are all for the static load case typical of members of a building-type structure. If other conditions exist, the specifications of AISC must be consulted. Special cases of columns and connections are discussed later.

Aluminum. When used in building-type structures carrying static loads, the Aluminum Association specifies design stresses for aluminum in its publication *Specifications for Aluminum Structures* (1). The complete specification includes a listing of the minimum properties of the various alloys typically used for structures, the design factors, and the formulas to be used for each loading case. Some of these are abstracted here to illustrate their use. For axial tension or tension in beams,

$$\sigma_d = \frac{s_y}{1.65} = 0.606s_y \tag{3-7}$$

or

$$\sigma_d = \frac{s_u}{1.95} = 0.513s_u \tag{3-8}$$

whichever is lower. These are similar to the values for structural steel.

Example Problem 3-1

A structural support for a machine will be subjected to a static tensile load of 16 kN. If it is to be made from AISI 1020 hot-rolled steel, what cross-sectional area of the support would be required?

Solution   The following steps can be used to solve this problem:

1. Compute the design stress for the steel.
2. Let the actual expected stress in the steel be equal to the design stress.
3. From the direct tensile stress formula [Equation (3-1)], solve for the area required.

   Begin with step 1 now.

You should have $\sigma_d = 103.5$ MPa. The AISI 1020 hot-rolled steel is a ductile material, as can be seen by the 25% elongation listed in Appendix A-13. Then the conditions described are for a ductile material under a static load and Equation (3-2) is used for the design stress. The yield strength of the steel can be found from Appendix A-13 to be 207 MPa. Then

$$\sigma_d = \frac{s_y}{N} = \frac{207 \text{ MPa}}{2} = 103.5 \text{ MPa}$$

Now do steps 2 and 3 and find the area required.

The area required is 154.6 mm², found as follows. Let

$$\sigma = \sigma_d = \frac{P}{A}$$

Solving for A gives

$$A = \frac{P}{\sigma_d} = \frac{16\ 000 \text{ N}}{103.5 \text{ N/mm}^2} = 154.6 \text{ mm}^2$$

Extending the problem, what dimensions would you specify for the tensile rod if it is to have a square cross section?

From $A = S^2$, the minimum allowable dimension for the side of the square cross section is the square root of the area required,

$$S = \sqrt{A} = \sqrt{154.6 \text{ mm}^2} = 12.4 \text{ mm}$$

Specifying a standard dimension greater than this minimum value, use $S = 14$ mm. This example problem is concluded.

Example Problem 3-2

A tensile member of a roof truss for a building is to carry an axial tensile load of 19 800 lb. It has been proposed to use a standard equal-leg structural steel angle for this application. Specify a suitable angle from Appendix A-5, made from ASTM A36 steel, according to the AISC code.

Solution   The solution of this problem is similar to that for Example Problem 3-1 except that the AISC code should be used for computing the design stress. So, following the procedure set up for that problem, complete step 1 now.

The design stress is 21 600 psi. According to the AISC code,

$$\sigma_d = 0.60 \, s_y$$

For ASTM A36 steel, $s_y = 36\ 000$ psi. Then

$$\sigma_d = 0.60(36\ 000 \text{ psi}) = 21\ 600 \text{ psi}$$

Now compute the required area for the tensile member.

The computed stress for this member would be found from $\sigma = P/A$. Letting $\sigma = \sigma_d$ and solving for the area gives

$$A = \frac{P}{\sigma_d} = \frac{19\ 800 \text{ lb}}{21\ 600 \text{ lb/in}^2} = 0.917 \text{ in}^2$$

Now select a standard structural steel angle from the tables in the Appendix.

The lightest equal-leg angle having a cross-sectional area greater than 0.917 in² is the L2 × 2 × ¼, which has an area of 0.938 in² and weighs 3.19 lb/ft.
This completes the example problem.

Example Problem 3-3

A machine element in a packaging machine is subjected to a tensile load of 36.6 kN, which will be repeated several thousand times over the life of the machine. The cross section of the element is 12 mm thick and 20 mm wide. Specify a suitable material from which to make the element.

Solution   From the data given, we can compute the actual expected stress in the element. Normally, this would be compared with the design stress for the material, but in this case we do not know the material. So the analysis can proceed by specifying the appropriate method of computing the design stress and then solving for the required material strength. From that we can specify a suitable material. The procedure is as follows:

1. Compute the actual expected stress in the element.

2. Specify the equation for the design stress.
3. Let the actual stress be equal to the design stress.
4. Solve for the required strength of the material.
5. Specify a suitable material having a strength greater than that required.

Complete step 1 now.

The stress is 152.5 MPa, found from $\sigma = P/A$.

$$\sigma = \frac{P}{A} = \frac{36\ 600\ \text{N}}{(12\ \text{mm})(20\ \text{mm})} = 152.5\ \text{N/mm}^2 = 152.5\ \text{MPa}$$

Now, what equation should be used for the design stress?

We should use $\sigma_d = s_u/8$ because the load will be repeated. Furthermore, we should select a ductile material for such a load.

Now compute the required ultimate strength of the material.

Solving for s_u from the design stress equation and letting the actual stress equal the design stress, we would find

$$s_u = 8\sigma_d = 8(152.5\ \text{MPa}) = 1220\ \text{MPa}$$

What material would you specify?

Several of the heat-treated steels from Appendix A-13 have an ultimate strength greater than 1220 MPa and could be used. Recalling that a highly ductile material is desired, let's specify AISI 4140 OQT 900, which has $s_u = 1289$ MPa and 15% elongation. This is the highest percent elongation of any material that would meet the strength requirement.

This example problem is concluded.

Example Problem 3-4

Figure 3-1 shows a design for the support for a heavy machine. The support will be loaded in axial compression and is to be made from gray cast iron, grade 20. Specify the allowable load on the support.

Solution   The equation for the compressive stress is $\sigma = P/A$. Note that this requires the assumption that the support is a *short* compression member, which appears to be justified in this case. The cross-sectional area for the support can be computed from the geometry given in Figure 3-1. But because the load is not known, the actual stress in the support cannot be computed. The approach to the problem involves the computation of the design stress for the cast iron under the given conditions. Then, letting the design stress equal P/A, we can solve for the allowable load, P.

How can we compute the design stress?

Because the support is to be made from gray cast iron, a brittle material, and

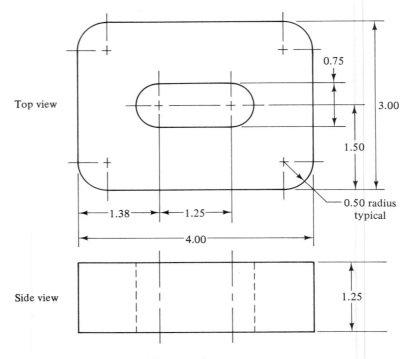

Top view

Side view

0.75

3.00

1.50

0.50 radius
typical

1.38

1.25

4.00

1.25

Dimensions in inches

Figure 3-1   Machine support for Example Problem 3-4.

because the load is a static load, the design equation is

$$\sigma_d = \frac{s_u}{N} = \frac{s_u}{6}$$

Now complete the computation of the design stress.

You should have $\sigma_d = 13\ 300$ psi. From Appendix A-16, the ultimate compressive strength of grade 20 cast iron is 80 000 psi. Then

$$\sigma_d = \frac{s_u}{6} = \frac{80\ 000\ \text{psi}}{6} = 13\ 300\ \text{psi}$$

Now, letting $\sigma_d = P/A$, the allowable load, P, is

$$P = A\sigma_d$$

Compute the area, A, and the allowable load, P, now.

The area is computed from the geometry of the support shown in Figure 3-1. Remember to use the *cross-sectional area* in the horizontal plane. It is a composite area computed by subtracting the slot and the corner fillets from the 3.00 by 4.00 in. rectangle. The result is $A = 10.41$ in^2.

Rectangle: $A_R = (3.00 \text{ in.})(4.00 \text{ in.}) = 12.00 \text{ in}^2$

Slot: $A_S = (0.75)(1.25) + \dfrac{\pi(0.75)^2}{4} = 1.38 \text{ in}^2$

The area of each fillet can be computed by the difference between the area of a square with sides equal to the radius of the corner (0.50 in.) and a quarter circle of the same radius. Then

Fillet: $A_F = r^2 - \frac{1}{4}(\pi r^2)$

$A_F = (0.50)^2 - \frac{1}{4}[\pi(0.50)^2] = 0.0537 \text{ in}^2$

Then the total area is

$A = A_R - A_S - 4A_F = 12.00 - 1.38 - 4(0.0537) = 10.41 \text{ in}^2$

We now have the data needed to compute the allowable load. Do that now.

The allowable load is 138 500 lb, found from

$P = A\sigma_d = (10.41 \text{ in}^2)(13\ 300 \text{ lb/in}^2) = 138\ 500 \text{ lb}$

This completes the example problem.

Example Problem 3-5

Figure 3-2 shows a piece of manufacturing equipment called a C-frame press used to pressform sheet-metal products. The ram is driven down with a large force, closing the dies and forming the part. The pressforming action causes the open end of the press to

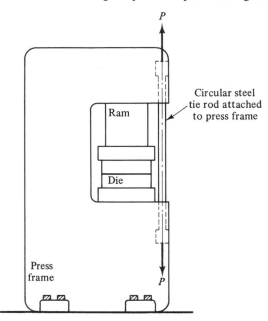

Figure 3-2   C-frame press for Example Problem 3-5.

tend to expand, an undesirable action if the deformation is excessive. As an aid in limiting the expansion, tie rods are installed across the open part of the C-frame and tightened under a high tensile load.

One proposed design for the tie rods calls for them to be made from AISI 5160 steel, OQT 1300, with a diameter of 2.00 in. During operation, a peak tensile force of 40 000 lb is applied to the rods. As the pressforming occurs, a moderate shock loading is experienced by the rods. Determine if this proposed design is acceptable.

Solution The following steps can be used to solve this problem:

1. Compute the design stress for the rod material for the loading conditions.
2. Compute the actual expected stress in the rod.
3. Compare the stresses from steps 1 and 2. If the actual stress is less than the design stress, the design is satisfactory. If not, some redesign is called for.

To start, we must identify the method of computing the design stress for this situation. One of the factors in this decision is whether the material is ductile or brittle. What is it in this case?

The material, AISI 5160 OQT 1300, is listed in Appendix A-13 and it has a percent elongation of 23%. Therefore, it is ductile. Now complete the specification of the method of computing the design stress.

Because of the shock loading on the tie rod, we should use Equation (3-3) with $N = 12$. Now compute the design stress.

You should have $\sigma_d = 9583$ psi. From Appendix A-13, the ultimate strength of the steel is $s_u = 115\ 000$ psi. Then the design stress is

$$\sigma_d = \frac{s_u}{12} = \frac{115\ 000\ \text{psi}}{12} = 9583\ \text{psi}$$

Now compute the actual expected stress in the rod.

The actual expected stress is approximately 12 700 psi. The rod is subjected to direct tensile stress, computed from Equation (3-1), $\sigma = P/A$. The cross-sectional area of the rod is

$$A = \frac{\pi D^2}{4} = \frac{\pi (2.00\ \text{in})^2}{4} = 3.14\ \text{in}^2$$

Then the stress in the rod is

$$\sigma = \frac{P}{A} = \frac{40\ 000\ \text{lb}}{3.14\ \text{in}^2} = 12\ 700\ \text{psi}$$

Is this acceptable?

No. Because the actual expected stress is greater than the design stress, it is not acceptable.

Let's extend the objective of the problem to propose a redesign that would be satisfactory for the given conditions of load. What could be changed?

The two obvious changes are the strength of the material and the diameter of the rod. Considering each change separately, what would be the required diameter of the rod if it were to be made from the same material, AISI 5160 OQT 1300?

The required diameter is 2.30 in. This value is found by letting the actual expected stress in the rod equal the design stress, 9583 psi, then solving for the required area of the cross section of the rod to limit the stress to this amount. Let

$$\sigma = \sigma_d = 9583 \text{ psi} = \frac{P}{A}$$

Then the area required is

$$A = \frac{P}{\sigma_d} = \frac{40\ 000 \text{ lb}}{9583 \text{ lb/in}^2} = 4.17 \text{ in}^2$$

But for the round circular cross section, $A = \pi D^2/4$. Solving for the diameter, D, gives

$$D = \sqrt{\frac{4A}{\pi}} = \sqrt{\frac{4(4.17 \text{ in}^2)}{\pi}} = 2.30 \text{ in.}$$

Now, suppose that we do not really want to increase the diameter above 2.00 in. Specify a suitable material which would be satisfactory for that size of rod. The first step is to compute the required strength of the material. How can we do that?

From the equation for design stress, we can solve for the required tensile strength, s_u, to give a design stress of 12 700 psi, the actual expected stress for the 2.00-in.-diameter rod. Remember that we want to keep the design factor at 12.

The required $s_u = 152\ 400$ psi, found as follows. Let

$$\sigma_d = 12\ 700 \text{ psi} = \frac{s_u}{12}$$

Then the required tensile strength is

$$s_u = 12(\sigma_d) = 12(12\ 700 \text{ psi}) = 152\ 400 \text{ psi}$$

Now specify a suitable material.

Several possible materials can be found in Appendix A-13, including

AISI 5160 OQT 1000; $s_u = 172\ 500$ psi (by interpolation between the values given for tempering temperatures of 900 and 1100)
AISI 4140 OQT 1000; $s_u = 167\ 000$ psi (by interpolation)
AISI 1141 OQT 800: $s_u = 169\ 500$ psi (by interpolation)

To choose among these possible materials, other factors such as cost, machinability, and availability should be considered.

This concludes the example problem.

3-6 DESIGN SHEAR STRESS

When members are subjected to shear stresses, design must be based on the *design shear stress*, τ_d. Two forms of design shear stress are used:

$$\tau_d = \frac{s_{ys}}{N} \qquad \text{based on the yield strength in shear} \qquad (3\text{-}9)$$

or

$$\tau_d = \frac{s_{us}}{N} \qquad \text{based on the ultimate shear strength} \qquad (3\text{-}10)$$

Of course, if the values of the yield strength in shear and the ultimate shear strength are available, they can be used in the design stress equations. For example, Appendix A-17 lists typical values for the ultimate shear strength for aluminum alloys. Appendix A-16 lists values for some cast irons.

But unfortunately, such values are frequently not reported and it is necessary to rely on estimates. For the yield strength in shear, a frequently used estimate is

$$s_{ys} = \frac{s_y}{2} = 0.5s_y \qquad (3\text{-}11)$$

This value is taken from the observation of a typical tensile test in which the shear stress is one-half of the direct tensile stress. This phenomenon, related to the *maximum shear stress theory of failure*, is somewhat conservative and will be discussed further in Chapter 10.

The Aluminum Association (1), in a table of mechanical properties for aluminum alloys, uses $s_{ys} = 0.58s_y$. This is consistent with a different theory called the *distortion energy theory of failure*, and is a fairly accurate predictor of failure. But in this book, we use Equation (3-11).

The data for aluminum alloys from Appendix A-17 show the ratio of the ultimate strength in shear to the tensile ultimate strength ranges from 0.54 to 0.69 and averages 0.62. Thus it is desirable to acquire specific test data for the material to be used whenever possible.

For occasions when published data are not available, Deutschman et al. (3) give the following guidelines:

$$s_{us} = 0.82s_u \qquad \text{for steel}$$

$$s_{us} = 0.90s_u \qquad \text{for malleable iron and copper alloys}$$

$$s_{us} = 1.30s_u \qquad \text{for gray cast iron}$$

$$s_{us} = 0.65s_u \qquad \text{for aluminum alloys}$$

Where necessary to make estimates of s_{us} in problems in this book, we use this list of factors.

The values to be used for the design factor, N, in Equations (3-9) and (3-10) can be the same as those used for normal stresses as reported in Table 3-2 and Appendix A-20.

Example Problem 3-6

Figure 3-3 shows a boat propeller mounted on a shaft with a cylindrical drive pin inserted through the hub and the shaft. The torque required to drive the propeller is 1575 lb · in. and the shaft is 3.00 in. in diameter inside the hub. Usually, the torque is steady and it is desired to design the pin to be safe for this condition.

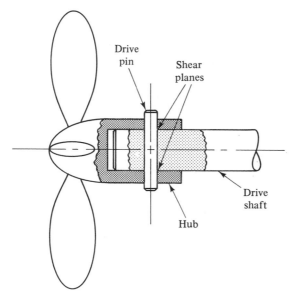

Figure 3-3 Propeller drive pin for Example Problem 3-6.

Solution   The following is a list of the major steps in this design problem.

1. Specify a suitable material for the pin.
2. Compute the design stress.
3. Compute the force exerted on the pin.
4. Solve for the required cross-sectional area of the pin.
5. Solve for the required diameter of the pin.
6. Specify a convenient size.

Step 1 requires a design decision. What properties would you recommend for the material?

It would be desirable to use a material with a fairly high strength for the pin so that the diameter is not excessively large. Also, because of the probability of shock loading from time to time, a ductile metal is preferred. Several materials meet these criteria, but let's choose AISI 1020 cold-drawn steel.

Now complete step 2.

The type of stress that will be developed in the pin is direct shear stress. Because the design is to be based on a steady torque, the design stress should be computed from Equation (3-9) with a design factor of approximately 2. Complete the calculation for the design stress now.

The design shear stress is 12 750 psi. Using $s_{ys} = 0.50s_y$ and $s_y = 51\ 000$ psi from Appendix A-13 yields

$$\tau_d = \frac{s_{ys}}{N} = \frac{0.50s_y}{2} = \frac{0.50(51\ 000\ \text{psi})}{2} = 12\ 750\ \text{psi}$$

Step 3 requires the computation of the force on the pin which tends to shear it at the interface between the shaft and the inside of the hub. Compute this now.

Figure 3-4 shows the pin passing through the hub and the shaft with two equal forces, F, acting on the pin. These two forces form a couple that must equal the applied torque of 1575 lb·in. That is,

$$F \cdot D = T = 1575\ \text{lb·in.}$$

where $D = 3.00$ in., the shaft diameter. Solving for F gives

$$F = \frac{T}{D} = \frac{1575\ \text{lb·in.}}{3.00\ \text{in.}} = 525\ \text{lb}$$

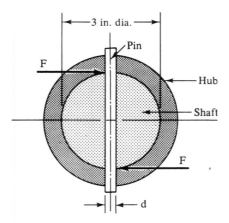

Figure 3-4 Cross section through propeller hub and shaft.

Step 4 calls for computing the required cross-sectional area for the pin. Do that now.

The required area is 0.041 in². Letting the design shear stress equal the computed shear stress, F/A, we can solve for A.

$$A = \frac{F}{\tau_d} = \frac{525 \text{ lb}}{12\ 750 \text{ lb/in}^2} = 0.041 \text{ in}^2$$

Now compute the required diameter for the pin.

The minimum acceptable diameter is 0.228 in. From

$$A = 0.041 \text{ in}^2 = \frac{\pi d^2}{4}$$

we can solve for d.

$$d = \sqrt{\frac{4A}{\pi}} = \sqrt{\frac{4(0.041 \text{ in}^2)}{\pi}} = 0.228 \text{ in}$$

Now specify a standard convenient size for the pin.

From Appendix A-2, a standard size is $\frac{1}{4}$ in. or 0.250 in. This completes the problem.

3-7 DESIGN BEARING STRESS

Bearing stress is a localized phenomenon created when two load-carrying parts are placed in contact. The stress condition is actually a compressive stress, but because of the localized nature of the stress, different allowable stresses are used.

Steel. According to the AISC, the allowable bearing stress in steel for flat surfaces or on the projected area of pins in reamed, drilled, or bored holes is

$$\sigma_{bd} = 0.90 s_y \tag{3-12}$$

When rollers or rockers are used to support a beam or other load-carrying member to allow for expansion of the member, the bearing stress is dependent on the diameter of the roller or rocker, d, and its length, L. The stress is inherently very high because the load is carried on only a small rectangular area. Theoretically, the contact between the flat surface and the roller is simply a line; but because of the elasticity of the materials, the actual area is rectangular. Instead of specifying an allowable bearing stress, the AISC standard allows the computation of the allowable bearing load, W_b, from

$$W_b = \frac{s_y - 13}{20}(0.66dL) \tag{3-13}$$

where s_y is in ksi, d and L are in inches, and W_b is in kips.

Example Problem 3-7

A short beam, shown in Figure 3-5, is made from a rectangular steel bar, 1.25 in. thick and 4.50 in. high. At each end, a length of 2.00 in. rests on a steel plate. If both the bar

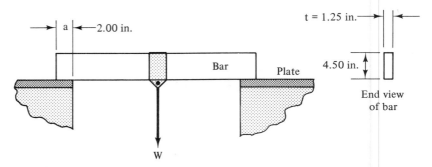

Figure 3-5 Beam for Example Problem 3-7.

and the plate are made from ASTM A36 structural steel, compute the maximum allowable load, W, which could be carried by the beam, based only on the bearing stress at the supports.

Solution The bearing stress at each support can be computed from the reaction force, R, divided by the bearing area, as described in Chapter 1.

$$\sigma_b = \frac{R}{A_b}$$

In this problem, the bearing area is known to be

$$A_b = ta = (1.25 \text{ in.})(2.00 \text{ in.}) = 2.50 \text{ in}^2$$

Also, the allowable bearing stress can be computed from Equation (3-12). Using $s_y = 36\ 000$ psi for A36 steel gives us

$$\sigma_{bd} = 0.90s_y = 0.90(36\ 000 \text{ psi}) = 32\ 400 \text{ psi}$$

Now we can solve for the allowable reaction.

$$R = A_b\sigma_{bd} = (2.50 \text{ in}^2)(32\ 400 \text{ lb/in}^2) = 81\ 000 \text{ lb}$$

Because the reaction is one-half of the total load, W,

$$W = 2R = 2(81\ 000 \text{ lb}) = 162\ 000 \text{ lb}$$

This is a very large load and the design of the beam itself may limit the load to a somewhat lesser value (see Chapters 8 and 9).

Example Problem 3-8

Figure 3-6 shows an alternative proposal for supporting the bar described in Example Problem 3-7. One end of the bar rests on a steel roller having a diameter of 2.00 in. The roller is made from AISI 1040 CD steel. For this arrangement, compute the allowable load, W.

Solution Equation (3-13) applies in this case.

$$W_b = \frac{s_y - 13}{20}(0.66dL) \tag{3-13}$$

The yield strength to be used is the smaller value for any material subjected to the bearing stress, in this case the ASTM A36 steel having a yield strength of 36 ksi. Using $d = 2.00$ in. (the diameter of the roller) and $L = 1.25$ in. (the thickness of the bar), the

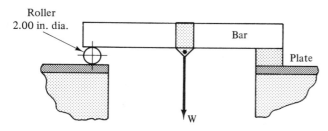

Figure 3-6 Beam for Example Problem 3-8.

allowable bearing load is

$$W_b = \frac{36 - 13}{20}(0.66)(2.00)(1.25) = 1.90 \text{ kips}$$

This would be the allowable reaction at each support. The total load is

$$W = 2W_b = 2(1.90 \text{ kips}) = 3.80 \text{ kips}$$

Note that this is significantly lower than the allowable load for the flat surfaces as found in Example Problem 3-7.

Aluminum.   The Aluminum Association (1) bases the allowable bearing stresses on aluminum alloys for flat surfaces and pins on the *bearing yield strength*.

$$\sigma_{bd} = \frac{\sigma_{by}}{2.48} \tag{3-14}$$

The minimum values for bearing yield strength are listed in Reference 1. But many references, including the appendix tables in this book, do not include these data. An analysis of the data shows that for most aluminum alloys, the bearing yield strength is approximately 1.60 times larger than the tensile yield strength. Then Equation (3-14) can be restated as

$$\sigma_{bd} = \frac{1.60s_y}{2.48} = 0.65s_y \tag{3-15}$$

We will use this form for bearing design stress in this book.

Example Problem 3-9

A rectangular bar is used as a hanger as shown in Figure 3-7. Compute the allowable load on the basis of bearing stress at the pin connection if the bar and the clevis members are made from 6061-T4 aluminum. The pin is to be made from a stronger material.

Solution   For cylindrical pins in close-fitting holes, the bearing stress is based on the *projected* area in bearing, found from the diameter of the pin times the length over which the load is distributed.

$$\sigma_b = \frac{F}{A_b} = \frac{F}{dL}$$

In this problem, the load is to be determined, the bearing area can be computed from the given dimensions, and the allowable bearing stress can be computed from Equation (3-15). Let's start by computing the allowable bearing stress.

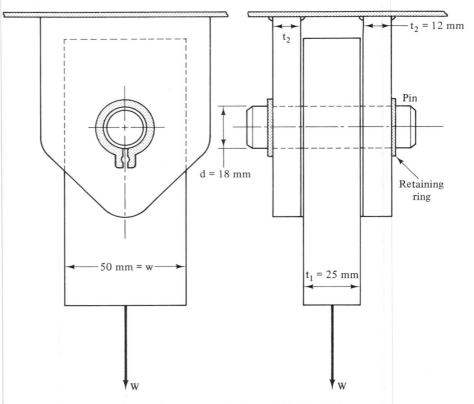

Figure 3-7 Hanger for Example Problem 3-9.

You should have σ_{bd} = 94.3 MPa. Using s_y = 145 MPa for the 6061-T4 aluminum in Equation (3-15) yields

$$\sigma_{bd} = 0.65s_y = 0.65(145 \text{ MPa}) = 94.3 \text{ MPa}$$

Now compute the bearing area.

The critical bearing area is in either of the two clevis members. Each takes one-half of the applied load, but their thickness is less than one-half that of the bar itself. Then the bearing area on one clevis member is

$$A_b = dt_2 = (18 \text{ mm})(12 \text{ mm}) = 216 \text{ mm}^2$$

We can now solve for the allowable load, W.

You should have W = 40.74 kN for the allowable load. Here is how that was found. Letting $\sigma_{bd} = \sigma_b$ gives us

$$\sigma_{bd} = \sigma_b = \frac{W/2}{A_b} = \frac{W}{2A_b}$$

Now, solving for W, we obtain

$$W = 2A_b\sigma_{bd} = 2(216 \text{ mm}^2)(94.3 \text{ N/mm}^2) = 40\ 740 \text{ N} = 40.74 \text{ kN}$$

This is a large force, and the other failure modes for the hanger would have to be analyzed.

This example problem is completed.

Masonry. The lower part of a support system is often made from concrete, brick, or stone. Loads transferred to such supports usually require consideration of bearing stresses because the strengths of these materials are relatively low compared with metals. It should be noted that the actual strengths of the materials should be used whenever possible because of the wide variation of properties. Also, some building codes list allowable bearing stresses for certain kinds of masonry.

In the absence of specific data, the AISC (2) recommends the allowable bearing stresses shown in Table 3-3.

TABLE 3-3 ALLOWABLE BEARING STRESSES ON MASONRY

Material	Allowable bearing stress	
	psi	MPa
Sandstone and limestone	400	2.76
Brick in cement mortar	250	1.72
Concrete: On full area of support	$0.35\sigma_c$	
(σ_c = Specified strength of concrete)		
σ_c = 1500 psi	525	3.62
σ_c = 2000 psi	700	4.83
σ_c = 2500 psi	875	6.03
σ_c = 3000 psi	1050	7.24
Concrete: On less than full area of support		
$\sigma_{bd} = 0.35\sigma_c\sqrt{A_2/A_1}$		
A_1 = Bearing area		
A_2 = Full area of support		
But maximum $\sigma_{bd} = 0.7\sigma_c$		

Soils. The masonry or concrete supports are often placed on soils to transfer the loads directly to the earth. *Marks' Standard Handbook for Mechanical Engineers* (5) lists the values for the safe bearing capacity of soils as shown in Table 3-4. Variations should be expected and test data should be obtained where possible.

Example Problem 3-10

Figure 1-42 shows a column resting on a foundation and carrying a load of 26 000 lb. Determine if the bearing stresses are acceptable for the concrete and the soil. The concrete has a specified strength of 2000 psi and the soil is compact gravel.

TABLE 3-4 SAFE BEARING CAPACITY OF SOILS

	Safe bearing capacity	
Nature of soil	psi	kPa
Solid hard rock	350	2400
Shale or medium rock	140	960
Soft rock	70	480
Hard clay or compact gravel	55	380
Soft clay or loose sand	15	100

Solution   The solution requires the calculation of the actual bearing stress at each critical surface and the allowable bearing stress for the materials involved. Then a comparison can be made to judge acceptability.

Let's start with the bearing stress on the top of the concrete foundation. Compute the bearing stress exerted on the concrete by the steel plate at the base of the column.

The stress is 180 psi, found from

$$\sigma_b = \frac{P}{A_b} = \frac{26\ 000\ \text{lb}}{(12\ \text{in.})^2} = 180\ \text{psi}$$

Now compute the allowable bearing stress for the concrete.

The allowable bearing stress is 1050 psi. Note that the bearing area is less than the full area of the concrete support. Then, from Table 3-3, the allowable bearing stress is

$$\sigma_{bd} = 0.35\sigma_c \sqrt{\frac{A_2}{A_1}}$$

The pertinent data are

$$\sigma_c = 2000\ \text{psi}$$
$$A_1 = (12\ \text{in.})^2 = 144\ \text{in}^2$$
$$A_2 = (18\ \text{in.})^2 = 324\ \text{in}^2$$

Then

$$\sigma_{bd} = 0.35(2000\ \text{psi})\sqrt{\frac{324}{144}} = 1050\ \text{psi}$$

Thus the bearing stress is acceptable.

Now compute the bearing stress on the gravel at the base of the foundation.

You should have $\sigma_b = 20.1$ psi, found from

$$\sigma_b = \frac{P}{A_b} = \frac{26\ 000\ \text{lb}}{(36\ \text{in.})^2} = 20.1\ \text{psi}$$

Is this acceptable?

Yes, from Table 3-4, the allowable bearing stress for compact gravel is 55 psi. This completes the example problem.

3-8 STRESS CONCENTRATION FACTORS

In defining the method for computing stress due to a direct tensile or compressive load on a member, it was emphasized that the member must have a uniform cross section in order for the equation $\sigma = P/A$ to be valid. The reason for this restriction is that wherever a change in the geometry of a loaded member occurs, the actual stress developed is higher than would be predicted by the standard equation. This phenomenon is called *stress concentration* because detailed studies reveal that localized high stresses appear to concentrate around sections where geometry changes occur.

An example of a member where a stress concentration occurs is shown in Figure 3-8(a). A circular bar is to be used as the cylinder rod in an automated packaging

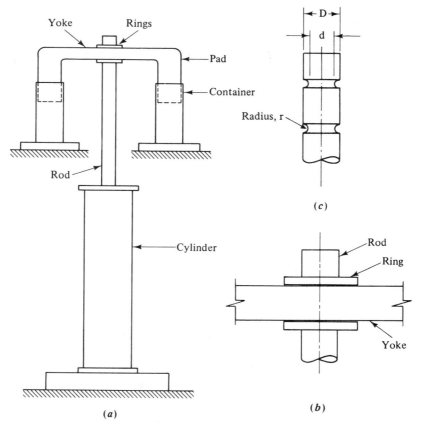

Figure 3-8 Illustration of a grooved rod in tension. (a) Overall view of packaging system (b) Connection of the yoke to the rod (c) Detail of the grooved rod

system. As the rod pulls down under the influence of the hydraulic cylinder, two pads attached to the rod compress a bulk granular chemical into containers. A maximum force of 2700 lb is exerted on the rod. The end of the rod has two circumferential grooves cut into it in order to use rings to secure the pads in position on the rod. Therefore, the rod is subjected to a tensile stress. A larger drawing of the end of the rod and a detailed drawing of the grooves are shown in parts (b) and (c) of the figure.

The stress in the rod at the location of the grooves is much higher than that which would be calculated by the axial stress formula, even when the smallest area at the bottom of the groove is used. To account for the higher stress, a modified stress formula is used whenever a change in geometry occurs. This formula is

$$\sigma = \frac{K_t P}{A} \tag{3-16}$$

where K_t is the *stress concentration factor*. The value of K_t represents the factor by which the actual stress is higher than the nominal stress computed by the standard formulas. In the present example it is related to direct tensile stress. However, stress concentration factors are used for other kinds of stresses, such as bending and torsion, which will be discussed in later chapters.

The actual geometry has a large effect on the magnitude of K_t. In the case of the grooved rod, the ratio of the major rod diameter to the smaller diameter at the root of the groove is important. Also, the sharpness of the radius at the bottom of the groove affects K_t. Data for stress concentration factors are usually presented in graphical form such as Appendix A-21. Thus, for any geometry, the value of K_t can be found.

It is most important to determine the basis on which the nominal stress is to be calculated. Normally, the smaller dimensions at a section where a change in geometry occurs are used in calculating the nominal stress. For example, the area at the root of the grooves in the rod in Figure 3-8 would be used. However, this is not universally true. In all cases where the method of computing the nominal stress is not stated, it should be assumed that the smaller section is used, in order to be on the conservative side.

Example Problem 3-11

The rod shown in Figure 3-8 is subjected to a tensile force of 2700 lb. The detailed dimensions of the rod and grooves are:

$$D = 0.50 \text{ in.}$$
$$d = 0.44 \text{ in.}$$
$$r = 0.04 \text{ in.}$$

Compute the maximum tensile stress in the rod.
Solution   Equation (3-16) will be used with $P = 2700$ lb. The area at the root of the grooves is

$$A = \frac{\pi d^2}{4} = \frac{\pi (0.44 \text{ in.})^2}{4} = 0.152 \text{ in}^2$$

To determine K_t, the ratios of D/d and r/d are needed.

$$\frac{D}{d} = \frac{0.50}{0.44} = 1.136$$

$$\frac{r}{d} = \frac{0.04}{0.44} = 0.09$$

Using Appendix A-21-1, $K_t = 2.25$. Then

$$\sigma = \frac{K_t P}{A}$$

$$= \frac{(2.25)(2700 \text{ lb})}{0.152 \text{ in}^2} = 39\ 950 \text{ lb/in}^2$$

Notice that neglecting the stress concentration and using $\sigma = P/A$ would have produced a calculated stress of only 17 800 lb/in^2. Stress concentrations can be very critical in design.

Other situations where stress concentrations occur are described in Appendix A-21, along with the graphs for determining the value of K_t. It is expected that all problem solutions will consider the effect of stress concentrations where they exist.

REFERENCES

1. Aluminum Association, *Specifications for Aluminum Structures*, Washington, D.C., 1982.
2. American Institute of Steel Construction, *Manual of Steel Construction*, 8th ed., Chicago, 1980.
3. Deutschman, A. D., W. J. Michels, and C. E. Wilson, *Machine Design Theory and Practice*, Macmillan, New York, 1975.
4. Juvinall, R. C., *Fundamentals of Machine Component Design*, Wiley, New York, 1983.
5. Baumeister, T., Editor-in-Chief, *Marks' Standard Handbook for Mechanical Engineers*, 8th ed., McGraw-Hill, New York, 1978.
6. Mott, R. L., *Machine Elements in Mechanical Design*, Charles E. Merrill, Columbus, Ohio, 1985.
7. Shigley, J. E., *Mechanical Engineering Design*, 4th ed., McGraw-Hill, New York, 1984.
8. Spotts, M. F., *Design of Machine Elements*, 6th ed., Prentice-Hall, Englewood Cliffs, N.J., 1985.

PROBLEMS

3-1. Specify a suitable aluminum alloy for a round bar having a diameter of 10 mm subjected to a static direct tensile force of 8.50 kN.

3-2. A rectangular bar having cross-sectional dimensions of 10 mm by 30 mm is subjected to a direct tensile force of 20.0 kN. If the force is to be repeated many times, specify a suitable steel material.

3-3. A link in a mechanism for an automated packaging machine is subjected to a direct tensile force of 1720 lb, repeated many times. The link is square, 0.40 in. on a side. Specify a suitable steel for the link.

3-4. A circular steel rod $\frac{3}{8}$ in. in diameter supports a heater assembly and carries a static tensile load of 1850 lb. Specify a suitable structural steel for the rod.

3-5. A tension member in a wood roof truss is to carry a static tensile force of 5200 lb. It has been proposed to use a standard 2 × 4 made from southern pine, No. 2 grade. Would this be acceptable?

3-6. For the data in Problem 3-5, suggest an alternative design that would be safe for the given loading. A different size member or a different material may be specified.

3-7. A guy wire for an antenna tower is to be aluminum, having an allowable stress of 12 000 psi. If the expected maximum load on the wire is 6400 lb, determine the required diameter of the wire.

3-8. A hopper having a mass of 1150 kg is designed to hold a load of bulk salt having a mass of 6350 kg. The hopper is to be suspended by four rectangular straps, each carrying one-fourth of the load. Steel plate with a thickness of 8.0 mm is to be used to make the straps. What should be the width in order to limit the stress to 70 MPa?

3-9. A shelf is being designed to hold crates having a total mass of 1840 kg. Two support rods like that shown in Figure 3-9 will hold the shelf. Assume that the center of gravity of the crates is at the middle of the shelf. Specify the required diameter of the circular rods to limit the stress to 110 MPa.

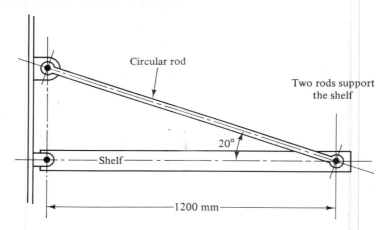

Figure 3-9 Shelf support rods for Problem 3-9.

3-10. A concrete column base is circular, 8.0 in. in diameter, and carries a static direct compressive load of 70 000 lb. Specify the required rated strength of the concrete according to the recommendations in Section 2-10.

3-11. Three short wood blocks made from standard 4 × 4 posts support a machine weighing 29 500 lb and share the load equally. Specify a suitable type of wood for the blocks.

3-12. A circular pier to support a column is to be made of concrete having a rated strength of 3000 psi (20.7 MPa). Specify a suitable diameter for the pier if it is to carry a direct compressive load of 1.50 MN.

3-13. An aluminum ring has an outside diameter of 12.0 mm and an inside diameter of 10 mm. If the ring is short and is made of 2014-T6, compute the force required to produce ultimate compressive failure in the ring. Assume that s_u is the same in both tension and compression.

3-14. A wooden cube 40 mm on a side is made from No. 2 hemlock. Compute the allowable compressive force that could be applied to the cube either parallel or perpendicular to the grain.

3-15. A round bar of ASTM A242 structural steel is to be used as a tension member to stiffen a frame. If a maximum static load of 4000 lb is expected, specify a suitable diameter for the rod.

3-16. A portion of a casting made of ASTM A48, grade 20 gray cast iron has the shape shown in Figure 1-28 and is subjected to compressive force in line with the centroidal axis of the section. If the member is short and carries a load of 52 000 lb, compute the stress in the section and the design factor.

3-17. A part for a truck suspension system is to carry a compressive load of 135 kN with the possibility of shock loading. Malleable iron ASTM A220 grade 45008 is to be used. The cross section is to be rectangular with the long dimension twice the short dimension. Specify suitable dimensions for the part.

3-18. A rectangular plastic link in an office printer is to be made from glass-filled acetal copolymer (see Appendix A-19). It is to carry a tensile force of 110 lb. Space limitations permit a maximum thickness for the link of 0.20 in. Specify a suitable width of the link if a design factor of 8 based on the tensile strength of the plastic is expected.

3-19. Figure 1-29 shows the cross section of a short compression member that is to carry a static load of 640 kN. Specify a suitable material for the member.

3-20. Figure 1-21 shows a bar carrying several static loads. If the bar is made from ASTM A36 structural steel, is it safe?

3-21. In Figure 1-26, specify a suitable aluminum alloy for the member AB if the load is to be repeated many times. Consider only the square part near the middle of the member.

3-22. Figure 3-10 shows the design of the lower end of the member AB shown in Figure 1-26. Use the load shown in Figure 1-26 and assume that it is static. Member 1 is made from aluminum alloy 6061-T4; member 2 is made from aluminum alloy 2014-T4; pin 3 is made from aluminum alloy 2014-T6. Perform the following analyses.

(**a**) Evaluate member 1 for safety in tension in the area of the pin holes.

(**b**) Evaluate member 1 for safety in bearing at the pin.

(**c**) Evaluate member 2 for safety in bearing at the pin.

(**d**) Evaluate the pin 3 for safety in bearing.

(**e**) Evaluate the pin 3 for safety in shear.

3-23. A standard steel angle, L2 × 2 × $\frac{1}{4}$, serves as a tension member in a truss carrying a static load. If the angle is made from ASTM A36 structural steel, compute the allowable tensile load based on the AISC specifications.

3-24. A machine weighs 90 kN and rests on four legs. Two concrete supports serve as the foundation. The load is symmetrical, as shown in Figure 3-11. Evaluate the proposed design with regard to the safety of the steel plate, the concrete foundation, and the soil for bearing.

3-25. A column for a building is to carry 160 kN. If the column is to be supported on a square concrete foundation that sits on soft rock, determine the required dimensions of the foundation block.

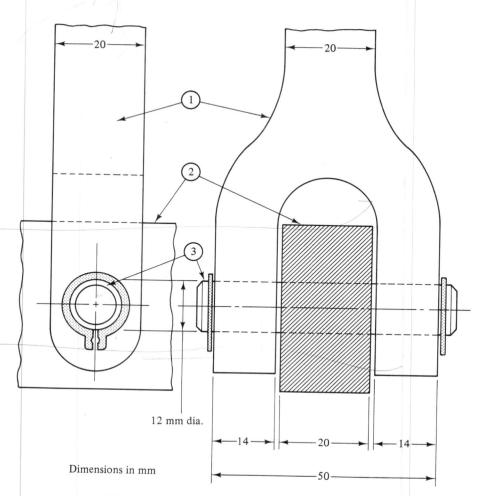

Figure 3-10 Clevis for beam support for Problem 3-22.

3-26. A base for moving heavy machinery is designed as shown in Figure 3-12. How much load could the base support based on the bearing capacity of the steel plate if it is 1.25 in. thick and made from ASTM A36 structural steel?

3-27. Repeat Problem 3-26 but use ASTM A242 high-strength low-alloy steel plate.

3-28. Figure 3-13 shows an alternative design for the machinery moving base described in Problem 3-26. Compute the allowable load for this design if it rests on (a) ASTM A36 steel or (b) ASTM A242 steel.

3-29. Two $4 \times 2 \times \frac{1}{4}$ steel beams are used to support a machine weighing a total of 4500 lb. Each foot of the machine transfers one-fourth of the load to the 2-in. top surface of the beam over a length of 1.25 inches. Compute the bearing stress at the foot and compare it with the allowable for ASTM A500 structural steel, grade A.

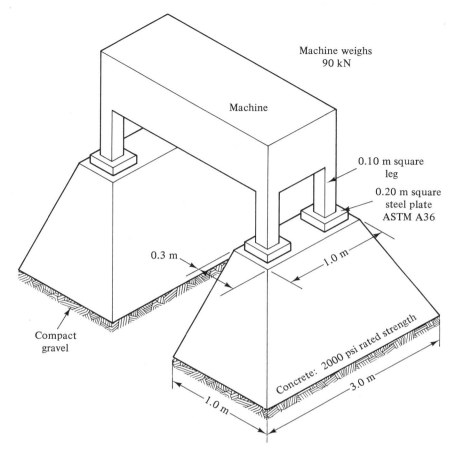

Machine weighs
90 kN

Machine

0.10 m square
leg

0.20 m square
steel plate
ASTM A36

0.3 m

1.0 m

Compact
gravel

Concrete: 2000 psi rated strength

3.0 m

1.0 m

Figure 3-11 Machine supports for Problem 3-24.

3-30. One end of a beam is supported on a rocker having a radius of 200 mm and a width of 150 mm. If the rocker and the plate on which it rests are made of ASTM A36 structural steel, specify the maximum allowable reaction at this end of the beam.

3-31. A gear transmits 550 N · m of torque to a circular shaft 2.00 in. in diameter. A square key 0.50 in. on a side connects the shaft to the hub of the gear as shown in Figure 1-34. The key is made from AISI 1020 cold-drawn steel. Determine the required length of the key, L, to be safe for shear and bearing. Use a design factor of 2.0 based on yield for shear and the AISC allowable bearing stress.

3-32. A support for a beam is made as shown in Figure 3-14. Determine the required thickness of the projecting ledge a if the maximum shear stress is to be 6000 psi. The load on the support is 21 000 lb.

3-33. A section of pipe is supported by a saddle-like structure which, in turn, is supported on two steel pins, as illustrated in Figure 3-15. If the load on the saddle is 42 000 lb, determine the required diameter and length of the pins. Use AISI 1040 cold-drawn steel.

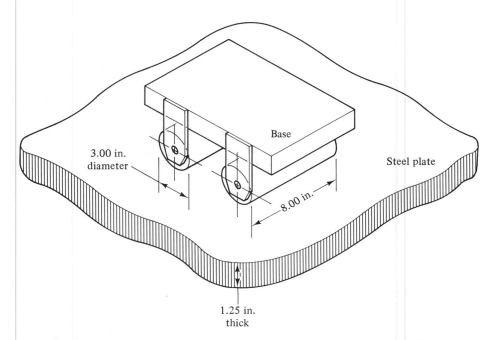

Figure 3-12 Roller base for moving machinery for Problems 3-26 and 3-27.

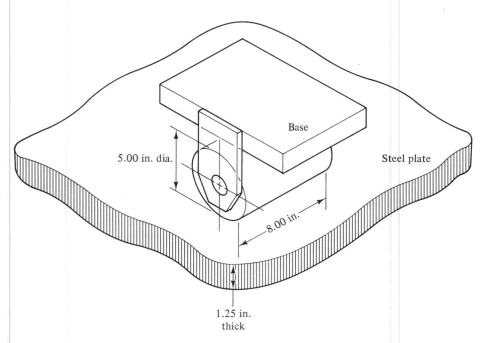

Figure 3-13 Roller base for moving machinery for Problem 3-28.

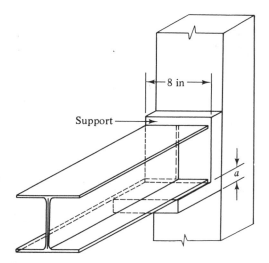

Figure 3-14 Beam support for Problem 3-32.

Figure 3-15 Pipe saddle for Problem 3-33.

Lab. **3-34.** The lower control arm on an automotive suspension system is connected to the frame by a round steel pin 16 mm in diameter. Two sides of the arm transfer loads from the frame to the arm, as sketched in Figure 3-16. How much shear force could the pin withstand if it is made of AISI 1040 cold-drawn steel and a design factor of 6 based on the yield strength in shear is desired?

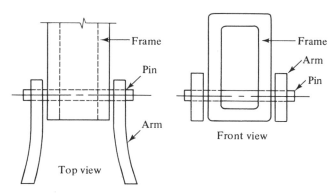

Figure 3-16 Pin for automotive suspension system for Problem 3-34.

3-35. A centrifuge is used to separate liquids according to their densities using centrifugal force. Figure 1-19 illustrates one arm of a centrifuge having a bucket at its end to hold the liquid. In operation, the bucket and the liquid have a mass of 0.40 kg. The centrifugal force has the magnitude in newtons of

$$F = 0.010\ 97 \cdot m \cdot R \cdot n^2$$

where m = rotating mass of bucket and liquid (kg)
R = radius to center of mass (meters)
n = rotational speed (rpm)

The centrifugal force places the pin holding the bucket in direct shear. Compute the stress in the pin due to a rotational speed of 3000 rpm. Then, specify a suitable steel for the pin, considering the load to be repeated.

3-36. A circular punch is used to punch a 20.0-mm-diameter hole in a sheet of AISI 1020 hot-rolled steel having a thickness of 8.0 mm. Compute the force required to punch out the slug.

3-37. Repeat Problem 3-36, but the material is aluminum 6061-T4.

3-38. Repeat Problem 3-36, but the material is C14500 copper, hard.

3-39. Repeat Problem 3-36, but the material is AISI 430 cold-worked stainless steel.

3-40. Determine the force required to punch a slug the shape shown in Figure 1-32 from a sheet of AISI 1020 hot-rolled steel having a thickness of 5.0 mm.

3-41. Determine the force required to punch a slug the shape shown in Figure 1-33 from a sheet of aluminum 3003-H18 having a thickness of 0.194 in.

3-42. A notch is made in a piece of wood, as shown in Figure 1-31, to support an external load of 1800 lb. Compute the shear stress in the wood. Is the notch safe? (See Appendix A-18.)

3-43. Compute the force required to shear a straight edge of a sheet of AISI 1040 cold-drawn steel having a thickness of 0.105 in. The length of the edge is 7.50 in.

3-44. Repeat Problem 3-43 but the material is AISI 5160 OQT 700 steel.

3-45. Repeat Problem 3-43 but the material is AISI 301 cold-worked stainless steel.

3-46. Repeat Problem 3-43 but the material is C26000 brass, hard.

3-47. Repeat Problem 3-43, but the material is aluminum 5154-H32.

3-48. For the lever shown in Figure 3-17, called a *bellcrank*, compute the required diameter of the pin A if the load is repeated and the pin is made from C17200 beryllium copper, hard. The loads are repeated many times.

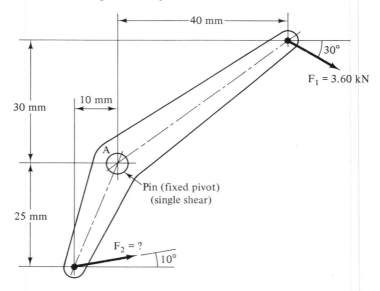

Figure 3-17   Bell crank for Problem 3-48.

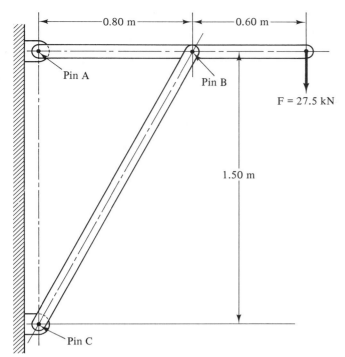

Figure 3-18 Pin-connected structure for Problem 3-49.

3-49. For the structure shown in Figure 3-18 determine the required diameter of each pin if it is made from AISI 1020 cold-drawn steel. Each pin is in double shear and the load is static.

3-50. For the structure shown in Figure 1-24a determine the required diameter of each pin if it is made from ASTM A572 high-strength low-alloy columbium-vanadium structural steel, grade 50. Each pin is in double shear and the load is static.

3-51. A pry bar as shown in Figure 3-19 is used to provide a large mechanical advantage for lifting heavy machines. An operator can exert a force of 280 lb on the handle. Compute the lifting force and the shear stress on the wheel axle.

3-52. Figure 3-20 illustrates a type of engineering chain used for conveying applications. All components are made from AISI 1040 cold-drawn steel. Evaluate the allowable tensile force (repeated) on the chain with respect to:
(a) Shear of the pin
(b) Bearing of the pin on the side plates
(c) Tension in the side plates.

3-53. Figure 3-21 shows an anvil for an impact hammer held in a fixture by a circular pin. If the force is to be 500 lb, specify a suitable diameter for the steel pin if it is to be made from AISI 1040 WQT 900.

3-54. Figure 3-22 shows a steel flange forged integrally with the shaft, which is to be loaded in torsion. Eight bolts serve to couple the flange to a mating flange. Assume that each bolt carries an equal load. Compute the maximum permissible torque on the coupling if the shear stress in the bolts is not to exceed 6000 psi.

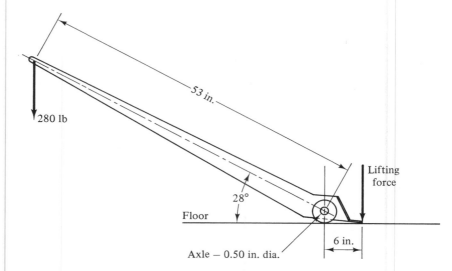

Figure 3-19 Pry bar for Problem 3-51.

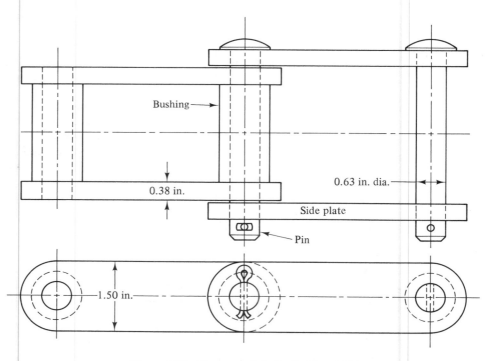

Figure 3-20 Conveyor chain for Problem 3-52.

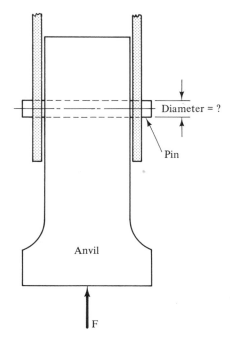

Diameter = ?

Pin

Anvil

F

Figure 3-21 Impact hammer for Problem 3-53.

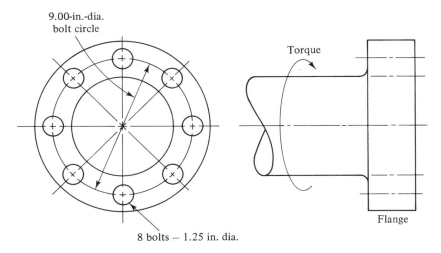

9.00-in.-dia. bolt circle

Torque

Flange

8 bolts — 1.25 in. dia.

Figure 3-22 Flange coupling for Problem 3-54.

3-55. Figure 3-23 shows a piston, connecting rod, and crank from an engine. The connecting rod is subjected to a repeated axial load of 3000 lb during operation. Specify a suitable size for the square cross section of the connecting rod if it is to be made from AISI 5160 OQT 1100 steel. Consider the stress concentration at the lower end of the rod.

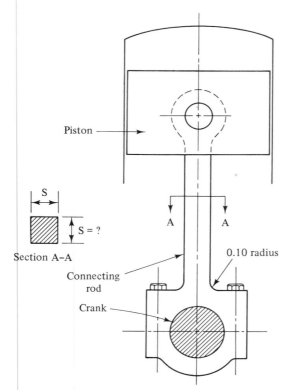

Piston

S

S = ?

Section A–A

Connecting
rod

0.10 radius

Crank

A A

Figure 3-23 Connecting rod for Problem 3-55.

3-56. A valve stem in an automotive engine is subjected to an axial tensile load of 900 N due to the valve spring, as shown in Figure 3-24. Compute the maximum stress in the stem at the place where the spring force acts against the shoulder.

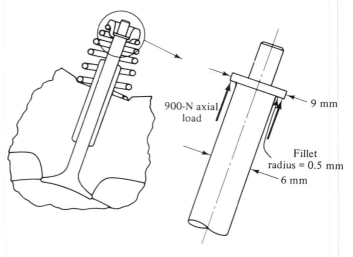

900-N axial
load

9 mm

Fillet
radius = 0.5 mm

6 mm

Figure 3-24 Valve stem for Problem 3-56.

3-57. A round shaft has two grooves in which rings are placed to retain a gear in position, as shown in Figure 3-25. If the shaft is subjected to an axial tensile force of 36 kN, compute the maximum tensile stress in the shaft.

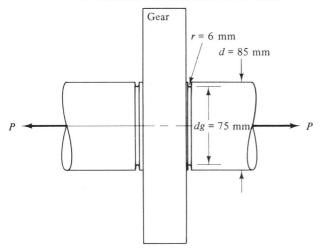

Figure 3-25 Shaft for Problem 3-57.

3-58. Figure 3-26 shows the proposed design for a tensile rod. The larger diameter is known, $D = 1.00$ in., along with the hole diameter, $a = 0.50$ in. It has also been decided that the stress concentration factor at the fillet is to be 1.7. The smaller diameter, d, and the fillet radius, r, are to be specified such that the stress at the fillet is the same as that at the hole.

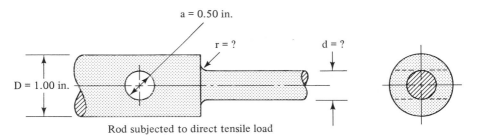

Rod subjected to direct tensile load

Figure 3-26 Tensile rod for Problem 3-58.

4

Deformation and Thermal Stress

4-1 OBJECTIVES OF THIS CHAPTER

The study of strength of materials involves both the determination of stresses in load-carrying members and the deflection or deformation of those members. In general terms, this requires the study of both *stress* and *strain*, as defined in Chapter 1. The material presented in Chapters 1, 2, and 3 enabled you to compute the magnitude of stresses created in members subjected to direct axial forces, either tensile or compressive. This chapter extends your knowledge of such members to include deformation.

Two kinds of deformation are presented in this chapter, *elastic deformation* due to the application of external loads, and *thermal deformation* due to changes in temperature. When a material is heated it tends to expand, and when it is cooled it tends to contract. If these thermal deformations are permitted to occur without restraint, no stresses are created. But if the member is held in such a way as to restrain movement, stresses are developed. Such stresses are called *thermal stresses*.

The principles of elastic deformation can also be used to solve some fairly complex problems in which members made of more than one material are subjected to loads. Such members are often *statically indeterminate*; that is, the internal forces and stresses cannot be found by simple equations of statics. We will show the combined use of statics, stress analysis, and elastic deformation analysis to solve such problems.

After completing this chapter, you should be able to:

1. Compute the amount of elastic deformation for a member carrying an axial tensile or compressive load.
2. Design axially loaded members to limit their deformation to a set value.
3. Define *coefficient of thermal expansion*, and select the appropriate value for use in computing thermal deformation.
4. Compute the amount of thermal deformation for a member subjected to changes in temperature when the deformation is unrestrained.
5. Compute the thermal stress resulting from the restraint of a member subjected to changes in temperature.
6. Compute the stress in components of a composite structure having members made of more than one material subjected to axial loads.

4-2 ELASTIC DEFORMATION IN TENSION AND COMPRESSION MEMBERS

Deformation refers to some change in the dimensions of a load-carrying member. Being able to compute the magnitude of deformation is important in the design of precision mechanisms, machine tools, building structures, and machine structures.

An example of where deformation is important is shown in Figure 3-2. The tie rods are subjected to tension when in operation. Since they contribute to the rigidity of the press, the amount that they deform under load is something the designer needs to be able to determine.

To develop the relationship from which deformation can be computed for members subjected to axial tension or compression, some concepts from Chapter 1 must be reviewed. *Strain* is defined as the ratio of the total deformation to the original length of a member. Using the symbols ϵ for strain, δ for total deformation, and L for length, the formula for strain becomes

$$\epsilon = \frac{\delta}{L} \tag{4-1}$$

The stiffness of a material is a function of its modulus of elasticity E, defined as

$$E = \frac{\text{stress}}{\text{strain}} = \frac{\sigma}{\epsilon} \tag{4-2}$$

Solving for strain gives

$$\epsilon = \frac{\sigma}{E} \tag{4-3}$$

Now Equations (4-1) and (4-3) can be equated:

$$\frac{\delta}{L} = \frac{\sigma}{E} \qquad (4\text{-}4)$$

Solving for deformation gives

$$\delta = \frac{\sigma L}{E} \qquad (4\text{-}5)$$

Since this formula applies to members which are subjected to either direct tensile or compressive forces, the direct-stress formula can be used to compute the stress σ. That is, $\sigma = P/A$, where P is the applied load and A is the cross-sectional area of the member. Substituting this into Equation (4-5) gives

$$\delta = \frac{\sigma L}{E} = \frac{PL}{AE} \qquad (4\text{-}6)$$

Equation (4-6) can be used to compute the total deformation of any load-carrying member, provided that it meets the conditions defined for direct tensile and compressive stress. That is, the member must be straight and have a constant cross section; the material must be homogeneous; the load must be directly axial; and the stress must be below the proportional limit of the material.

Example Problem 4-1

The tie rods in the press in Figure 3-2 are made of steel and have a diameter of 2.00 in. The length of each rod before installation is 68.5 in. If each rod is subjected to an axial tensile load of 40 000 lb, compute the deformation of each rod.

Solution   Equation (4-6) will be used to compute the deformation, in this case an elongation, or stretching, of the rod. The conditions on Equation (4-6) as described above are met if the stress is below the proportional limit.

$$\sigma = \frac{P}{A}$$

$$P = 40\ 000\ \text{lb}$$

$$A = \frac{\pi D^2}{4} = \frac{\pi (2.0\ \text{in.})^2}{4} = 3.14\ \text{in}^2$$

Then

$$\sigma = \frac{40\ 000\ \text{lb}}{3.14\ \text{in}^2} = 12\ 700\ \text{psi}$$

This level of stress is well below the proportional limit of any steel.
Now from Equation (4-6),

$$\delta = \frac{PL}{AE}$$

We know $P = 40\ 000\ \text{lb}$, $L = 68.5\ \text{in.}$, and $A = 3.14\ \text{in}^2$. From the notes for Appendix A-13, it is found that $E = 30 \times 10^6$ psi is approximately correct for any carbon or alloy steel. Then

$$\delta = \frac{PL}{AE} = \frac{(40\ 000\ \text{lb})(68.5\ \text{in.})}{(3.14\ \text{in}^2)(30 \times 10^6\ \text{lb/in}^2)} = 0.029\ \text{in.}$$

Example Problem 4-2

A large pendulum is composed of a 10.0-kg ball suspended by an aluminum wire having a diameter of 1.00 mm and a length of 6.30 m. The aluminum is the alloy 7075-T6. Compute the elongation of the wire due to the weight of the 10-kg ball.

Solution   To use Equation (4-6) to compute the elongation, the stress in the wire must be below the proportional limit for the 7075-T6 aluminum alloy.

$$\sigma = \frac{P}{A}$$

The force P is equal to the weight of the 10.0-kg ball.

$$P = m \cdot g = 10.0\ \text{kg} \cdot 9.81\ \text{m/s}^2 = 98.1\ \text{N}$$

$$A = \frac{\pi D^2}{4} = \frac{\pi (1.00\ \text{mm})^2}{4} = 0.785\ \text{mm}^2$$

Then

$$\sigma = \frac{P}{A} = \frac{98.1\ \text{N}}{0.785\ \text{mm}^2} = 125\ \text{N/mm}^2 = 125\ \text{MPa}$$

The proportional limit for 7075-T6 aluminum alloy would be close to its yield strength. From Appendix A-17, the yield strength is 503 MPa. Therefore, the stress is well below the proportional limit.

In this case it is most convenient to use Equation (4-5) to compute the deformation since stress is already known.

$$\delta = \frac{\sigma L}{E}$$

Also from Appendix A-17, $E = 72$ GPa for 7075-T6 aluminum.
Then

$$\delta = \frac{\sigma L}{E} = \frac{(125\ \text{MPa})(6.30 \text{m})}{72\ \text{GPa}} = \frac{(125 \times 10^6\ \text{Pa})(6.30\ \text{m})}{72 \times 10^9\ \text{Pa}}$$

$$= 10.9 \times 10^{-3}\ \text{m} = 10.9\ \text{mm}$$

Example Problem 4-3

A tension link in a machine must have a length of 610 mm and will be subjected to a maximum axial load of 3000 N. It has been proposed that the link be made of steel and that it have a square cross section. Determine the required dimensions of the link if the elongation under load must not exceed 0.05 mm.

Solution   The objective of the solution is to determine the dimensions of the square link. Thus the area is unknown. Then in Equation (4-6),

$$\delta = \frac{PL}{AE}$$

we can solve for A.

$$A = \frac{PL}{E\delta}$$

we let $P = 3000$ N, $L = 610$ mm, and $\delta = 0.05$ mm. To keep consistent units, it is convenient to express E in the units of N/m^2. Appendix A-13 gives $E = 207$ GPa. Then we will use $E = 207 \times 10^9$ N/m^2. Now

$$A = \frac{PL}{E\delta} = \frac{(3000 \text{ N})(610 \text{ mm})}{(207 \times 10^9 \text{ N/m}^2)(0.05 \text{ mm})} = 176.8 \times 10^{-6} \text{ m}^2$$

Converting to mm^2 yields

$$A = 176.8 \times 10^{-6} \text{ m}^2 \times \frac{(10^3 \text{ mm})^2}{\text{m}^2} = 176.8 \text{ mm}^2$$

If we call each side of the square cross section of the link d,

$$A = d^2$$

and

$$d = \sqrt{A} = \sqrt{176.8 \text{ mm}^2} = 13.3 \text{ mm}$$

Therefore, the link must have cross-sectional dimensions of at least 13.3 mm by 13.3 mm.

Example Problem 4-4

Figure 1-22 shows a steel pipe being used as a bracket to support equipment through cables attached as shown. The forces are $F_1 = 8000$ lb and $F_2 = 2500$ lb. Select the smallest standard schedule 40 steel pipe that will limit the stress to no more than 18 000 psi. Then for the pipe selected, determine the total downward deflection of point C at the bottom of the pipe as the loads are applied.

Solution The maximum load in the pipe will occur between points A and B, where the pipe is subjected to the sum of F_2 and the vertical components of both F_1 forces. That is,

$$P_{A-B} = F_2 + 2 F_1 \cos 30°$$

$$= 2500 \text{ lb} + 2(8000 \text{ lb}) \cos 30°$$

$$= 16\ 400 \text{ lb}$$

Then, letting $\sigma = 18\ 000$ psi, the required cross-sectional area of the metal in the pipe is

$$A = \frac{P}{\sigma} = \frac{16\ 400 \text{ lb}}{18\ 000 \text{ lb/in}^2} = 0.911 \text{ in}^2$$

From Appendix A-12, listing the properties of steel pipe, the standard size with the next larger cross-sectional area is the 2-in. schedule 40 pipe with $A = 1.075$ in^2.

In computing the deflection of point C, it is necessary to observe that the force in the pipe is 16 400 lb between A and B but only 2500 lb between B and C. Therefore, the deformations of the two segments must be computed separately.

$$\delta_{A-B} = \left(\frac{PL}{AE}\right)_{A-B} = \frac{(16\ 400 \text{ lb})(48 \text{ in.})}{(1.075 \text{ in}^2)(30 \times 10^6 \text{ lb/in}^2)} = 0.024 \text{ in.}$$

$$\delta_{B-C} = \left(\frac{PL}{AE}\right)_{B-C} = \frac{(2500 \text{ lb})(36 \text{ in.})}{(1.075 \text{ in}^2)(30 \times 10^6 \text{ lb/in}^2)} = 0.003 \text{ in.}$$

Then

$$\delta_C = \delta_{A-B} + \delta_{B-C} = 0.027 \text{ in.}$$

Point C moves downward 0.027 in. as the loads are applied to the 2-in. schedule 40 pipe.

4-3 DEFORMATION DUE TO TEMPERATURE CHANGES

A machine or a structure could undergo deformation or be subjected to stress by changes in temperature in addition to the application of loads. Bridge members and other structural components see temperatures as low as $-30°F$ ($-34°C$) to as high as $110°F$ ($43°C$) in some areas. Vehicles and machinery operating outside experience similar temperature variations. Frequently, a machine part will start at room temperature and then become quite hot as the machine operates. Examples are parts of engines, furnaces, metal-cutting machines, rolling mills, plastics molding and extrusion equipment, food-processing equipment, air compressors, hydraulic and pneumatic devices, and high-speed automation equipment.

As a metal part is heated, it tends to expand. If the expansion is unrestrained, the dimensions of the part will grow but no stress will be developed in the metal. However, in some cases the part is restrained, preventing the change in dimensions. Under such circumstances, stresses will occur.

Different materials expand at different rates when subjected to temperature changes. The property of a material that indicates its tendency to change dimensions with a change in temperature is its coefficient of thermal expansion. The lowercase Greek letter α denotes this coefficient. It is a measure of the change in length of a material per unit length for a 1-degree change in temperature. The units, then, for α in the U.S. Customary system would be

$$\text{in./(in.} \cdot °F) \quad \text{or} \quad 1/°F \quad \text{or} \quad °F^{-1}$$

In SI units, α would be

$$\text{m/(m} \cdot °C) \quad \text{or} \quad \text{mm/(mm} \cdot °C) \quad \text{or} \quad 1/°C \quad \text{or} \quad °C^{-1}$$

For use in computations, the last form of each unit type is most convenient. However, the first form will help you remember the physical meaning of the term.

It follows from the definition of the coefficient of thermal expansion that the change in length δ of a member can be computed from the equation

$$\delta = \alpha \cdot L \cdot \Delta t \tag{4-7}$$

where L = original length of the member
Δt = change in temperature

Table 4-1 gives values for the coefficient of thermal expansion for several common materials. The actual values vary somewhat with temperature. However, the

TABLE 4-1 COEFFICIENTS OF THERMAL EXPANSION, α

Material	α	
	$°F^{-1}$	$°C^{-1}$
Steel, AISI		
1020	6.5×10^{-6}	11.7×10^{-6}
1040	6.3×10^{-6}	11.3×10^{-6}
4140	6.2×10^{-6}	11.2×10^{-6}
Structural steel	6.5×10^{-6}	11.7×10^{-6}
Gray cast iron	6.0×10^{-6}	10.8×10^{-6}
Stainless steel		
AISI 301	9.4×10^{-6}	16.9×10^{-6}
AISI 430	5.8×10^{-6}	10.4×10^{-6}
AISI 501	6.2×10^{-6}	11.2×10^{-6}
Aluminum alloys		
2014	12.8×10^{-6}	23.0×10^{-6}
6061	13.0×10^{-6}	23.4×10^{-6}
7075	12.9×10^{-6}	23.2×10^{-6}
Brass, C26000	11.1×10^{-6}	20.0×10^{-6}
Bronze, C22000	10.2×10^{-6}	18.4×10^{-6}
Copper, C14500	9.9×10^{-6}	17.8×10^{-6}
Magnesium, ASTM AZ63A-T6	14.0×10^{-6}	25.2×10^{-6}
Titanium, Ti-6A1-4V	5.3×10^{-6}	9.5×10^{-6}
Plate glass	5.0×10^{-6}	9.0×10^{-6}
Wood (pine)	3.0×10^{-6}	5.4×10^{-6}
Concrete	6.0×10^{-6}	10.8×10^{-6}

numbers in Table 4-1 are approximately average values over the range of temperature from 0°C (32°F) to 100°C (212°F).

Example Problem 4-5

A rod made from AISI 1040 steel is used as a link in a steering mechanism of a large truck. If its nominal length is 56 in., compute its change in length as the temperature changes from −30°F to 110°F.

Solution Using Equation (4-7), we find the change in length to be

$$\delta = \alpha \cdot L \cdot \Delta t$$

From Table 4-1 we find $\alpha = 6.3 \times 10^{-6}°F^{-1}$. The term Δt is the change in temperature; thus

$$\Delta t = 110°F - (-30°F) = 140°F$$

Then

$$\delta = (6.3 \times 10^{-6}°F^{-1})(56 \text{ in.})(140°F) = 0.049 \text{ in.}$$

Example Problem 4-6

A pushrod in the valve mechanism of an automotive engine has a nominal length of 203 mm. If the rod is made of AISI 4140 steel, compute the elongation due to a temperature change from −20°C to 140°C.

Solution   Again using Equation (4-7), we have

$$\delta = \alpha \cdot L \cdot \Delta t$$

From Table 4-1, we find $\alpha = 11.2 \times 10^{-6}°C^{-1}$. The change in temperature is

$$\Delta t = 140°C - (-20°C) = 160°C$$

Then

$$\delta = (11.2 \times 10^{-6}°C^{-1})(203 \text{ mm})(160°C) = 0.364 \text{ mm}$$

It would be important to accommodate this expansion in the design of the valve mechanism.

Example Problem 4-7

An aluminum frame of 6061 alloy for a window is 4.350 m long and holds a piece of plate glass 4.347 m long when the temperature is 35°C. At what temperature would the aluminum and glass be the same length?

Solution   The temperature would have to decrease in order for the aluminum and glass to reach the same length, since aluminum contracts at a greater rate than glass. As the temperature decreases, the change in temperature Δt would be the same for both the aluminum and the glass. After the temperature change, the length of the aluminum would be

$$L_{a2} = L_{a1} - \alpha_a \cdot L_{a1} \cdot \Delta t$$

where the subscript a refers to the aluminum, 1 refers to the initial condition, and 2 refers to the final condition. The length of the glass would be

$$L_{g2} = L_{g1} - \alpha_g \cdot L_{g1} \cdot \Delta t$$

But when the glass and the aluminum have the same length,

$$L_{a2} = L_{g2}$$

Then

$$L_{a1} - \alpha_a \cdot L_{a1} \cdot \Delta t = L_{g1} - \alpha_g \cdot L_{g1} \cdot \Delta t$$

Solving for Δt gives

$$\Delta t = \frac{L_{a1} - L_{g1}}{\alpha_a \cdot L_{a1} - \alpha_g \cdot L_{g1}}$$

Then from Table 4-1 we find that $\alpha_a = 23.4 \times 10^{-6}°C^{-1}$ and $\alpha_g = 9.0 \times 10^{-6}°C^{-1}$.

$$\Delta t = \frac{4.350 \text{ m} - 4.347 \text{ m}}{(23.4 \times 10^{-6}°C^{-1})(4.350 \text{ m}) - (9.0 \times 10^{-6}°C^{-1})(4.347 \text{ m})}$$

$$= \frac{0.003}{(0.000102) - (0.000039)} °C = 48°C$$

If the temperature decreases by 48°C from the original temperature of 35°C, the resulting temperature would be −13°C. Since this is well within the possible ambient temperature for a building, a dangerous condition could be created by this window. The window frame and glass would contract without stress until a temperature of −13°C was reached. If the temperature continued to decrease, the frame would contract faster than the glass and would generate stress in the glass. Of course, if the stress is great enough, the glass would fracture, possibly causing injury. The window should be reworked so there is a larger difference in size between the glass and the aluminum frame.

Stresses due to thermal expansion and contraction when members are restrained is discussed in the next section.

4-4 THERMAL STRESS

In the preceding section, parts which were subjected to changes in temperature were unrestrained, so that they could grow or contract freely. If the parts were held in such a way that deformation was resisted, stresses would be developed.

Consider a steel structural member in a furnace that is heated while the members to which it is attached are kept at a lower temperature. Assuming the ideal case, the supports would be considered rigid and immovable. Thus all expansion of the steel member would be prevented.

If the steel part were allowed to expand, it would elongate by an amount $\delta = \alpha \cdot L \cdot \Delta t$. But since it is restrained, this represents the apparent total strain in the steel. Then the unit strain would be

$$\epsilon = \frac{\delta}{L} = \frac{\alpha \cdot L \cdot \Delta t}{L} = \alpha(\Delta t) \tag{4-8}$$

The resulting stress in the part can be found from

$$\sigma = E\epsilon$$

or   (4-9)

$$\sigma = E\alpha(\Delta t)$$

Example Problem 4-8

A steel structural member in a furnace is made from AISI 1020 steel and undergoes an increase in temperature of 95°F while being held rigid at its ends. Compute the resulting stress in the steel.

Solution   Using Equation (4-9), we have

$$\sigma = E\alpha(\Delta t)$$

From Table 4-1 we find $\alpha = 6.5 \times 10^{-6}{}^\circ F^{-1}$. For steel,

$$E = 30 \times 10^{-6} \text{ psi}$$

Then

$$\sigma = (30 \times 10^6 \text{ psi})(6.5 \times 10^{-6}{}^\circ F^{-1})(95°F) = 18\ 500 \text{ psi}$$

Example Problem 4-9

An aluminum rod of alloy 2014-T6 in a machine is held at its ends while being cooled from 95°C. At what temperature would be the tensile stress in the rod equal half of the yield strength of the aluminum if it is originally at zero stress?

Solution In Equation (4-9), we can solve for the change in temperature Δt.

$$\sigma = E\alpha(\Delta t)$$

$$\Delta t = \frac{\sigma}{E\alpha}$$

From Appendix A-17, for aluminum alloy 2014-T6, $s_y = 414$ MPa and $E = 73$ GPa. From Table 4-1, $\alpha = 23.0 \times 10^{-6}\,°C^{-1}$. Then

$$\sigma = \frac{s_y}{2} = \frac{414 \text{ MPa}}{2} = 207 \text{ MPa}$$

and

$$\Delta t = \frac{\sigma}{E\alpha} = \frac{207 \text{ MPa}}{(73 \text{ GPa})(23.0 \times 10^{-6}\,°C^{-1})}$$

$$= \frac{207 \times 10^{-6} \text{ Pa}}{(73 \times 10^{9} \text{ Pa})(23.0 \times 10^{-6}\,°C^{-1})} = 123°C$$

Since the rod had zero stress when its temperature was 95°C, the temperature at which the stress would be 207 MPa would be

$$t = 95°C - 123°C = -28°C$$

4-5 MEMBERS MADE OF MORE THAN ONE MATERIAL

When two or more materials in a load-carrying member share the load, a special analysis is required to determine what portion of the load each material takes. Consideration of the elastic properties of the materials is required.

Figure 4-1 shows a steel pipe filled with concrete and used to support part of a large structure. The load is distributed evenly across the top of the support. It is desired to determine the stress in both the steel and the concrete.

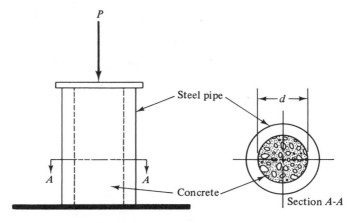

Figure 4-1 Steel and concrete post.

Two concepts must be understood in deriving the solution to this problem.

1. The total load P is shared by the steel and the concrete such that $P = P_s + P_c$.
2. Under the compressive load P, the composite support deforms and the two materials deform in equal amounts. That is, $\delta_s = \delta_c$.

Now since the steel and the concrete were originally the same length,

$$\frac{\delta_s}{L} = \frac{\delta_c}{L}$$

But

$$\frac{\delta_s}{L} = \epsilon_s \quad \text{and} \quad \frac{\delta_c}{L} = \epsilon_c$$

Also,

$$\epsilon_s = \frac{\sigma_s}{E_s} \quad \text{and} \quad \epsilon_c = \frac{\sigma_c}{E_c}$$

Then

$$\frac{\sigma_s}{E_s} = \frac{\sigma_c}{E_c}$$

Solving for σ_s yields

$$\sigma_s = \frac{\sigma_c E_s}{E_c} \tag{4-10}$$

This equation gives the relationship between the two stresses.
Now consider the loads,

$$P_s + P_c = P$$

Since both materials are subjected to axial stress,

$$P_s = \sigma_s A_s \quad \text{and} \quad P_c = \sigma_c A_c$$

where A_s and A_c are the areas of the steel and concrete, respectively. Then

$$\sigma_s A_s + \sigma_c A_c = P \tag{4-11}$$

Substituting Equation (4-10) into Equation (4-11) gives

$$\frac{A_s \sigma_c E_s}{E_c} + \sigma_c A_c = P$$

Now, solving for σ_c gives

$$\sigma_c = \frac{P E_c}{A_s E_s + A_c E_c} \tag{4-12}$$

Equations (4-10) and (4-12) can now be used to compute the stresses in the steel and the concrete. Of course, these equations can be used for any other composite member of two materials by substituting the appropriate values for area and modulus of elasticity.

Example Problem 4-10

For the support shown in Figure 4-1, the pipe is a standard 6-in. schedule 40 steel pipe completely filled with concrete. If the load P is 155 000 lb, compute the stress in the concrete and the steel. For the concrete use $E = 2 \times 10^6$ psi, and for steel use $E = 30 \times 10^6$ psi.

Solution   Equations (4-10) and (4-12) can be used. The areas of the steel pipe and the concrete are required. From Appendix A-12, the cross-sectional area of metal in a 6-in. schedule 40 pipe is 5.581 in². Its inside diameter d is 6.065 in. Then the area of the concrete is

$$A_c = \frac{\pi d^2}{4} = \frac{\pi (6.065 \text{ in.})^2}{4} = 28.89 \text{ in}^2$$

Then in Equation (4-12),

$$\sigma_c = \frac{(155\ 000 \text{ lb})(2 \times 10^6 \text{ psi})}{(5.581 \text{ in}^2)(30 \times 10^6 \text{ psi}) + (28.89 \text{ in}^2)(2 \times 10^6 \text{ psi})}$$

$$= 1376 \text{ psi}$$

Using Equation (4-10) gives

$$\sigma_s = \frac{\sigma_c E_s}{E_c} = \frac{(1376 \text{ psi})(30 \times 10^6 \text{ psi})}{2 \times 10^6 \text{ psi}} = 20\ 600 \text{ psi}$$

PROBLEMS

4-1. A post of No. 2 grade hemlock is $3\frac{1}{2}$ in. square and 6.0 ft long. How much would it be shortened when it is loaded in compression up to its allowable load parallel to the grain?

4-2. Determine the elongation of a strip of plastic 0.75 mm thick by 12 mm wide by 375 mm long if it is subjected to a load of 90 N and is made of (a) glass-reinforced ABS or (b) phenolic (see Appendix A-19).

4-3. A hollow aluminum cylinder made of 2014-T4 has an outside diameter of 2.50 in. and a wall thickness of 0.085 in. Its length is 14.5 in. What axial compressive force would cause the cylinder to shorten by 0.005 in.? What is the resulting stress in the aluminum?

4-4. A square bar 0.25 in. on a side is found in a stock bin, and it is not known if it is aluminum or magnesium. Discuss two ways in which you could determine what it is.

4-5. A tensile member is being designed for a car. It must withstand a repeated load of 3500 N and not elongate more than 0.12 mm in its 630-mm length. Use a design factor of 8 based on ultimate strength, and compute the required diameter of a round rod to satisfy these requirements using (a) AISI 1020 hot-rolled steel, (b) AISI 4140 OQT 700 steel, and (c) aluminum alloy 6061-T6. Compare the mass of the three options.

4-6. A steel bolt has a diameter of 12.0 mm in the unthreaded portion. Determine the elongation in a length of 220 mm if a force of 17.0 kN is applied.

4-7. In an aircraft structure, a rod is designed to be 1.25 m long and have a square cross section 8.0 mm on a side. Determine the amount of elongation which would occur if it is made of (a) titanium Ti-6Al-4V and (b) AISI 501 OQT 1000 stainless steel. The load is 5000 N.

4-8. A tension member in a welded steel truss is 13.0 ft long and subjected to a force of 35 000 lb. Choose an equal-leg angle made of ASTM A36 steel that will limit the stress to 21 600 psi. Then compute the elongation in the angle due the force. Use $E = 29.0 \times 10^6$ psi for structural steel.

4-9. A link in a mechanism is subjected alternately to a tensile load of 450 lb and a compressive load of 50 lb. Compute the elongation and compression of the link if it is a rectangular steel bar $\frac{1}{4}$ in. by $\frac{1}{8}$ in. in cross section and 8.40 in. long.

4-10. A round steel bar having a cross section of 0.50 in^2 is attached at the top and is subjected to three axial forces as shown in Figure 4-2. Determine the deflection of the free end caused by these forces.

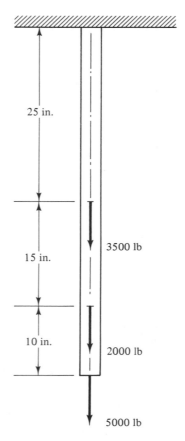

25 in.

15 in.

3500 lb

10 in.

2000 lb

5000 lb

Figure 4-2 Bar in axial tension for Problem 4-10.

4-11. A tension link in a mechanism is made of a hollow aluminum tube with an outside diameter of 1.250 in. and an inside diameter of 1.126 in. The tube is 36.0 in. long and is made from 6061-T6 aluminum. What force would be required to produce a deflection of the bar of 0.050 in.? Would the stress produced by the force just found be safe if the load is applied repeatedly?

Deformation and Thermal Stress   Chap. 4

4-12. A tension member in a frame is subjected to a static load of 2500 lb. Specify a suitable aluminum alloy that would be safe if the member is a tube with an outside diameter of 0.750 in. and an inside diameter of 0.563 in. Then compute the elongation of the member if its original length is 8.75 ft.

4-13. A hollow aluminum tube, 40 mm long, is used as a spacer in a machine and is subjected to an axial compressive force of 18.2 kN. The tube has an outside diameter of 56.0 mm and an inside diameter of 48.0 mm. Compute the deflection of the tube and the resulting compressive stress.

4-14. A steel guy wire is 0.375 in. in diameter and 135 ft long. Compute the stress in the wire and its deflection when subjected to a tensile force of 1600 lb.

4-15. Compute the total elongation of the bar shown in Figure 1-20 if it is made from titanium Ti-6A1-4V.

4-16. During a tensile test of a metal bar it was found by measurement that it elongated 0.023 in. when subjected to an axial load of 10 000 lb. The bar had an original length of 10.000 in. and a diameter of 0.75 in. Compute the modulus of elasticity of the metal. What kind of metal was it probably made from?

4-17. The bar shown in Figure 1-21 carries three loads. Compute the deflection of point D relative to point A. The bar is made from polycarbonate plastic.

4-18. A column is made from a concrete base, 6.00 in. in diameter, supporting a standard hollow square steel tube, $4 \times 4 \times \frac{1}{2}$. The concrete base is 3.0 ft long and the steel tube is 8.60 ft long. Assuming the column does not buckle, compute the amount by which the total column would be shortened by an axial compressive load of 96 000 lb. Use $E_c = 2 \times 10^6$ psi.

4-19. A length of 14-gauge copper electrical wire (C14500), 10.5 ft long, is rigidly attached to a beam at its top. If the wire has a diameter of 0.064 in., how much would it stretch if a person weighing 120 lb hung from the bottom? How much would it stretch if the person weighed 200 lb?

4-20. A steel measuring tape used by carpenters is 25 ft long and is made from strip steel having a cross section of 0.006 in. by 0.750 in. Compute how much the tape would elongate if a tensile force of 25 lb is applied. What is the resulting stress in the steel?

4-21. A wooden post is made from a standard 4×4 (Appendix A-4). If it is made from No. 2 grade southern pine, compute the axial compressive load it could carry before reaching its allowable stress in compression parallel to the grain. Then, if the post is 10.75 ft long, compute the amount that it would be shortened under that load.

4-22. Ductile iron, ASTM A536, grade 60-40-18, is formed into a hollow square shape, 200 mm outside dimension and 150 mm inside dimension. Compute the load that would produce an axial compressive stress in the iron of 200 MPa. Then, for that load, compute the amount that the member would be shortened from its original length of 1.80 m.

4-23. A brass wire (C26000) has a diameter of 3.00 mm and is initially 3.600 m long. At this condition, the lower end, with a plate for applying a load, is 6.0 mm from the floor. How many kilograms of lead would have to be added to the plate to just cause it to touch the floor? What would be the stress in the wire at that time?

4-24. Compute the elongation of the square bar AB in Figure 1-26 if it is 1.25 m long and made from aluminum 6016-T6.

4-25. A concrete slab in a highway is 80 ft long. Determine the change in length of the slab if the temperature changes from $-30°F$ to $+110°F$.

4-26. A steel rail for a railroad siding is 12.0 m long. Determine the change in length of the rail if the temperature changes from $-34°C$ to $+43°C$.

4-27. Determine the stress that would result in the rail described in Problem 4-26 if it were completely restrained from expanding.

4-28. The pushrods that actuate the valves on a six-cylinder engine are AISI 1040 steel and are 625 mm long and 8.0 mm in diameter. Calculate the change in length of the rods if their temperature varies from $-40°C$ to $+116°C$ and the expansion is unrestrained.

4-29. If the pushrods described in Problem 4-28 were installed with zero clearance with other parts of the valve mechanism at 25°C, compute the following:
(a) The clearance between parts at $-40°C$.
(b) The stress in the rod due to a temperatue rise to 116°C.
Assume that mating parts are rigid.

4-30. A bridge deck is made as one continuous concrete slab to 140 ft long at 30°F. Determine the required width of expansion joints at the ends of the bridge if no stress is to be developed when the temperature varies from $+30°F$ to $+110°F$.

4-31. For the bridge deck in Problem 4-30, what stress would be developed if only 0.25 in. expansion joints were used at each end and the supports were rigid? For concrete use $E = 2 \times 10^6$ psi.

4-32. For the bridge deck in Problem 4-30, assume that the deck is to be just in contact with its support at the temperature of 110°F. If the deck is to be installed when the temperature is 60°F, what should the gap be between the deck and its supports?

4-33. A ring of AISI 301 stainless steel is to be placed on a shaft having a temperature of 20°C and a diameter of 55.200 mm. The inside diameter of the ring is 55.100 mm. To what temperature must the ring be heated to make it 55.300 mm in diameter and thus allow it to be slipped onto the shaft?

4-34. When the ring of Problem 4-33 is placed on the shaft and then cooled back to 20°C, what tensile stress will be developed in the ring?

4-35. A heat exchanger is made by arranging several brass (C26000) tubes inside a stainless steel (AISI 430) shell. Initially, when the temperature is 10°C, the tubes are 4.20 m long and the shell is 4.50 m long. Determine how much each will elongate when heated to 85°C.

4-36. In Alaska, a 40-ft section of steel pipe may see a variation in temperature from $-50°F$ when it is at ambient temperature to $+140°F$ when it is carrying heated oil. Compute the change in the length of the pipe under these conditons.

4-37. A square bar of magnesium is 30 mm on a side and 250.0 mm long at 20°C. It is placed between two rigid supports set 250.1 mm apart. The bar is then heated to 70°C while the supports do not move. Compute the resulting stress in the bar.

4-38. A square rod, 8.0 mm on a side, is made from AISI 1040 cold-drawn steel and has a length of 175 mm. It is placed snugly between two unmoving supports with no stress in the rod. Then the temperature is increased by 90°C. What is the final stress in the rod?

4-39. An aluminum bar (6061-T4) has a length of 10.500 in. at 75°F. It is placed between rigid walls spaced 10.505 in. apart. The temperature of the bar is then raised to 400°F. Describe what would happen to the bar.

4-40. A straight carpenter's level rests on top of two bars 24.00 in. apart and is perfectly level at a temperature of 65°F. One bar is made of 6061-T6 aluminum and the other is made of titanium Ti-6A1-4V. The bars are initially 30.00 in. long. What would be the angle of tilt of the level if the temperature is increased to 212°F?

4-41. When manufactured, a steel (AISI 1040) measuring tape was exactly 25.000 ft long at a temperature of 68°F. Compute the error that would result if the tape is used at $-15°F$.

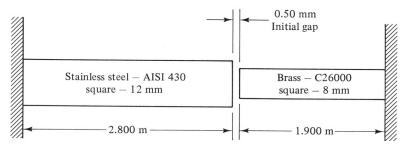

Figure 4-3 Problem 4-42.

4-42. Figure 4-3 shows two bars of different materials separated by 0.50 mm when the temperature is 20°C. At what temperature would they touch?

4-43. A stainless steel (AISI 301) wire is stretched between rigid supports so that a stress of 40 MPa is induced in the wire at a temperature of 20°C. What would be the stress at a temperature of −15°C?

4-44. For the conditions described in Problem 4-43, at what temperature would the stress in the wire be zero?

4-45. A short post is made by welding steel plates into a square, as shown in Figure 4-4 and then filling the area inside with concrete. Compute the stress in the steel and in the concrete if $b = 150$ mm, $t = 10$ mm, $E_s = 207$ GPa, $E_c = 13.8$ GPa, and the post carries an axial load of 1.40 MN.

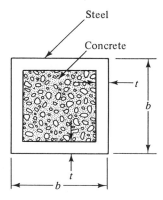

Figure 4-4 Post for Problems 4-45 and 4-46.

4-46. A short post is made by welding steel plates into a square, as shown in Figure 4-4 and then filling the area inside with concrete. The steel has an allowable stress of 21 600 psi, and the concrete has an allowable stress of 2000 psi. If $b = 6.00$ in. and $t = \frac{3}{8}$ in., compute the allowable axial load on the post. Use $E_s = 30 \times 10^6$ psi and $E_c = 2 \times 10^6$ psi

4-47. A short post is being designed to support an axial compressive load of 500 000 lb. It is to be made by welding $\frac{1}{2}$-in. thick plates of A36 steel into a square and filling the area inside with concrete, as shown in Figure 4-4. It is required to determine the dimension of the side of the post b in order to limit the stress in the steel to no more than 21 600 psi and in the concrete to no more than 2000 psi. Use $E_c = 2 \times 10^6$ psi and $E_s = 30 \times 10^6$ psi.

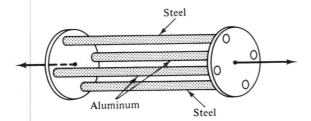

Figure 4-5   Problem 4-48.

4-48. Two disks are connected by four rods as shown in Figure 4-5. All rods are 6.0 mm in diameter and have the same length. Two rods are steel ($E = 207$ GPa), and two are aluminum ($E = 69$ GPa). Compute the stress in each rod when an axial force of 11.3 kN is applied to the disks.

4-49. An array of three wires is used to suspend a casting having a mass of 2265 kg in such a way that the wires are symmetrically loaded (see Figure 4-6). The outer two wires are AISI 430 stainless steel, cold worked. The middle wire is hard beryllium copper, C17200. All three wires have the same diameter and length. Determine the required diameter of the wires if none is to be stressed beyond one-half of its yield strength.

4-50. Figure 4-7 shows a load being applied to an inner cylindrical member which is initially

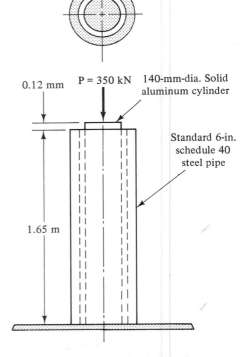

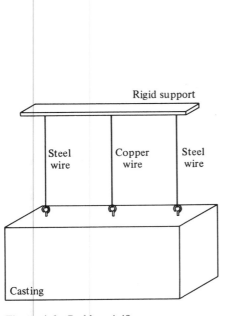

Figure 4-6   Problem 4-49.

Figure 4-7   Aluminum bar in a steel pipe under an axial compression load for Problem 4-50.

Deformation and Thermal Stress   Chap. 4

0.12 mm longer than a second concentric hollow pipe. What would be the stress in both members if a total load of 350 kN is applied?

4-51. Figure 4-8 shows an aluminum cylinder being capped by two end plates which are held in position with four steel tie rods. A clamping force is created by tightening the nuts on the ends of the tie rods. Compute the stress in the cylinder and the tie rods if the nuts are turned one full turn from the hand-tight condition.

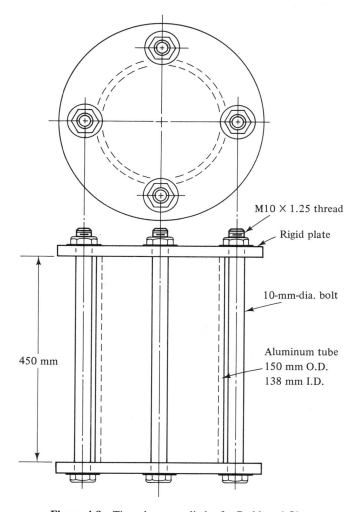

M10 × 1.25 thread

Rigid plate

10-mm-dia. bolt

Aluminum tube
150 mm O.D.
138 mm I.D.

450 mm

Figure 4-8 Tie rods on a cylinder for Problem 4-51.

4-52. A column for a building is made by encasing a W6 × 15 wide-flange shape in concrete as shown in Figure 4-9. The concrete helps to protect the steel from the heat of a fire and also shares in carrying the load. What stress would be produced in the steel and the concrete by a total load of 50 kips?

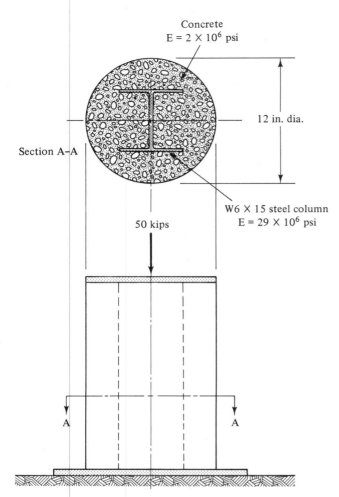

Concrete
E = 2 × 10⁶ psi

12 in. dia.

Section A–A

50 kips

W6 × 15 steel column
E = 29 × 10⁶ psi

A   A

Figure 4-9 Steel column encased in concrete for Problem 4-52.

5

Torsional Shear Stress and Torsional Deflection

5-1 OBJECTIVES OF THIS CHAPTER

Torsion refers to the loading of a member that tends to twist it. Such a load is called a *torque*, *twisting moment*, or *couple*. When a torque is applied to a member, such as a round shaft, *shearing stress* is developed within the shaft and a *torsional deflection* is created, resulting in an angle of twist of one end of the shaft relative to the other.

After completing this chapter, you should be able to:

1. Define *torque* and compute the magnitude of torque exerted on a member subjected to torsional loading.
2. Define the relationship among the three critical variables involved in power transmission: power, torque, and rotational speed.
3. Manipulate the units for power, torque, and rotational speed in both the SI metric system and the U.S. Customary system.
4. Compute the maximum shear stress in a member subjected to torsional loading.
5. Define the *polar moment of inertia* and compute its value for solid and hollow round shafts.
6. Compute the shear stress at any point within a member loaded in torsion.
7. Specify a suitable design shear stress for a member loaded in torsion.

8. Define the *polar section modulus* and compute its value for solid and hollow round shafts.

9. Determine the required diameter of a shaft to carry a given torque safely.

10. Compare the design of solid and hollow shafts on the basis of the mass of the shafts required to carry a certain torque while limiting the torsional shear stress to a certain design value.

11. Apply stress concentration factors to members in torsion.

12. Compute the angle of twist of a member loaded in torsion.

13. Define the *shear modulus of elasticity*.

14. Discuss the method of computing torsional shear stress and torsional deflection for members with noncircular cross sections.

15. Describe the general shapes of members having relatively high torsional stiffness.

5-2 TORQUE, POWER, AND ROTATIONAL SPEED

A necessary task in approaching the calculation of torsional shear stress and deflection is the understanding of the concept of *torque* and the relationship among the three critical variables involved in power transmission: *torque*, *power*, and *rotational speed*.

Figure 5-1 shows a socket wrench with an extension shaft being used to tighten a bolt. The *torque*, applied to both the bolt and the extension shaft, is the product of the applied force and the distance from the line of action of the force to the axis of the bolt. That is,

$$\text{torque} = T = F \cdot d \qquad (5\text{-}1)$$

Thus torque is expressed in the units of *force times distance*, which is $N \cdot m$ in the SI metric system and $lb \cdot in$ or $lb \cdot ft$ in the U.S. Customary system.

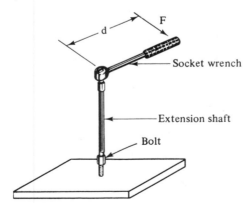

Figure 5-1 Wrench applying a torque to a bolt.

Example Problem 5-1

For the wrench in Figure 5-1, compute the magnitude of the torque applied to the bolt if a force of 50 N is exerted at a point 250 mm out from the axis of the socket.

Solution Using Equation (5-1) yields

$$T = F \cdot d = (50 \text{ N})(250 \text{ mm})\left(\frac{1 \text{ m}}{1000 \text{ mm}}\right) = 12.5 \text{ N} \cdot \text{m}$$

Figure 5-2 shows a drive system for a boat. Power developed by the engine flows through the transmission and the drive shaft to the propeller, where it drives the boat forward. The crankshaft inside the engine, the various power transmission shafts in the transmission, and the drive shaft are all subjected to torsion. The magnitude of the torque in a power transmission shaft is dependent on the amount of power it carries and on the speed of rotation, according to the following relation

power = torque · rotational speed

$$P = T \cdot n \tag{5-2}$$

This is a very useful relationship because if any two values, P, n, or T, are known, the third can be computed.

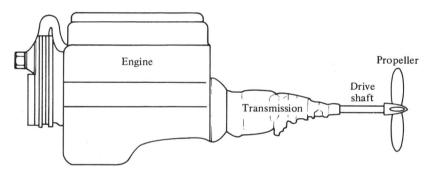

Figure 5-2 Drive system for a boat.

Careful attention must be paid to units when working with torque, power, and rotational speed. Appropriate units in the SI metric system and the U.S. Customary system are reviewed next.

SI metric system. Power is defined as the rate of transferring energy. In the SI metric system, the *joule* is the standard unit for energy and it is equivalent to the N · m, the standard unit for torque. That is,

$$1.0 \text{ J} = 1.0 \text{ N} \cdot \text{m}$$

Then power is defined as

$$\text{power} = \frac{\text{energy}}{\text{time}} = \frac{\text{joule}}{\text{second}} = \frac{\text{J}}{\text{s}} = \frac{\text{N} \cdot \text{m}}{\text{s}} \tag{5-3}$$

But 1.0 J/s is defined to be 1.0 watt (1.0 W). The watt is a rather small unit of power, so the kilowatt (kW; 1000 watts) is often used.

The standard unit for rotational speed in the SI metric system is *radians per second*, rad/s. Frequently, however, rotational speed is expressed in revolutions per minute, rpm. The conversion required is illustrated below, converting 1750 rpm to rad/s.

$$n = \frac{1750 \text{ rev}}{\text{min}} \cdot \frac{2\pi \text{ rad}}{\text{rev}} \cdot \frac{1 \text{ min}}{60 \text{ s}} = 183 \text{ rad/s}$$

When using n in rad/s in Equation (5-2), the radian is considered to be *no unit at all*, as illustrated in the following example problem.

Example Problem 5-2

The drive shaft for the boat shown in Figure 5-2 transmits 95 kW of power while rotating at 525 rpm. Compute the torque in the drive shaft.

Solution First, let's convert the rotational speed to rad/s.

$$n = \frac{525 \text{ rev}}{\text{min}} \cdot \frac{2\pi \text{ rad}}{\text{rev}} \cdot \frac{1 \text{ min}}{60 \text{ s}} = 55.0 \text{ rad/s}$$

Note that 95 kW is equivalent to 95 000 W, and

$$1.0 \text{ W} = 1.0 \text{ N} \cdot \text{m/s}$$

Then

$$\text{power} = P = 95\ 000 \text{ N} \cdot \text{m/s}$$

Now, solving for torque from Equation (5-2) gives

$$T = \frac{P}{n} = \frac{95\ 000 \text{ N} \cdot \text{m}}{\text{s}} \cdot \frac{1}{55.0 \text{ rad/s}} = 1727 \text{ N} \cdot \text{m}$$

Here the radian unit is considered no unit at all.

U.S. Customary units. Typical units for torque, power, and rotational speed in the U.S. Customary unit system are

$$T = \text{torque (lb} \cdot \text{in.)}$$

$$n = \text{rotational speed (rpm)}$$

$$P = \text{power (horsepower, hp)}$$

Note that $1.0 \text{ hp} = 6600 \text{ lb} \cdot \text{in./s}$. Then the unit conversions required to ensure consistent units are

$$\text{power} = T\,(\text{lb} \cdot \text{in.}) \cdot n \left(\frac{\text{rev}}{\text{min}}\right) \cdot \frac{1 \text{ min}}{60 \text{ s}} \cdot \frac{2\pi \text{ rad}}{\text{rev}} \cdot \frac{1 \text{ hp}}{6600 \text{ lb} \cdot \text{in./s}}$$

or

$$\text{power} = \frac{Tn}{63\ 000} \tag{5-4}$$

Example Problem 5-3

Compute the power being transmitted by a shaft if it is developing a torque of 15 000 lb · in. and rotating at 525 rpm.

Solution   Using Equation (5-4), the power in horsepower is

$$P = \frac{Tn}{63\ 000}$$

when T is in lb $\cdot$ in. and n is in rpm.

$$P = \frac{(15\ 000)(525)}{63\ 000} = 125 \text{ hp}$$

5-3 TORSIONAL SHEAR STRESS IN MEMBERS WITH CIRCULAR CROSS SECTIONS

When a member is subjected to an externally applied torque, an internal resisting torque must be developed in the material from which the member is made. The internal resisting torque is the result of stresses developed in the material.

Figure 5-3(a) shows a circular bar subjected to a torque, T. Section N would be rotated relative to section M as shown. If an element on the surface of the bar were isolated, it would be subjected to shearing forces on the sides parallel to cross sections M and N, as shown. These shearing forces result in shear stresses on the element. For the stress element to be in equilibrium, equal shearing stresses must exist on the top and bottom faces of the element. The following discussion develops the equation from which the magnitude of the maximum shear stress can be computed, along with the shear stress at any point within the bar.

Considering two cross sections M and N, at different places on the bar, section N would be rotated through an angle θ relative to section M. The fibers of the material would undergo a strain which would be maximum at the outside surface of the bar and varying linearly with radial position to zero at the center of the bar. Because, for elastic materials obeying Hooke's law, stress is proportional to strain, the maximum stress would also occur at the outside the bar as shown in Figure 5-3(b). The linear variation of stress, τ, with radial position in the cross section, r, is also shown. Then, by proportion using similar triangles,

$$\frac{\tau}{r} = \frac{\tau_{\max}}{c} \tag{5-5}$$

Then the shear stress at any radius can be expressed as a function of the maximum shear stress at the outside of the shaft,

$$\tau = \tau_{\max} \cdot \frac{r}{c} \tag{5-6}$$

It should be noted that the shear stress τ acts uniformly on a small ring-shaped area, dA, of the shaft, as illustrated in Figure 5-3(c). Now since force equals stress times area, the force on the area dA is

$$dF = \underbrace{\tau}_{} \, dA = \underbrace{\tau_{\max} \frac{r}{c}}_{\text{stress}} \cdot \underbrace{dA}_{\text{area}}$$

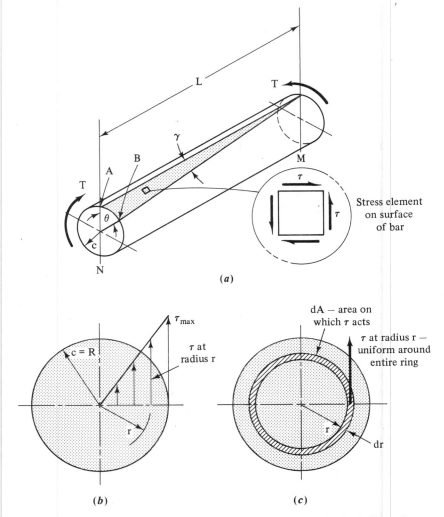

Figure 5-3 Torsional shear stress on a circular bar. (a) Circular bar loaded in torsion (b) Distribution of shear stress on a cross section of the bar (c) Shear stress τ at radius r acting on the area dA.

The next step is to consider that the torque dT developed by this force is the product of dF and the radial distance to dA. Then

$$dT = dF \cdot r = \underbrace{\tau_{\max} \frac{r}{c} dA}_{\text{force}} \cdot \underbrace{r}_{\text{radius}} = \tau_{\max} \frac{r^2}{c} dA$$

This equation is the internal resisting torque developed on the small area dA. The total torque on the entire area would be the sum of all the individual torques on all areas of the cross section. The process of summing is accomplished by the mathematical technique of integration, illustrated as follows:

$$T = \int_A dT = \int_A \tau_{max} \frac{r^2}{c} \, dA$$

In the process of integration, constant terms such as τ_{max} and c can be brought outside the integral sign, leaving

$$T = \frac{\tau_{max}}{c} \int_A r^2 \, dA \qquad (5\text{-}7)$$

In mechanics, the term $\int r^2 \, dA$ is given the name *polar moment of inertia* and is identified by the symbol J. Equation (5-7) can then be written

$$T = \tau_{max} \frac{J}{c}$$

or

$$\boxed{\tau_{max} = \frac{Tc}{J}} \qquad (5\text{-}8)$$

The method of evaluating J is developed in the next section.

Equation (5-8) can be used to compute the maximum shear stress on a circular bar subjected to torsion. The maximum shear stress occurs anywhere on the outer surface of the bar. Equation (5-6) can be used to compute the shear stress at any point within the bar.

5-4 POLAR MOMENT OF INERTIA

Two different kinds of members can be analyzed for torsional shear stress using Equation (5-8). They are solid circular bars and hollow circular tubes. The difference between the two is in the evaluation of J.

Solid circular bars. Refer to Figure 5-3 showing a solid circular cross section. To evaluate J from

$$J = \int r^2 \, dA$$

it must be seen that dA is the area of a small ring located at a distance r from the center of the section and having a thickness dr.

For a small magnitude of dr, the area is that of a strip having a length equal to the circumference of the ring times the thickness.

thickness of the ring

$$dA = 2\pi r \, dr$$

circumference of a ring at the radius r

Then the polar moment of inertia for the entire cross section can be found by integrating from $r = 0$ at the center of the bar to $r = R$ at the outer surface.

$$J = \int_0^R r^2 \, dA = \int_0^R r^2 (2\pi r) \, dr = \int_0^R 2\pi r^3 \, dr = \frac{2\pi R^4}{4} = \frac{\pi R^4}{2}$$

It is usually more convenient to use diameter rather than radius. Then since $R = D/2$,

$$\boxed{J = \frac{\pi(D/2)^4}{2} = \frac{\pi D^4}{32}}$$ (5-9)

Hollow circular tube. A similar process can be used for finding J for a hollow tube (Refer to Figure 5-4).

$$J = \int r^2 \, dA$$

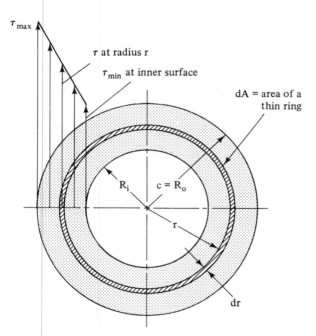

τ_{max}

τ at radius r

τ_{min} at inner surface

dA = area of a thin ring

R_i $c = R_o$

r

dr

Figure 5-4 Notation for variables used to derive J for a hollow round shaft.

Again, $dA = 2\pi r \, dr$. But for the hollow tube, r varies only from R_i to R_o. Then

$$J = \int_{R_i}^{R_o} r^2 (2\pi r) \, dr = 2\pi \int_{R_i}^{R_o} r^3 \, dr = \frac{2\pi(R_o^4 - R_i^4)}{4}$$

$$J = \frac{\pi}{2}(R_o^4 - R_i^4)$$

Substituting $R_o = D_o/2$ and $R_i = D_i/2$ gives

Torsional Shear Stress and Torsional Deflection   Chap. 5

$$J = \frac{\pi}{32}(D_o^4 - D_i^4)$$ (5-10)

This is the equation for the polar moment of inertia for a hollow circular tube.

Summary of Relationships for Torsional Shear Stresses in Circular Members

Maximum shear stress:

$$\tau_{max} = \frac{Tc}{J}$$ (5-8)

occurs at the outer surface of the bar or tube, where c is the radius of the bar or tube.

Shear stress at any radial position r:

$$\tau = \tau_{max}\frac{r}{c} = \frac{Tr}{J}$$ (5-6)

Polar moment of inertia for solid round bars:

$$J = \frac{\pi D^4}{32}$$ (5-9)

Polar moment of inertia for hollow tubes:

$$J = \frac{\pi}{32}(D_o^4 - D_i^4)$$ (5-10)

Example Problem 5-4

For the socket wrench extension shown in Figure 5-1, compute the maximum torsional shear stress in the central portion, where the diameter is 9.5 mm. The applied torque is 10 N·m.

Solution   Using Equations (5-8) and (5-9) yields

$$\tau_{max} = \frac{Tc}{J}$$

where $T = 10$ N·m

$$c = \frac{D}{2} = \frac{9.5 \text{ mm}}{2} = 4.75 \text{ mm}$$

$$J = \frac{\pi D^4}{32} = \frac{\pi(9.5 \text{ mm})^4}{32} = 800 \text{ mm}^4$$

Then

$$\tau_{max} = \frac{(10 \text{ N·m})(4.75 \text{ mm})}{800 \text{ mm}^4} \times \frac{10^3 \text{ mm}}{\text{m}} = 59.4 \text{ N/mm}^2$$

$$\tau_{max} = 59.4 \text{ MPa}$$

Example Problem 5-5

For the propeller drive shaft of Figure 5-2, compute the torsional shear stress when it is transmitting a torque of 1.76 kN · m. The shaft is a hollow tube having an outside diameter of 60 mm and an inside diameter of 40 mm. Find the stress at both the outer and inner surfaces.

Solution Using Equations (5-8), (5-10), and (5-6),

$$\tau_{max} = \frac{Tc}{J} \quad \text{at the outer surface}$$

$$J = \frac{\pi}{32}(D_o^4 - D_i^4) = \frac{\pi}{32}(60^4 - 40^4) \text{ mm}^4 = 1.02 \times 10^6 \text{ mm}^4$$

$$c = \frac{D_o}{2} = \frac{60 \text{ mm}}{2} = 30 \text{ mm}$$

$$T = 1.76 \text{ kN} \cdot \text{m} = 1.76 \times 10^3 \text{ N} \cdot \text{m}$$

Then

$$\tau_{max} = \frac{Tc}{J} = \frac{(1.76 \times 10^3 \text{ N} \cdot \text{m})(30 \text{ mm})}{1.02 \times 10^6 \text{ mm}^4} \times \frac{10^3 \text{ mm}}{\text{m}}$$

$$= 51.8 \text{ N/mm}^2 = 51.8 \text{ MPa}$$

At the inner surface,

$$\tau = \tau_{max}\frac{r}{c}$$

and

$$r = \frac{D_i}{2} = \frac{40 \text{ mm}}{2} = 20 \text{ mm}$$

Then

$$\tau = 51.8 \text{ MPa} \times \frac{20 \text{ mm}}{30 \text{ mm}} = 34.5 \text{ MPa}$$

Example Problem 5-6

Calculate the torsional shear stress that would be developed in a solid circular shaft having a diameter of 1.25 in. if it is transmitting 125 hp while rotating at 525 rpm.

Solution The first steps in a problem involving power transmission by shafts is to determine the torque developed in the shaft. Since this problem is stated with U.S. Customary units, Equation (5-4) can be used.

$$\text{Power} = P = \frac{Tn}{63\ 000}$$

Solving for the torque T gives

$$T = \frac{63\ 000P}{n}$$

Recall that this equation will give the value of torque directly in lb · in. when P is in horsepower and n is in rpm. Then

$$T = \frac{63\,000(125)}{525} = 15\,000 \text{ lb} \cdot \text{in.}$$

Now the torsional shear stress can be computed using Equation (5-8).

$$\tau_{\max} = \frac{Tc}{J}$$

where

$$c = \text{radius of shaft} = \frac{D}{2} = \frac{1.25 \text{ in.}}{2} = 0.625 \text{ in.}$$

$$J = \frac{\pi D^4}{32} = \frac{\pi(1.25 \text{ in.})^4}{32} = 0.240 \text{ in}^4$$

Then

$$\tau_{\max} = \frac{Tc}{J} = \frac{(15\,000 \text{ lb} \cdot \text{in.})(0.625 \text{ in.})}{0.240 \text{ in}^4} = 39\,100 \text{ psi}$$

5-5 DESIGN OF CIRCULAR MEMBERS UNDER TORSION

In a design problem, the loading on a member is known, and it is required to determine the geometry of the member to ensure that it will carry the loads safely. Material selection and the determination of design stresses are integral parts of the design process. *The techniques developed in this section are for circular members only, subjected only to torsion.* Of course, both solid and hollow circular members are covered. Torsion in noncircular members is covered in a later section of this chapter. The combination of torsion with bending and axial loads is presented in a later chapter.

The basic torsional shear stress equation, Equation (5-8), was expressed as

$$\tau_{\max} = \frac{Tc}{J} \tag{5-8}$$

In design, we can substitute a certain design stress τ_d for $\tau_{\max}$. As in the case of members subjected to direct shear stress and made of ductile materials, the design stress is related to the yield strength of the material in shear. That is,

$$\tau_d = \frac{S_{ys}}{N}$$

where N is the design factor chosen by the designer based on the manner of loading. Table 5-1 can be used as a guide to determine the value of N.

Where the data for S_{ys} are not available, the value can be estimated as $S_y/2$. This will give a reasonable, and usually conservative, estimate for ductile metals, especially steel. Then

$$\tau_d = \frac{S_{ys}}{N} = \frac{S_y}{2N} \tag{5-11}$$

TABLE 5-1 DESIGN FACTORS FOR DESIGN SHEAR STRESS IN EQUATION (5-11) FOR DUCTILE MATERIALS

Manner of loading	Design factor
Static load	2
Repeated load	4
Impact or shock	6

The torque T would be known in a design problem. Then, in Equation (5-8), only c and J are left to be determined. Notice that both c and J are properties of the geometry of the member which is being designed. For solid circular members (shafts), the geometry is completely defined by the diameter. It has been shown that

$$c = \frac{D}{2}$$

and

$$J = \frac{\pi D^4}{32}$$

It is now convenient to note that if the quotient J/c is formed, a simple expression involving D is obtained.

In the study of strength of materials, the term J/c is given the name *polar section modulus*, and the symbol Z_p is used to denote it.

$$Z_p = \frac{J}{c} = \frac{\pi D^4}{32} \cdot \frac{1}{D/2} = = \frac{\pi D^3}{16} \qquad (5\text{-}12)$$

Substituting Z_p for J/c in Equation (5-8) gives

$$\tau_{max} = \frac{T}{Z_p} \qquad (5\text{-}13)$$

To use this equation in design, we can let $\tau_{max} = \tau_d$ and then solve for Z_p.

$$Z_p = \frac{T}{\tau_d} \qquad (5\text{-}14)$$

Equation (5-14) will give the required value of the polar section modulus of a circular shaft to limit the torsional shear stress to τ_d when subjected to a torque T. Then Equation (5-12) can be used to find the required diameter of a solid circular shaft. Solving for D gives us

$$D = \sqrt[3]{\frac{16Z_p}{\pi}} \qquad (5\text{-}15)$$

If a hollow shaft is to be designed,

$$Z_p = \frac{J}{c} = \frac{\pi}{32}(D_o^4 - D_i^4) \cdot \frac{1}{D_o/2}$$

$$Z_p = \frac{\pi}{16}\frac{D_o^4 - D_i^4}{D_o} \qquad (5\text{-}16)$$

In this case, one of the diameters *or* the relationship between the two diameters would have to be specified in order to solve for the complete geometry of the hollow shaft.

Example Problem 5-7

The final drive to a conveyor that feeds coal into a railroad car is a shaft loaded in pure torsion and carrying 800 N·m of torque. Determine the required diameter of the shaft if it is to have a solid circular cross section. Specify a suitable steel for the shaft.

Solution In selecting a material, it should be noted that many alloys could be used. Because of the rough loading likely from a coal conveyor, a high ductility is desirable. Also, machinability would be important, as some machining would most likely be required on the shaft. Alloy AISI 1141 OQT 1300, from Appendix A-13, fits these requirements. The 1100 series steels generally have good machinability, and the 28 percent elongation indicates good ductility. The yield strength is 469 MPa.

A design factor of $N = 6$ would be reasonable based on yield strength since the shaft would likely be subjected to some shock or impact loading. Then

$$\tau_d = \frac{s_y}{2N} = \frac{469 \text{ MPa}}{2(6)} = 39.1 \text{ MPa}$$

Now, from Equation (5-14), the required polar section modulus of the shaft can be found.

$$Z_p = \frac{T}{\tau_d} = \frac{800 \text{ N·m}}{39.1 \text{ MPa}}$$

Restating the units in other forms gives

$$Z_p = \frac{T}{\tau_d} = \frac{800 \text{ N·m}}{39.1 \text{ N/mm}^2} \cdot \frac{10^3 \text{ mm}}{\text{m}} = 20.5 \times 10^3 \text{ mm}^3$$

Now the diameter can be found using Equation (5-15).

$$Z_p = \frac{J}{c} = \frac{\pi D^3}{16}$$

Then

$$D = \sqrt[3]{\frac{16Z_p}{\pi}} = \sqrt[3]{\frac{16(20.5 \times 10^3) \text{ mm}^3}{\pi}} = 47.1 \text{ mm}$$

A preferred nominal size of 50 mm should be specified.

Example Problem 5-8

An alternative design for the shaft described in Example Problem 5-7 would be to use a hollow tube for the shaft. Assume that a tube having an outside diameter of 60 mm is available in the same material as specified for the solid shaft (AISI 1141 OQT 1300). Compute what maximum inside diameter the tube can have which would result in the same stress in the steel as the 50-mm solid shaft.

Solution It was shown in Example Problem 5-7 that the 50-mm solid shaft would encounter a torsional shear stress less than 39.1 MPa. The actual stress in the shaft caused by the 800-N · m applied torque would be

$$\tau_{max} = \frac{T}{Z_p}$$

For the 50-mm-diameter shaft,

$$Z_p = \frac{\pi D^3}{16} = \frac{\pi (50)^3 \text{ mm}^3}{16} = 24.5 \times 10^3 \text{ mm}^3$$

Then

$$\tau_{max} = \frac{800 \text{ N} \cdot \text{m}}{24.5 \times 10^3 \text{ mm}^3} \times \frac{10^3 \text{ mm}}{1 \text{ m}} = 32.6 \text{ N/mm}^2 = 32.6 \text{ MPa}$$

Since torsional shear stress is inversely proportional to the polar section modulus, it is necessary that the hollow tube have the same value for Z_p as does the solid shaft if it is to experience the same stress. For the hollow tube,

$$Z_p = \frac{\pi}{16} \frac{D_o^4 - D_i^4}{D_o}$$

The outside diameter, D_o, is known to be 60 mm. We can then solve for the required inside diameter, D_i.

$$D_i = \left(D_o^4 - \frac{16 Z_p D_o}{\pi} \right)^{1/4}$$

Now, for $D_o = 60$ mm and $Z_p = 24.5 \times 10^3$ mm^3,

$$D_i = \left[(60)^4 - \frac{(16)(24.5 \times 10^3)(60)}{\pi} \right]^{1/4} \text{ mm} = 48.4 \text{ mm}$$

This is the maximum allowable inside diameter, since to use a larger value for D_i would result in a thinner tube and higher stress. Figure 5-5 shows a comparison of the two designs for the shaft, drawn full scale.

5-6 COMPARISON OF SOLID AND HOLLOW CIRCULAR MEMBERS

In many design situations, economy of material usage is a major criterion of performance for a product. In aerospace applications, every reduction in the mass of the aircraft or space vehicle allows increased payload. Automobiles achieve higher fuel economy when they are lighter. Also, since raw materials are purchased on a price per unit mass basis, a lighter part generally costs less.

Providing economy of material usage for load-carrying members requires that all the material in the member be stressed to a level approaching the safe design stress. Then every portion is carrying its share of the load.

Example Problems 5-7 and 5-8 can be used to illustrate this point. Recall that both designs shown in Figure 5-5 result in the same maximum torsional shear stress in the steel shaft. The hollow shaft is slightly larger in outside diameter, but it is the

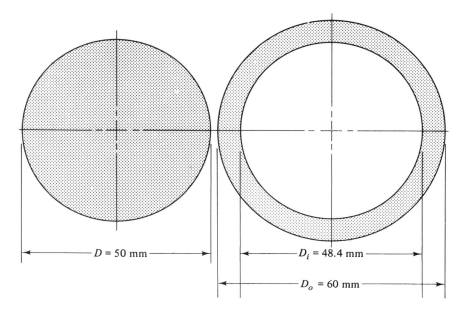

Figure 5-5 Comparison of solid and hollow shafts for Example Problem 5-8.

volume of metal which determines the mass of the shaft. Consider a length of shaft 1.0 m long. For the solid shaft, the volume is the cross-sectional area times the length.

$$V_s = AL = \frac{\pi D^2}{4} L$$

$$= \frac{\pi (50 \text{ mm})^2}{4} \cdot 1.0 \text{ m} \cdot \frac{1 \text{ m}^2}{(10^3 \text{ mm})^2} = 1.96 \times 10^{-3} \text{ m}^3$$

The mass is the volume times the density. Appendix A-13 gives the density of steel to be 7680 kg/m³. Then the mass of the solid shaft is

$$M_s = 1.96 \times 10^{-3} \text{ m}^3 \cdot 7680 \text{ kg/m}^3 = 15.1 \text{ kg}$$

Now for the hollow shaft the volume is

$$V_H = AL = \frac{\pi}{4} (D_o^2 - D_i^2)(L)$$

$$= \frac{\pi}{4} (60^2 - 48.4^2) \text{mm}^2 (1.0 \text{ m}) \cdot \frac{1 \text{ m}^2}{(10^3 \text{ mm})^2}$$

$$= 0.988 \times 10^{-3} \text{ m}^3$$

The mass of the hollow shaft is

$$M_H = 0.988 \times 10^{-3} \text{ m}^3 \cdot 7680 \text{ kg/m}^3 = 7.58 \text{ kg}$$

Thus it can be seen that the hollow shaft has almost exactly *one-half the mass* of the solid shaft, even though both are subjected to the same stress level for a given applied torque. Why?

The reason for the hollow shaft being lighter is that a greater portion of its material is being stressed to a higher level than in the solid shaft. Figure 5-3 shows a sketch of the stress distribution in the solid shaft. The maximum stress, 32.6 MPa, occurs at the outer surface. The stress then varies linearly with the radius for other points within the shaft to *zero* at the center. From this it can be seen that the material near the middle of the shaft is not being used efficiently.

Contrast this with the sketch of the hollow shaft in Figure 5-4. Again the stress at the outer surface is the maximum, 32.6 MPa. The stress at the inner surface of the hollow shaft can be found from Equation (5-6).

$$\tau = \tau_{max} \frac{r}{c}$$

At the inner surface, $r = R_i = D_i/2 = 48.4$ mm/2 $= 24.2$ mm. Also, $c = R_o = D_o/2 = 60$ mm/2 $= 30$ mm. Then

$$\tau = 32.6 \text{ MPa} \frac{24.2}{30} = 26.3 \text{ MPa}$$

The stress at points between the inner and outer surfaces varies linearly with the radius to each point. Thus it can be seen that all of the material in the hollow shaft shown in Figure 5-4 is being stressed to a fairly high but safe level. This illustrates why the hollow section requires less material.

Of course, the specific data used in the illustration above cannot be generalized to all problems. However, it can be said that for torsional loading of circular members, a hollow section can be designed which is lighter than a solid section while subjecting the material to the same maximum torsional shear stress.

5-7 STRESS CONCENTRATIONS IN TORSIONALLY LOADED MEMBERS

Changes in the cross section of a member loaded in torsion cause the local stress near the changes to be higher than would be predicted by using the torsional shear stress formula. The actual level of stress in such cases is determined experimentally. Then a *stress concentration factor* is determined which allows the stress in similar designs to be computed from the relationship

$$\tau = K_t \tau_{nom} \tag{5-17}$$

The term τ_{nom} is the nominal stress due to torsion which would be developed in the parts if the stress concentration were not present. Thus the standard torsional shear stress formulas [Equations (5-8) and (5-13)] can be used to compute the nominal stress. The value of K_t is a factor by which the actual stress is greater than the nominal stress.

Several cases in which stress concentrations occur are illustrated in Figure 5-6 and in the following appendix charts:

Appendix A-21-3: Grooved Round Bar in Torsion
Appendix A-21-4: Stepped Round Bar in Torsion
Appendix A-21-6: Shafts with Keyseats
Appendix A-21-8: Round Bar with Transverse Hole in Torsion

Grooved round bar. Round-bottomed grooves are cut into round bars for the purpose of installing seals or for distributing lubricating oil around a shaft. The stress concentration factor is dependent on the ratio of the shaft diameter to the diameter of the groove and on the ratio of the groove radius to the groove diameter. The groove is cut with a tool having a rounded nose to produce the round-bottomed groove. The radius should be as large as possible to minimize the stress concentration factor. Note that the nominal stress is based on the diameter *at the base of the groove.*

Stepped round bar. Shafts are often made with two or more diameters, resulting in a stepped shaft like that shown in Appendix A-21-4. The face of the step provides a convenient means of locating one side of an element mounted on the shaft, such as a bearing, gear, pulley, or chain sprocket. Care should be exercised in defining the radius at the bottom of the step, called the *fillet radius.* Sharp corners are to be avoided, as they cause extremely high stress concentration factors. The radius should be as large as possible while being compatible with the elements mounted on the shaft.

Retaining rings seated in grooves cut into the shaft are often used to locate machine elements as shown in Figure 5-6. The grooves are typically flat bottomed with small radii at the sides. Some designers treat such grooves as two steps on the shaft close together and use the stepped shaft chart (Appendix A-21-4) to determine the stress concentration factor. Because of the small radius at the base of the groove, the relative radius is often quite small, resulting in high values of K_t off the chart. In such cases a value of $K_t = 3.0$ is sometimes used.

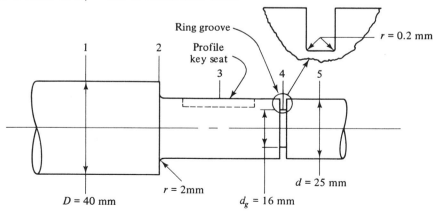

Figure 5-6 Shaft with stress concentrations.

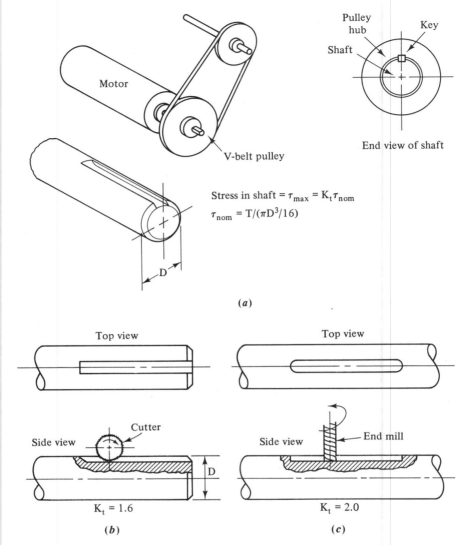

Stress in shaft $= \tau_{max} = K_t \tau_{nom}$

$\tau_{nom} = T/(\pi D^3/16)$

(a)

$K_t = 1.6$

(b)

$K_t = 2.0$

(c)

Figure 5-7 Stress concentration factors for keyseats. (a) Typical application (b) Sled runner type key seat made with a circular milling cutter (c) Profile type key seat made with an end mill

Shafts with keyseats. Power transmission elements typically transmit torque to and from shafts through keys fitted into keyseats cut into the shaft as shown in Figure 5-7. The V-belt pulley mounted on the end of the motor shaft shown is an example. Two types of keyseats are in frequent use: the *sled-runner* and the *profile* keyseats.

A circular milling cutter having a thickness equal to the width of the keyseat is used to cut the sled-runner keyseat, typically on the end of a shaft as shown in Figure

5-7(b). As the cutter ends its cut, it leaves a gentle radius as shown in the side view, resulting in $K_t = 1.6$ as a design value.

A profile keyseat is cut with an end mill having a diameter equal to the width of the keyseat. Usually used at a location away from the ends of the shaft, it leaves a square corner at the ends of the keyseat when viewed from the side as shown in Figure 5-7(c). This is more severe than the sled runner and a value of $K_t = 2.0$ is used. Note that the stress concentration factors account for both the removal of material from the shaft and the change in geometry.

Round bar with transverse hole. One purpose for drilling a hole in the shaft is to insert a pin through a hole in the hub of a machine element and through the shaft to locate it in position and to transmit torque. The chart in Appendix A-21-8 shows two curves for stress concentration factor. One is based on the nominal stress for the *gross cross section* and accounts for the removal of material and the change in geometry. The other is based on the nominal stress for the *net cross section* and accounts only for the change in geometry. The polar section modulus for the net section must be computed, often a difficult task. Therefore, the gross section is easier to use.

The use of stress concentration factors is illustrated in the following example problem.

Example Problem 5-9

Figure 5-6 shows a portion of a shaft on which a pulley is mounted. The pulley rests against a shoulder at its left face. A ring in a groove locates the pulley on its right face. A profile keyseat is used at the pulley. Compute the stress in the shaft at the sections 1, 2, 3, 4, and 5. Assume that the torque is 20 N · m at all sections. Then, using a design factor $N = 4$ based on yield strength, specify a suitable material for the shaft.

Solution The stress at *section 1* is computed for the larger 40-mm-diameter portion of the shaft with no stress concentration. Then

$$\tau = \frac{T}{Z_p}$$

$$Z_p = \frac{\pi D^3}{16} = \frac{\pi(40 \text{ mm})^3}{16} = 12\ 570 \text{ mm}^3$$

$$\tau = \frac{20 \text{ N} \cdot \text{m}}{12\ 570 \text{ mm}^3} \cdot \frac{10^3 \text{ mm}}{\text{m}} = 1.59\frac{\text{N}}{\text{mm}^2} = 1.59 \text{ MPa}$$

At *section 2*, the stress concentration due to the shoulder fillet must be considered (see Appendix A-21-4).

$$\tau = K_t \tau_{\text{nom}}$$

$$\tau_{\text{nom}} = \frac{T}{\pi d^3/16} = \frac{20 \text{ N} \cdot \text{m}}{[\pi(25)^3/16] \text{ mm}^3} \cdot \frac{10^3 \text{ mm}}{\text{m}}$$

$$= 6.52 \text{ N/mm}^2 = 6.52 \text{ MPa}$$

The value of K_t depends on the ratios D/d and r/d.

$$\frac{D}{d} = \frac{40 \text{ mm}}{25 \text{ mm}} = 1.60$$

$$\frac{r}{d} = \frac{2 \text{ mm}}{25 \text{ mm}} = 0.08$$

Then from Appendix A-21-4, $K_t = 1.45$. Then

$$\tau = (1.45)(6.52 \text{ MPa}) = 9.45 \text{ MPa}$$

At *section 3*, the profile-type keyseat presents a stress concentration factor of 2.0. The nominal stress is the same as that computed at the shoulder fillet. Then

$$\tau = K_t \tau_{\text{nom}} = (2.0)(6.52 \text{ MPa}) = 13.04 \text{ MPa}$$

Section 4 is the location of the ring groove. Here the nominal stress is computed on the basis of the root diameter of the groove.

$$\tau_{\text{nom}} = \frac{T}{\pi d_g^3/16} = \frac{20 \text{ N} \cdot \text{m}}{[\pi(16)^3/16] \text{ mm}^3} \cdot \frac{10^3 \text{ mm}}{\text{m}} = 24.9 \frac{\text{N}}{\text{mm}^2}$$

$$= 24.9 \text{ MPa}$$

The value of K_t depends on d/d_g and r/d_g.

$$\frac{d}{d_g} = \frac{25 \text{ mm}}{16 \text{ mm}} = 1.56$$

$$\frac{r}{d_g} = \frac{0.2 \text{ mm}}{16 \text{ mm}} = 0.013$$

Referring to Appendix A-21-4, the stress concentration factor is off the chart. This is the type of case for which $K_t = 3.0$ is reasonable.

$$\tau = K_t \tau_{\text{nom}} = (3.0)(24.9 \text{ MPa}) = 74.7 \text{ MPa}$$

Section 5 is in the smaller portion of the shaft, where no stress concentration occurs. then

$$\tau = \frac{T}{Z_p} = \frac{T}{\pi d^3/16}$$

Notice that this is identical to the nominal stress computed for sections 2 and 3. Then at section 5,

$$\tau = 6.52 \text{ MPa}$$

It can be seen that there is a large variation in the magnitude of the torsional shear stress at the various points around the pulley mounted on the shaft. The maximum computed stress occurs at section 4, where the ring groove is cut into the shaft. There, both the smallest diameter and the largest value of K_t occur. The selection of a material must be based on the stress at section 4. Letting that stress, 74.7 MPa, be the design stress in Equation (5-11), the required yield strength for the material is

$$s_y = \tau_d \cdot 2N = 74.7 \text{ MPa } (2)(4) = 598 \text{ MPa}$$

From Appendix A-13, two suitable steels for this requirement are AISI 1040 WQT 900 and AISI 4140 OQT 1300. Both have adequate strength and a high ductility as measured by the percent elongation. Certainly, other alloys and heat treatments could be used.

5-8 TWISTING—ELASTIC TORSIONAL DEFORMATION

Stiffness in addition to strength is an important design consideration for torsionally loaded members. The measure of torsional stiffness is the angle of twist of one part of a shaft relative to another part when a certain torque is applied.

In mechanical power transmission applications, excessive twisting of a shaft may cause vibration problems which would result in noise and improper synchronization of moving parts. One guideline for torsional stiffness is related to the desired degree of precision as listed in Table 5-2 (See Ref. 1 and 3).

TABLE 5-2 RECOMMENDED TORSIONAL STIFFNESS: ANGLE OF TWIST PER UNIT LENGTH

Application	Torsional deflection	
	deg/in.	rad/m
General machine part	1×10^{-3} to 1×10^{-2}	6.9×10^{-4} to 6.9×10^{-3}
Moderate precision	2×10^{-5} to 4×10^{-4}	1.4×10^{-5} to 2.7×10^{-4}
High precision	1×10^{-6} to 2×10^{-5}	6.9×10^{-7} to 1.4×10^{-5}

In structural design, load-carrying members are sometimes loaded in torsion as well as tension or bending. The rigidity of the structure then depends on the torsional stiffness of the components. Any load applied off from the axis of a member and transverse to the axis will produce torsion. This section will discuss twisting of circular members, both solid and hollow. Noncircular sections will be covered in a later section. It is very important to note that the behavior of an open-section shape such as a channel or angle is much different from that of a closed section such as a pipe or rectangular tube. In general, the open sections have very low torsional stiffness.

To aid in the development of the relationship for computing the angle of twist of a circular member, consider the shaft shown in Figure 5-3. One end of the shaft is held fixed while a torque T is applied to the other end. Under these conditions the shaft will twist between the two ends through an angle θ.

The derivation of the angle-of-twist formula depends on some basic assumptions about the behavior of a circular member when subjected to torsion. As the torque is applied, an element along the outer surface of the member, which was initially straight, rotates through a small angle γ (gamma). Likewise, a radius of the member in a cross section rotates through a small angle θ. In Figure 5-3, the rotations γ and θ are both related to the arc length AB on the surface of the bar. From geometry, for small angles, the arc length is the product of the angle in radians and the radius from the center of the rotation. Therefore, the arc length AB can be expressed as either

$$AB = \gamma L \qquad (5\text{-}18)$$

or

$$AB = \theta c \qquad (5\text{-}19)$$

where c is the outside radius of the bar. These two expressions for the arc length AB can be equated to each other,

$$\gamma L = \theta c$$

Solving for γ gives

$$\gamma = \frac{\theta c}{L} \qquad (5\text{-}20)$$

The angle γ is a measure of the maximum shearing strain in an element on the outer surface of the bar. It was discussed in Chapter 1 that the shearing strain, γ, is related to the shearing stress, τ, by the modulus of elasticity in shear, G. That was expressed as Equation (1-7),

$$G = \frac{\tau}{\gamma} \qquad (1\text{-}7)$$

At the outer surface, then,

$$\tau = G\gamma$$

But the torsional shear stress formula [Equation (5-8)] states

$$\tau = \frac{Tc}{J}$$

Equating these two expressions for γ gives

$$G\gamma = \frac{Tc}{J}$$

Now, substituting from Equation (5-20) for γ, we obtain

$$\frac{G\theta c}{L} = \frac{Tc}{J}$$

We can now cancel c and solve for θ:

$$\boxed{\theta = \frac{TL}{JG}} \qquad (5\text{-}21)$$

The resulting angle of twist, θ, is in radians. When consistent units are used for all terms in the calculation, all units will cancel leaving a dimensionless number. This should be interpreted as the angle, θ, in radians.

Equation (5-21) can be used to compute the angle of twist of one section of a circular bar, either solid or hollow, with respect to another section where L is the distance between them, provided that the torque, T, the polar moment of inertia, J, and the shear modulus of elasticity, G, are the same over the entire length, L. If any of these factors vary in a given problem, the bar can be subdivided into segments over which they are constant to compute angles of rotation for those segments. Then the computed angles can be combined algebraically to get the total angle of twist. This principle, called *superposition*, will be illustrated in example problems.

The shear modulus of elasticity, G is a measure of the torsional stiffness of the material of the bar. Table 5-3 gives values for G for selected materials.

TABLE 5-3 SHEAR MODULUS OF ELASTICITY, G

Material	Shear modulus, G	
	GPa	psi
Plain carbon and alloy steels	80	11.5×10^6
Stainless steel type 304	69	10.0×10^6
Aluminum 6061-T6	26	3.75×10^6
Beryllium copper	48	7.0×10^6
Magnesium	17	2.4×10^6
Titanium alloy	43	6.2×10^6

Example Problem 5-10

Determine the angle of twist between two sections 250 mm apart in a steel rod having a diameter of 10 mm when a torque of 15 N·m is applied. Figure 5-3 shows a sketch of the arrangement.

Solution   Equation (5-21) can be used directly.

$$\theta = \frac{TL}{JG}$$

For steel, $G = 80$ GPa $= 80 \times 10^9$ N/m^2. The value of J is

$$J = \frac{\pi D^4}{32} = \frac{\pi (10 \text{ mm})^4}{32} = 982 \text{ mm}^4$$

Then

$$\theta = \frac{TL}{JG} = \frac{(15 \text{ N} \cdot \text{m})(250 \text{ mm})}{(982 \text{ mm}^4)(80 \times 10^9 \text{ N/m}^2)} \times \frac{(10^3 \text{ mm})^3}{1 \text{ m}^3} = 0.048 \text{ rad}$$

Note that all units cancel. Expressing the angle in degrees,

$$\theta = 0.048 \text{ rad} \times \frac{180 \text{ deg}}{\pi \text{ rad}} = 2.73 \text{ deg}$$

Example Problem 5-11

Determine the required diameter of a round shaft made of aluminum alloy 6061-T6 if it is to twist not more than 0.08 deg in 1 ft of length when a torque of 75 lb·in. is applied.

Solution   Equation (5-21) must be solved for J since that factor is related to the unknown diameter of the shaft and all other factors are specified.

$$\theta = \frac{TL}{JG}$$

$$J = \frac{TL}{\theta G}$$

The angle of twist must be expressed in radians.

$$\theta = 0.08 \text{ deg} \times \frac{\pi \text{ rad}}{180 \text{ deg}} = 0.0014 \text{ rad}$$

From Table 5-3, $G = 3.75 \times 10^6$ psi. Then

$$J = \frac{TL}{\theta G} = \frac{(75 \text{ lb} \cdot \text{in.})(12 \text{ in.})}{(0.0014)(3.75 \times 10^6 \text{ lb/in.}^2)} = 0.171 \text{ in}^4$$

Now since $J = \pi D^4/32$,

$$D = \left(\frac{32 J}{\pi}\right)^{1/4} = \left[\frac{(32)(0.171 \text{ in}^4)}{\pi}\right]^{1/4} = 1.15 \text{ in.}$$

Example Problem 5-12

Figure 5-8 shows a steel rod to which three disks are attached. The rod is fixed against rotation at its left end, but free to rotate in a bearing at its right end. Each disk is 300 mm in diameter. Downward forces act at the outer surfaces of the disks so that torques are applied to the rod. Determine the angle of twist of the section A relative to the fixed section E.

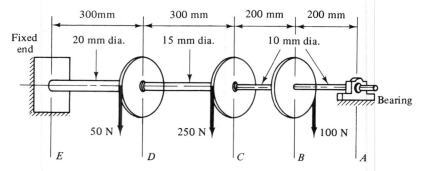

Figure 5-8 Rod for Example Problem 5-12.

Solution Obviously the diameter of the rod and the magnitude of the applied torque vary along the length of the rod. However, for each segment AB, BC, CD, and DE the conditions are constant. Then Equation (5-21) will be applied separately to each segment.

The first thing to be done is to determine the torque applied to the shaft at the points B, C, and D due to the loads on the disks. Since torque is the product of force times distance, we have:

Torque on disk B, clockwise:

$$T_B = (100 \text{ N})(150 \text{ mm}) = 15\,000 \text{ N} \cdot \text{mm} = 15 \text{ N} \cdot \text{m}$$

Torque on disk C, counterclockwise:

$$T_C = (250 \text{ N})(150 \text{ mm}) = 37\,500 \text{ N} \cdot \text{mm} = 37.5 \text{ N} \cdot \text{m}$$

Torque on disk D, counterclockwise:

$$T_D = (50 \text{ N})(150 \text{ mm}) = 7500 \text{ N} \cdot \text{mm} = 7.5 \text{ N} \cdot \text{m}$$

Now, to compute the angle of twist in the rod, the level of internal torque *in each segment of the rod* must be determined. Here, the use of the free-body approach can be helpful. Start first at the right end of the rod from the bearing A up to *but not including* B. If the rod were cut at a section to the right of B, there would be no internal torque because no torque is applied to that segment. Therefore T_{AB}, the torque in the rod from A to B is zero.

For segment BC, cut the shaft just to the right of C and observe that the internal torque in the rod must be 15 N·m to balance the applied torque at B. Then $T_{BC} = 15$ N·m.

For any section in the segment CD, the torque in the rod would be the difference between T_C and T_B. Then

$$T_{CD} = T_C - T_B = 37.5 \text{ N} \cdot \text{m} - 15 \text{ N} \cdot \text{m} = 22.5 \text{ N} \cdot \text{m}$$

Between D and E in the rod, the torque is the resultant of all the applied torques at D, C, and B.

$$T_{DE} = T_D + T_C - T_B = 7.5 \text{ N} \cdot \text{m} + 37.5 \text{ N} \cdot \text{m} - 15 \text{ N} \cdot \text{m} = 30 \text{ N} \cdot \text{m}$$

The fixed support at E must be capable of providing a reaction torque of 30 N·m to maintain the rod in equilibrium.

In summary, the distribution of torque in the rod can be shown in graphical form as in Figure 5-9. Notice that the applied torques T_B, T_C, and T_D are the *changes* in torque which occur at B, C, and D but that they are not the magnitudes of the torque *in the rod* at those points.

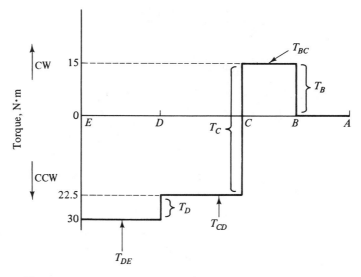

Figure 5-9 Torque distribution in rod for Example Problem 5-12.

Now the computation for angle of twist can be made.

Segment AB

$$\theta_{AB} = T_{AB} \frac{L}{JG}$$

Since $T_{AB} = 0$, $\theta_{AB} = 0$. There is no twisting of the rod between A and B.

Segment BC

$$\theta_{BC} = T_{BC} \frac{L}{JG}$$

We know $T_{BC} = 15$ N·m, $L = 200$ mm, and $G = 80$ GPa for steel. For the 10-mm-diameter rod,

$$J = \frac{\pi D_4}{32} = \frac{\pi (10 \text{ mm})^4}{32} = 982 \text{ mm}^4$$

Then

$$\theta_{BC} = \frac{(15 \text{ N·m})(200 \text{ mm})}{(982 \text{ mm}^4)(80 \times 10^9 \text{ N/m}^2)} \times \frac{(10^3)^3 \text{ mm}^3}{\text{m}^3} = 0.038 \text{ rad}$$

This means that section B is rotated 0.038 rad clockwise relative section C, since θ_{BC} is the total angle of twist in the segment BC.

Segment CD

$$\theta_{CD} = T_{CD} \frac{L}{JG}$$

Here $T_{CD} = 22.5$ N·m, $L = 300$ mm, and the rod diameter is 15 mm. Then

$$J = \frac{\pi D^4}{32} = \frac{\pi (15 \text{ mm})^4}{32} = 4970 \text{ mm}^4$$

$$\theta_{CD} = \frac{(22.5 \text{ N·m})(300 \text{ mm})}{(4970 \text{ mm}^4)(80 \times 10^9 \text{ N/m}^2)} \times \frac{(10^3)^3 \text{ mm}^3}{\text{m}^3} = 0.017 \text{ rad}$$

Section C is rotated 0.017 rad counterclockwise relative to section D.

Segment DE

$$\theta_{DE} = T_{DE} \frac{L}{JG}$$

Here $T_{DE} = 30$ N·m, $L = 300$ mm, and $D = 20$ mm. Then

$$J = \frac{\pi D^4}{32} = \frac{\pi (20 \text{ mm})^4}{32} = 15\,700 \text{ mm}^4$$

$$\theta_{DE} = \frac{(30 \text{ N·m})(300 \text{ mm})}{(15\,700 \text{ mm}^4)(80 \times 10^9 \text{ N/ mm}^2)} \times \frac{(10^3)^3 \text{ mm}^3}{\text{m}^3} = 0.007 \text{ rad}$$

Section D is rotated 0.007 rad counterclockwise relative to E.

Total Angle of Twist from E to A

$$\theta_{AE} = \theta_{AB} + \theta_{BC} - \theta_{CD} - \theta_{DE}$$

$$= 0 + 0.038 - 0.017 - 0.007 = 0.014 \text{ rad}$$

The angle of twist at any section can be found by referring to the graph in Figure 5-10. Notice that the angle varies linearly with position within any segment.

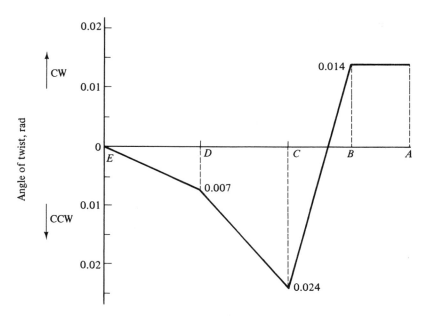

Figure 5-10 Angle of twist versus position on the rod for Example Problem 5-12.

5-9 TORSION IN NONCIRCULAR SECTIONS

The behavior of noncircular sections when subjected to torsion is vastly different from that of circular sections, for which the discussions earlier in this chapter applied. There is a large variety of shapes which can be imagined, and the analysis of stiffness and strength is different for each. The development of the relationships involved will not be done here. Compilations of the pertinent formulas occur in References 1 to 5 listed at the end of this chapter, and a few are given in this section.

Some generalizations can be made. Solid sections having the same cross-sectional area are stiffer when their shape more closely approaches a circle (see Figure 5-11). Conversely, a member made up of long, thin sections which do not form a closed, tube-like shape are very weak and flexible in torsion. Examples of flexible sections are common structural shapes such as wide flange beams, standard I-beams, channels, angles, and tees, as illustrated in Figure 5-12. Pipes, solid bars, and structural rectangular tubes have high rigidity, or stiffness (see Figure 5-13).

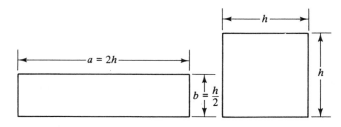

Figure 5-11 Comparison of stiffness for rectangle and square sections in torsion. Square is two times stiffer than rectangle even though both have the same area: $ab = h^2$.

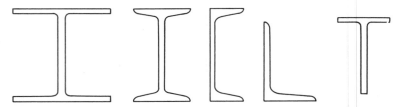

Figure 5-12 Torsionally flexible sections.

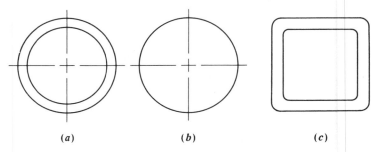

(*a*)   (*b*)   (*c*)

Figure 5-13 Torsionally stiff sections.

An interesting illustration of the lack of stiffness of open, thin sections is shown in Figure 5-14. The thin plate, (a) the angle, (b) and the channel (c) have the same thickness and cross-sectional area, and all have nearly the same torsional stiffness. Likewise, if the thin plate were formed into a circular shape, (d) but with a slit remaining, its stiffness would remain low. However, closing the tube completely as in Fig. 5-13(a) by welding or by drawing a seamless tube would produce a relatively stiff member. Understanding these comparisons is an aid to selecting a reasonable shape for members loaded in torsion.

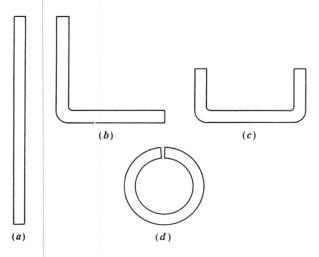

(*b*)   (*c*)

(*a*)   (*d*)

Figure 5-14 Sections having nearly equal (and low) torsional stiffness.

Cross-sectional shape	K = for use in $\theta = TL/GK$ Q = for use in $\tau = T/Q$	Black dot (●) denotes location of τ_{max}
Square 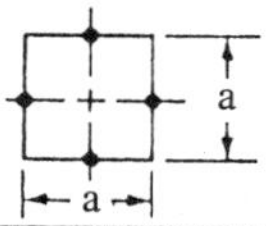	$K = 0.141a^4$ $Q = 0.208a^3$	τ_{max} at midpoint of each side
Rectangle 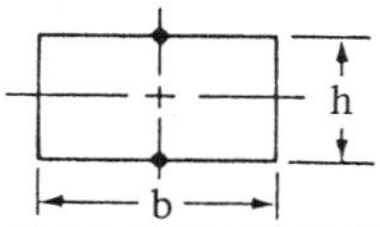	$K = bh^3\left[\dfrac{1}{3} - 0.21\dfrac{h}{b}\left(1 - \dfrac{(h/b)^4}{12}\right)\right]$ $Q = \dfrac{bh^2}{[3 + 1.8(h/b)]}$	(Approximate; within $\approx 5\%$) τ_{max} at midpoint of long sides
Triangle (equivalent) 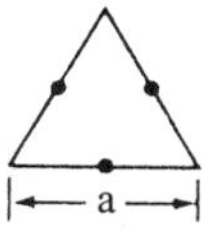	$K = 0.0217a^4$ $Q = 0.050a^3$	

Shaft with One Flat

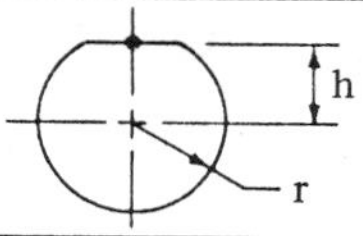

$K = C_1 r^4$

$Q = C_2 r^3$

h/r	0	0.2	0.4	0.6	0.8	1.0
C_1	0.30	0.51	0.78	1.06	1.37	1.57
C_2	0.35	0.51	0.70	0.92	1.18	1.57

Shaft with Two Flats

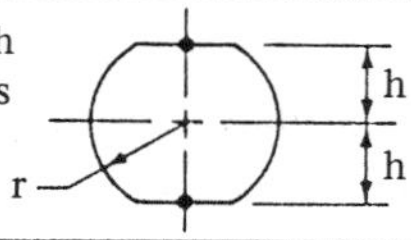

$K = C_3 r^4$

$Q = C_4 r^3$

h/r	0.5	0.6	0.7	0.8	0.9	1.0
C_3	0.44	0.67	0.93	1.19	1.39	1.57
C_4	0.47	0.60	0.81	1.02	1.25	1.57

Hollow Rectangle

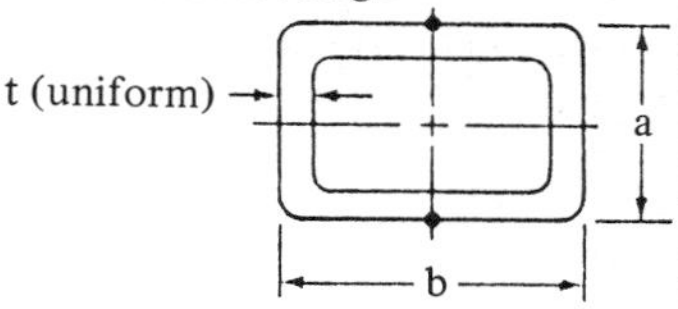

t (uniform) →

$$K = \frac{2t(a - t)^2(b - t)^2}{(a + b - 2t)}$$

$$Q = 2t(a - t)(b - t)$$

Gives average stress; good approximation of maximum stress if t is small

Inner corners should have generous fillets

Split Tube
 Mean radius (r)
 t (uniform)

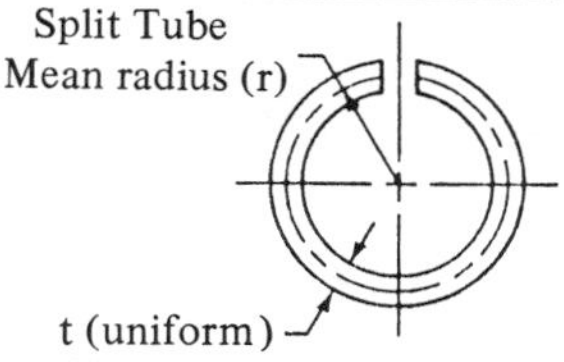

$$K = 2\pi r t^3/3$$

$$Q = \frac{4\pi^2 r^2 t^2}{(6\pi r + 1.8t)}$$

t must be small

Figure 5-15 Methods for determining values for K and Q for several types of cross sections. (Source: *Machine Elements in Mechanical Design*, Robert L. Mott, Copyright © Merrill Publishing Co., Columbus, Ohio. Reprinted by permission of the publisher)

Figure 5-15 shows seven cases of noncircular cross sections that are commonly encountered in machine design and structural analysis. The computation of the maximum shear stress and the angle of twist can be made by slightly modifying the formulas used for circular cross sections as given here.

$$\tau_{max} = \frac{T}{Q} \tag{5-22}$$

$$\theta = \frac{TL}{GK} \tag{5-23}$$

The term Q is analogous to the polar section modulus Z_p used for circular bars. The torsional stiffness is indicated by K, analogous to the polar moment of inertia J.

Note in Figure 5-15 that the points of maximum shear stress for the noncircular cross sections are indicated by a prominent black dot.

An example is now given which illustrates the high degree of flexibility of a slit tube in comparison with a closed tube.

Example Problem 5-13

A tube is made by forming a flat steel sheet, 4.0 mm thick, into the circular shape having an outside diameter of 90 mm. The final step is to weld the seam along the length of the tube. Figure 5-14(a), 5-14(d), and 5-13(a) show the stages of the process. Perform the following calculations to compare the behavior of the closed, welded tube with that of the open tube.

(a) Compute the torque that would create a stress of 10 MPa in the closed, welded tube.

(b) Compute the angle of twist of a 1.0-m length of the closed tube for the torque found in part (a).

(c) Compute the stress in the open tube for the torque found in part (a).

(d) Compute the angle of twist of a 1.0-m length of the open tube for the torque found in part (a).

(e) Compare the stress and deflection of the open tube with those of the closed tube.

Solution (a) The torque applied to the closed tube can be found by using Equation (5-8) and solving for T.

$$\tau_{max} = \frac{Tc}{J}$$

Then $\tag{5-8}$

$$T = \frac{\tau_{max} J}{c}$$

We can compute J from Equation (5-10),

$$J = \frac{\pi}{32} (D_o^4 - D_i^4) \tag{5-10}$$

Using $D_o = 90$ mm $= 0.09$ m and $D_i = 82$ mm $= 0.082$ m,

$$J = \frac{\pi}{32} (0.09^4 - 0.082^4) \text{ m}^4 = 2.00 \times 10^{-6} \text{ m}^4$$

Now, letting $\tau_{max} = 10$ MPa $= 10 \times 10^6$ N/m², we have

$$T = \frac{\tau_{max} J}{c} = \frac{(10 \times 10^6 \text{ N/m}^2)(2.00 \times 10^{-6} \text{ m}^4)}{0.045 \text{ m}} = 444 \text{ N} \cdot \text{m}$$

That is, a torque of 444 N · m applied to the closed welded tube would produce a maximum torsional shear stress of 10 MPa in the tube. Note that this is a very low stress level for steel.

(b) The angle of twist for the closed tube can be found from Equation (5-21). Using $T = 444$ N · m yields

$$\theta = \frac{TL}{GJ} = \frac{(444 \text{ N} \cdot \text{m})(1.0 \text{ m})}{(80 \times 10^9 \text{ N/m}^2)(2.00 \times 10^{-6} \text{ m}^4)} = 0.00278 \text{ rad}$$

Converting θ to degrees gives

$$\theta = 0.00278 \text{ rad} \frac{180 \text{ deg}}{\pi \text{ rad}} = 0.159 \text{ deg}$$

Again, note that this is a very small angle of twist.

(c) Equation (5-22) can be used to compute the maximum shear stress for the open tube before it is welded, treating it as a noncircular cross section. The formula for Q is given in Figure 5-15.

$$Q = \frac{4\pi^2 r^2 t^2}{6\pi r + 1.8t}$$

The mean radius is

$$r = \frac{D_o}{2} - \frac{t}{2} = \frac{90 \text{ mm}}{2} - \frac{4 \text{ mm}}{2} = 43 \text{ mm}$$

Then

$$Q = \frac{4\pi^2 (43)^2 (4)^2}{6\pi(43) + 1.8(4)} \text{ mm}^3 = 1428 \text{ mm}^3$$

The stress in the open tube is, then,

$$\tau_{max} = \frac{T}{Q} = \frac{444 \text{ N} \cdot \text{m}}{1428 \text{ mm}^3} \frac{1000 \text{ mm}}{\text{m}} = 311 \text{ MPa}$$

(d) The angle of twist for the open tube can be computed using Equation (5-23). Figure 5-15 gives us the formula for the torsional rigidity constant, K. Using $r = 43$ mm $= 0.043$ m and $t = 4$ mm $= 0.004$ m yields

$$K = \frac{2\pi r t^3}{3} = \frac{2\pi(0.043)(0.004)^3}{3} = 5.764 \times 10^{-9} \text{ m}^4$$

Then the angle of twist is

$$\theta = \frac{TL}{GK} = \frac{(444 \text{ N} \cdot \text{m})(1.0 \text{ m})}{(80 \times 10^9 \text{ N/m}^2)(5.764 \times 10^{-9} \text{ m}^4)} = 0.963 \text{ rad}$$

Converting θ to degrees, we obtain

$$\theta = 0.963 \text{ rad} \frac{180 \text{ deg}}{\pi \text{ rad}} = 55.2 \text{ deg}$$

(e) Comparing the torsional shear stress and the angle of twist for the closed and open tubes shows a remarkable difference; far greater stress and deformation for the open tube. We compute ratios:

$$\frac{\text{stress in open tube}}{\text{stress in closed tube}} = \frac{311 \text{ MPa}}{10 \text{ MPa}} = 31.1 \text{ times greater}$$

$$\frac{\text{twist of open tube}}{\text{twist of closed tube}} = \frac{55.2 \text{ deg}}{0.159 \text{ deg}} = 347 \text{ times greater}$$

In addition, the actual stress level computed for the open tube is probably greater than the allowable stress for many steels. Recall that the design shear stress is

$$\tau_d = \frac{0.5 s_y}{N}$$

Then the required yield strength of the material to be safe at a stress level of 311 MPa and a design factor of 2.0 is

$$s_y = \frac{N\tau_d}{0.5} = \frac{2(311 \text{ MPa})}{0.5} = 1244 \text{ MPa}$$

Only a few highly heat treated steels listed in Appendix A-13 have this level of yield strength. These comparisons should illustrate the importance of using closed sections for torsionally loaded members.

REFERENCES

1. BLODGETT, O. W., *Design of Weldments,* James F. Lincoln Arc Welding Foundation, Cleveland, Ohio, 1963.
2. BORESI, A. P., O. M. SIDEBOTTOM, F. B. SEELY, and J. O. SMITH, *Advanced Mechanics of Materials*, 3rd ed., New York, 1978.
3. MOTT, R. L., *Machine Elements in Mechanical Design*, Charles E. Merrill, Columbus, Ohio. 1985.
4. POPOV, E. P., *Mechanics of Materials*, 2nd ed. Prentice-Hall, Englewood Cliffs, N. J., 1978.
5. ROARK, R. J., and W. C. YOUNG, *Formulas for Stress and Strain*, 5th ed. McGraw-Hill, New York, 1975.

PROBLEMS

5-1. Compute the torsional shear stress which would be produced in a solid circular shaft having a diameter of 20 mm when subjected to a torque of 280 N · m.

5-2. A hollow shaft has an outside diameter of 35 mm and an inside diameter of 25 mm. Compute the torsional shear stress in the shaft when it is subjected to a torque of 560 N · m.

5-3. Compute the torsional shear stress in a shaft having a diameter of 1.25 in. when carrying a torque of 1550 lb · in.

5-4. A steel tube is used as a shaft carrying 5500 lb · in. of torque. The outside diameter is 1.75 in., and the wall thickness is $\frac{1}{8}$ in. Compute the torsional shear stress at the outside and the inside surfaces of the tube.

5-5. A movie projector drive mechanism is driven by a 0.08-kW motor whose shaft rotates at 180 rad/s. Compute the torsional shear stress in its 3.0-mm-diameter shaft.

5-6. The impeller of a fluid agitator rotates at 42 rads/s and requires 35 kW of power. Compute the torsional shear stress in the shaft which drives the impeller if it is hollow and has an outside diameter of 40 mm and an inside diameter of 25 mm.

5-7. A drive shaft for a milling machine transmits 15.0 hp at a speed of 240 rpm. Compute the torsional shear stress in the shaft if it is solid with a diameter of 1.44 in. Would the shaft be safe if the torque is applied with shock and it is made from AISI 4140 OQT 1300 steel?

5-8. Repeat Problem 5-7 if the shaft contains a profile keyseat.

5-9. Figure 5-16 shows the end of the vertical shaft for a rotary lawnmower. Compute the maximum torsional shear stress in the shaft if it is transmitting 7.5 hp to the blade when rotating 2200 rpm. Specify a suitable steel for the shaft.

5-10. Figure 5-17 shows a stepped shaft in torsion. The larger section also has a hole drilled through.

(a) Compute the maximum shear stress at the step for an applied torque of 7500 lb · in.

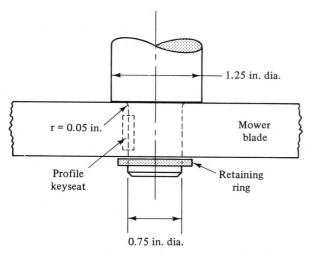

Figure 5-16 Shaft for Problem 5-9.

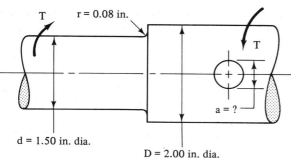

Figure 5-17 Shaft for Problem 5-10.

(b) Determine the largest hole that could be drilled in the shaft and still maintain the stress near the hole at or below that at the step.

5-11. Compute the torsional shear stress and the angle of twist in degrees in an aluminum tube, 600 mm long, having an inside diameter of 60 mm and an outside diameter of 80 mm when subjected to a steady torque of 4500 N · m. Then specify a suitable aluminum alloy for the tube.

5-12. Two designs for a shaft are being considered. Both have an outside diameter of 50 mm and are 600 mm long. One is solid but the other is hollow with an inside diameter of 40 mm. Both are made from steel. Compare the torsional shear stress, angle of twist, and the mass of the two designs if they are subjected to a torque of 850 N · m.

5-13. Determine the required inside and outside diameters for a hollow shaft to carry a torque of 1200 N · m with a maximum torsional shear stress of 45 MPa. Make the ratio of the outside diameter to the inside diameter approximately 1.25.

5-14. A gear drive shaft for a milling machine transmits 7.5 hp at a speed of 240 rpm. Compute the torsional shear stress in the 0.860-in.-diameter solid shaft.

5-15. The input shaft for the gear drive described in Problem 5-14 also transmits 7.5 hp, but rotates at 1140 rpm. Determine the required diameter of the input shaft to give it the same stress as the output shaft.

5-16. Determine the stress which would result in a $1\frac{1}{2}$-in. schedule 40 steel pipe if a plumber applies a force of 80 lb at the end of a wrench handle 18 in. long.

5-17. A rotating sign makes 1 rev every 5 s. In a high wind, a torque of 30 lb·ft is required to maintain rotation. Compute the power required to drive the sign. Also compute the stress in the final drive shaft if it is 0.60 in. in diameter. Specify a suitable steel for the shaft to provide a design factor of 4 based on yield strength in shear.

5-18 A short, cylindrical bar is welded to a rigid plate at one end, and then a torque is applied at the other. If the bar has a diameter of 15 mm and is made of AISI 1020 cold-drawn steel, compute the torque which must be applied to it to subject it to a stress equal to its yield strength in shear. Use $s_{ys} = s_y/2$.

5-19. A propeller drive shaft on a ship is to transmit 2500 hp at 75 rpm. It is to be made of AISI 1040 WQT 1300 steel. Use a design factor of 6 based on the yield strength in shear. The shaft is to be hollow, with the inside diameter equal to 0.80 times the outside diameter. Determine the required diameter of the shaft.

5-20. If the propeller shaft of Problem 5-19 was to be solid instead of hollow, determine the required diameter. Then compute the ratio of the weight of the solid shaft to that of the hollow shaft.

5-21. A power screwdriver uses a shaft with a diameter of 5.0 mm. What torque can be applied to the screwdriver if the limiting stress due to torsion is 80 MPa?

5-22. An extension for a socket wrench similar to that shown in Figure 5-1 has a diameter of 6.0 mm and a length of 250 mm. Compute the stress and angle of twist in the extension when a torque of 5.5 N · m is applied. The extension is steel.

5-23. Compute the angle of twist in a steel shaft 15 mm in diameter and 250 mm long when a torque of 240 N · m is applied.

5-24. Compute the angle of twist in an aluminum tube that has an outside diameter of 80 mm and an inside diameter of 60 mm when subjected to a torque of 2250 N · m. The tube is 1200 mm long.

5-25. A steel rod 8 ft long and 0.625 in. in diameter is used as a long wrench to unscrew a plug at the bottom of a pool of water. If it requires 40 lb · ft of torque to loosen the plug, compute the angle of twist of the rod.

5-26. For the rod described in Problem 5-25, what must the diameter be if only 2.0 deg of twist is desired when 40 lb · ft of torque is applied?

5-27. Compute the angle of twist of the free end relative to the fixed end of the steel bar shown in Figure 5-18.

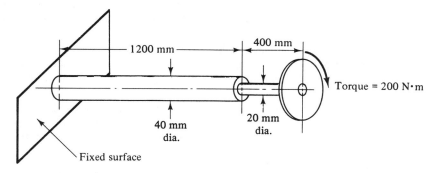

Figure 5-18 Bar for Problem 5-27.

5-28. A meter for measuring torque uses the angle of twist of a shaft to indicate torque. The shaft is to be 150 mm long and made of 6061-T6 aluminum alloy. Determine the required diameter of the shaft if it is desired to have a twist of 10.0 deg when a torque of 5.0 N · m is applied to the meter. For the shaft thus designed, compute the torsional shear stress and then compute the resulting design factor for the shaft. Is it satisfactory? If not, what would you do?

5-29. A beryllium copper wire having a diameter of 1.50 mm and a length of 40 mm is used as a small torsion bar in an instrument. Determine what angle of twist would result in the wire when it is stressed to 250 MPa.

5-30. A fuel line in a aircraft is made of a titanium alloy. The tubular line has an outside diameter of 18 mm and an inside diameter of 16 mm. Compute the stress in the tube if a length of 1.65 m must be twisted through an angle of 40 deg during installation. Determine the design factor based on the yield strength in shear if the tube is Ti-6A1-4V, aged.

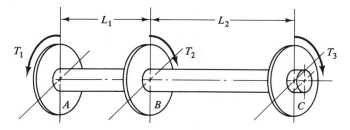

Figure 5-19 Shaft for Problem 5-31.

5-31. For the shaft shown in Figure 5-19 compute the angle of twist of pulleys *B* and *C* relative to *A*. The steel shaft has a diameter of 35 mm throughout its length. The torques are:

$T_1 = 1500$ N · m, $T_2 = 1000$ N · m, $T_3 = 500$ N · m. The lengths are: $L_1 = 500$ mm, $L_2 = 800$ mm.

5-32. A torsion bar in a light truck suspension is to be 820 mm long and made of steel. It is subjected to a torque of 1360 N · m and must be limited to 2.2 deg of twist. Determine the required diameter of the solid round bar. Then compute the stress in the bar.

5-33. A steel drive shaft for an automobile is a hollow tube 1525 mm long. Its outside diameter is 75 mm, and its inside diameter if 55 mm. If the shaft transmits 120 kW of power at a speed of 225 rad/s, compute the torsional shear stress in the shaft and the angle of twist of one end relative to the other.

5-34. A rear axle of an automobile is a solid steel shaft having the configuration shown in Figure 5-20. Considering the stress concentration due to the shoulder, compute the torsional shear stress in the axle when it rotates at 70.0 rad/s, transmitting 60 kW of power.

5-35. An output shaft from an automotive transmission has the configuration shown in Figure 5-21. If the shaft is transmitting 105 kW at 220 rad/s, compute the maximum torsional shear stress in the shaft. Account for the stress concentration at the place where the speedometer gear is located.

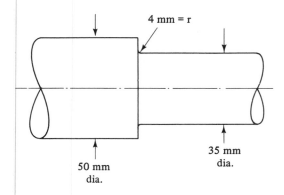

4 mm = r

35 mm dia.

50 mm dia.

Figure 5-20 Axle for Problem 5-34.

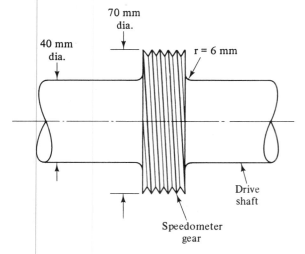

70 mm dia.

40 mm dia.

r = 6 mm

Drive shaft

Speedometer gear

Figure 5-21 Shaft for Problem 5-35.

The indicated figures for Problems 5-36 through 5-39 show portions of shafts from power transmission equipment. Compute the maximum repeated torque that could safely be applied to each shaft if it is to be made from AISI 1141 OQT 1100 steel.

5-36. Use Figure 5-22.

5-37. Use Figure 5-23.

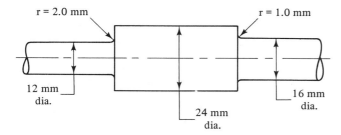

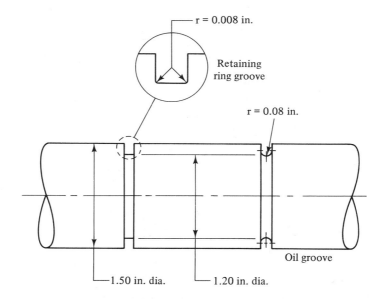

Figure 5-22 Shaft for Problem 5-36.

Figure 5-23 Shaft for Problem 5-37.

5-38. Use Figure 5-24.

5-39. Use Figure 5-25.

5-40. Compute the torque that would produce a torsional shear stress of 50 MPa in a square steel rod 20 mm on a side.

5-41. For the rod described in Problem 5-40 compute the angle of twist that would be produced by the torque found in the problem over a length of 1.80 m.

5-42. Compute the torque that would produce a torsional shear stress of 7500 psi in a square aluminum rod, 1.25 in. on a side.

5-43. For the rod described in Problem 5-42, compute the angle twist that would be produced by the torque found in the problem over a length of 48 in.

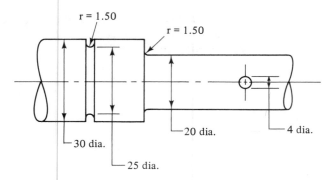

r = 1.50

r = 1.50

20 dia.

4 dia.

30 dia.

25 dia.

Dimensions in mm

Figure 5-24 Shaft for Problem 5-38.

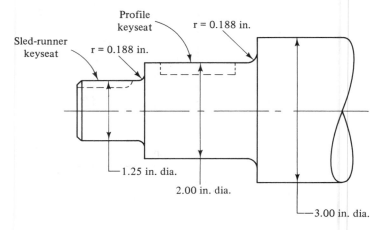

Profile
keyseat

r = 0.188 in.

Sled-runner
keyseat

r = 0.188 in.

1.25 in. dia.

2.00 in. dia.

3.00 in. dia.

Figure 5-25 Shaft for Problem 5-39.

5-44. Compute the torque that would produce a torsional shear stress of 7500 psi in a rectangular aluminum bar 1.25 in. thick by 3.00 in. wide.

5-45. For the bar described in Problem 5-44, compute the angle of twist that would be produced by the torque found in the problem over a length of 48 in.

5-46. An extruded aluminum bar is in the form of an equilateral triangle 30 mm on a side. What torque is required to cause an angle of twist in the bar of 0.80 deg over a length of 2.60 m?

5-47. For the triangular bar described in Problem 5-46, what stress would be developed in the bar when carrying the torque found in the problem?

5-48. As shown in Figure 5-26, a portion of a steel shaft 1.75 in. in diameter has a flat machined on one side such that the measurement from the flat to the opposite side is 1.50 in. Compute the torsional shear stress in both the circular section and the one with the flat when a torque of 850 lb · in. is applied.

5-49. For the steel shaft shown in Figure 5-26, compute the angle of twist of one end relative to the other if a torque of 850 lb · in. is applied uniformly along the length.

5-50. Repeat Problem 5-48 with all the data the same except that two flats are machined on the shaft resulting in a total measurement across the flats of 1.25 in.

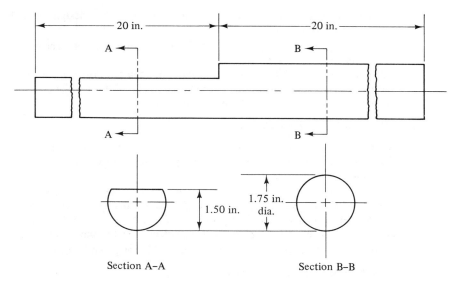

1.75 in. dia.

1.50 in.

Section A–A

Section B–B

Figure 5-26 Problems 5-48 and 5-49.

5-51. Repeat Problem 5-49 for a shaft having two flats resulting in a total measurement across the flats of 1.25 in.

5-52. A square stud 200 mm long and 8 mm on a side is made from titanium, Ti-6A1-4V, aged. What angle of twist will result when a wrench is applying a pure torque that causes the stress to equal the yield strength of the material in shear?

5-53. A standard square structural steel tube has cross-sectional dimensions of $4 \times 4 \times \frac{1}{4}$ in. and is 8.00 ft long. Compute the torque required to twist the tube 3.00 deg.

5-54. For the tube in Problem 5-53, compute the maximum stress in the tube when it is twisted 3.00 deg. Would this be safe if the tube is made from ASTM A501 structural steel and the load was static?

5-55. Repeat Problem 5-53 for a rectangular tube, $6 \times 4 \times \frac{1}{4}$.

5-56. Repeat Problem 5-54 for a rectangular tube, $6 \times 4 \times \frac{1}{4}$.

5-57. A standard 6-in. schedule 40 steel pipe has approximately the same cross-sectional area as a square tube, $6 \times 6 \times \frac{1}{4}$, and thus both would weigh about the same for a given length. If the same torque were applied to both, compare the resulting torsional shear stress and angle of twist for the two shapes.

COMPUTER PROGRAMMING ASSIGNMENTS

1. Given the need to design a solid circular shaft for a given torque, a given material yield strength, and a given design factor, compute the required diameter for the shaft.

ENHANCEMENTS TO ASSIGNMENT 1

(a) For a given power transmitted and speed of rotation, compute the applied torque.

(b) Include a table of materials from which the designer can select one. Then automatically look up the yield strength.

(c) Include the design factor table, Table 5-1. Then prompt the designer to specify the manner of loading only and determine the appropriate design factor from the table within the program.

2. Repeat assignment 1, except design a hollow circular shaft. Three possible solution procedures exist:

   (a) For a given outside diameter, compute the required inside diameter.
   (b) For a given inside diameter, compute the required outside diameter.
   (c) For a given ratio of D_i/D_o, find both D_i and D_o.

ENHANCEMENTS TO ASSIGNMENT 2

   (a) Compute the mass of the resulting design for a given length and material density.
   (b) If the computer has graphics capability, draw the resulting cross section and dimension it.

3. Enter the stress concentration factor curves into a program, allowing the automatic computaton of K_t for given factors such as fillet radius, diameter ratio, hole diameter, and so on. Any of the cases shown in Appendix A-21-3, A-21-4, or A-21-8 could be used. This program could be run by itself or made an enhancement of other stress analysis programs.

4. Compute the angle of twist from Equation (5-21) for given T, L, G, and J.

ENHANCEMENTS TO ASSIGNMENT 4

   (a) Compute J for given dimensions for the shaft, either solid or hollow.
   (b) Include a table of values for G, from Table 5-3, in the program.

5. Compute the required diameter of a solid circular shaft to limit the angle of twist to a specified amount.

6. Compute the angle of twist for one end of a multisection shaft relative to the other, similar to Example Problem 5-12. Allow the lengths, diameters, materials, and torques to be different in each section.

7. Write a program to compute the values for the effective section modulus, Q, and the torsional stiffness constant, K, from Figure 5-15 for any or all cases.

ENHANCEMENT TO ASSIGNMENT 7

Determine the equations for C_1, C_2, C_3, and C_4 in terms of the ratio, h/r, for shafts with flats. Use a curve-fitting routine.

6

Shearing Forces and Bending Moments in Beams

6-1 OBJECTIVES OF THIS CHAPTER

Much of the discussion in the next six chapters deals with beams. A *beam* is a member that carries loads transversely, that is, perpendicular to its long axis. Such loads cause *shearing stresses* to be developed in the beam and give the beam its characteristic *bent* shape, resulting in *bending stresses* as well.

To compute shearing stresses and bending stresses, it is necessary to be able to determine the magnitude of internal *shearing forces* and *bending moments* developed in beams due to a variety of loads. That is the primary objective of this chapter, after completion of which you should be able to:

1. Define the term *beam* and recognize when a load-carrying member is a beam.
2. Describe several kinds of beam loading patterns: *concentrated loads, uniformly distributed loads, linearly varying distributed loads,* and *concentrated moments.*
3. Describe several kinds of beams according to the manner of support: *simple beam, overhanging beam, cantilever,* and *composite beam* having more than one part.
4. Draw free-body diagrams for beams and parts of beams showing all external forces and reactions.

5. Compute the magnitude of reaction forces and moments and determine their directions.

6. Define *shearing force* and determine the magnitude of shearing force anywhere within a beam.

7. Draw free-body diagrams of *parts* of beams and show the internal shearing forces.

8. Draw complete shearing force diagrams for beams carrying a variety of loading patterns and with a variety of support conditions.

9. Define *bending moment* and determine the magnitude of bending moment anywhere within a beam.

10. Draw free-body diagrams of *parts* of beams and show the internal bending moments.

11. Draw complete bending moment diagrams for beams carrying a variety of loading patterns and with a variety of support conditions.

12. Use the *laws of beam diagrams* to relate the load, shear, and bending moment diagrams to each other and to draw the diagrams.

13. Draw free-body diagrams of parts of composite beams and structures and draw the shearing force and bending moment diagrams for each part.

14. Properly consider *concentrated moments* in the analysis of beams.

6-2 BEAM LOADING

A *beam* is a member which carries loads transversely, that is, perpendicular to its long axis. Figures 6-1 through 6-4 show several examples of beams. Figure 6-1(a) shows a beam, supported at its ends by rods, which carries four pipes in a utility tunnel supplying water, steam, compressed air, and natural gas to production processes in a factory. Such a beam is called a *simple beam* because of the manner of support. Note that the rods at the end provide only vertical supporting forces and supply these only at the ends of the beam. Also, the pipes on the beams are examples of *concentrated loads*. That is, the forces exerted on the beam act essentially at a single point or a very small part of the length of the beam. For purposes of analyzing this beam, the loads due to the pipes would be shown as arrows acting downward, as shown in Figure 6-1(b).

A second kind of loading is the *uniformly distributed load*, as shown in Figure 6-2. This shows a roof assembly made up of several beams running across the width of a structure. The roofing materials are then supported on the beams. Snow piled on top of the roof applies additional loading. Considering that the total roof load is shared evenly by the several beams, each beam can be considered to carry the uniformly distributed load as sketched in Figure 6-2(b). The load is usually described as being so much force per unit length. For example, the loading could be 20.3 kN/m or 1800 lb/ft.

An *overhanging beam* is one in which part of the loaded beam extends outside the supports. Figure 6-3 shows a part of a building structure in which the upper

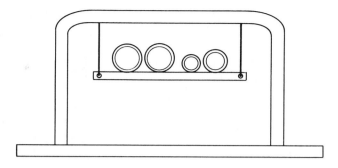

(a) Pictorial representation
of beam and loads

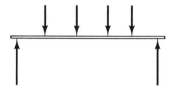

(b) Schematic representation
of beam and loads

Figure 6-1 Simple beam with concentrated loads.

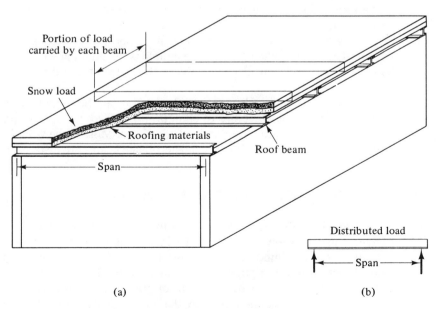

(a)

(b)

Figure 6-2 Simple beam with uniformly distributed load.

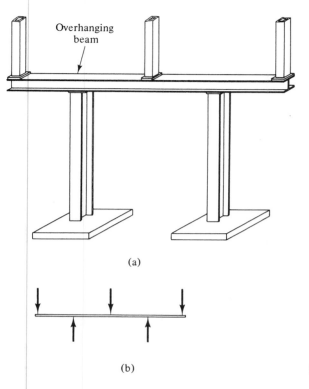

(a)

(b)

Figure 6-3 Overhanging beam.

columns apply loads to the beam supported by the two lower columns. Figure 6-3(b) shows the schematic diagram of the beam.

A *cantilever beam* is held fixed at the support and extends out to carry either concentrated or distributed loads. An example is the boom of a crane, such as the one shown in Figure 6-4. Part (b) of the figure shows the standard manner of sketching a cantilever beam.

For the four types of beams described so far, all the support reaction forces can be computed from the equations of equilibrium studied in statics or physics mechanics. Thus they are said to be *statically determinate*. Most of the work in this book will be for beams of this type. Statically indeterminate beams are discussed in Chapter 13.

6-3 BEAM SUPPORTS AND REACTIONS AT SUPPORTS

The first step in analyzing a beam in order to determine its safety under a given loading arrangement is to show completely the loads and support reactions on a free-body diagram. It is very important to be able to construct free-body diagrams from the physical picture or description of the loaded beam. This was done in each case for Figures 6-1 through 6-4, where part (b) of each figure is the free-body diagram.

After constructing the free-body diagram, it is necessary to compute the magnitude of all support reactions. It is assumed that the methods used to find reactions

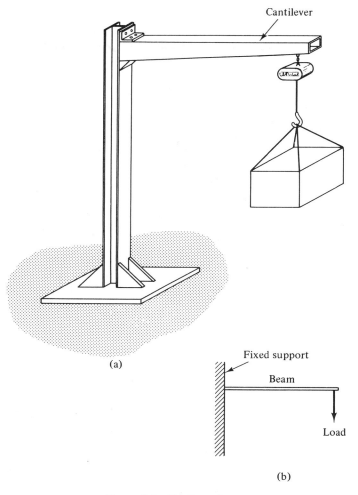

(a)

Fixed support

Beam

Load

(b)

Figure 6-4 Cantilever beam.

were studied previously. Therefore, only a few examples are shown here as a review and as an illustration of the techniques used throughout this book.

The following general procedure is recommended for solving for reactions on simple or overhanging beams.

1. Draw the free-body diagram.
2. Use the equilibrium equation $\Sigma M = 0$ by summing moments about the point of application of one support reaction. The resulting equation can then be solved for the other reaction.
3. Use $\Sigma M = 0$ by summing moments about the point of application of the second reaction to find the first reaction.
4. Use $\Sigma F = 0$ to check the accuracy of your calculations.

Example Problem 6-1

Figure 6-5 shows the free-body diagram for the beam carrying pipes, which was originally shown in Figure 6-1. Compute the reaction forces in the support rods.

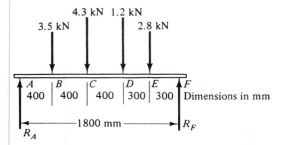

Figure 6-5   Beam loading.

Solution   Since the free-body diagram is given, step 2 of the solution procedure will now be applied. To find the reaction R_F, sum moments about point A.

$$\sum M_A = 0 = 3.5(400) + 4.3(800) + 1.2(1200) + 2.8(1500) - R_F(1800)$$

Note that all forces are in kilonewtons and distances in millimeters. Now solve for R_F.

$$R_F = \frac{3.5(400) + 4.3(800) + 1.2(1200) + 2.8(1500)}{1800} = 5.82 \text{ kN}$$

Now, to find R_A, sum moments about point F.

$$\sum M_F = 0 = 2.8(300) + 1.2(600) + 4.3(1000) + 3.5(1400) - R_A(1800)$$

$$R_A = \frac{2.8(300) + 1.2(600) + 4.3(1000) + 3.5(1400)}{1800} = 5.98 \text{ kN}$$

Now use $\Sigma F = 0$ for the vertical direction as a check.

Downward forces: $(3.5 + 4.3 + 1.2 + 2.8) \text{ kN} = 11.8 \text{ kN}$

Upward reactions: $(5.82 + 5.98) \text{ kN} = 11.8 \text{ kN}$ (check)

Remember to show the reaction forces R_A and R_F at their proper points on the beam.

Example Problem 6-2

Compute the reactions on the beam shown in Figure 6-6. Note that the distributed load covers only part of the beam.

Solution   For purposes of finding reactions, it is convenient to work with the resultant of a distributed load by considering that the total load acts at the centroid of the load. This is shown in Figure 6-6(b). Then the same procedure as before can be used to compute reactions.

$$\sum M_A = 0 = 22\ 000 \text{ lb (5 ft)} - R_C(12 \text{ ft})$$

$$R_C = \frac{22\ 000 \text{ lb (5 ft)}}{12 \text{ ft}} = 9167 \text{ lb}$$

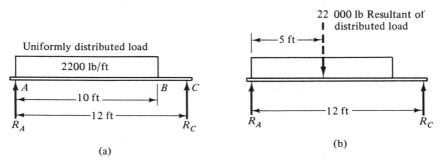

Figure 6-6 Beam loading.

$$\sum M_C = 0 = 22\ 000\ \text{lb}\,(7\ \text{ft}) - R_A(12\ \text{ft})$$

$$R_A = \frac{22\ 000\ \text{lb}\,(7\ \text{ft})}{12\ \text{ft}} = 12\ 833\ \text{lb}$$

Finally, as a check, in the vertical direction,

$$\sum F = 0$$

Downward forces: 22 000 lb

Upward forces: $R_A + R_C = 12\ 833 + 9167 = 22\ 000\ \text{lb}$ (check)

Example Problem 6-3

Compute the reactions for the overhanging beam shown in Figure 6-7.
Solution   First, summing moments about point B,

$$\sum M_B = 0 = 1000(200) - R_D(250) + 1200(400) - 800(100)$$

Notice that forces that tend to produce clockwise moments about B are considered positive in this calculation. Now solving for R_D gives

$$R_D = \frac{1000(200) + 1200(400) - 800(100)}{250} = 2400\ \text{N}$$

Summing moments about D will allow computation of R_B.

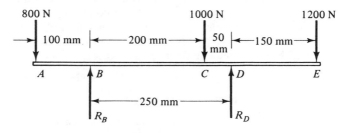

Figure 6-7 Beam loading.

$$\sum M_D = 0 = 1000(50) - R_B(250) + 800(350) - 1200(150)$$

$$R_B = \frac{1000(50) + 800(350) - 1200(150)}{250} = 600 \text{ N}$$

Check with $\sum F = 0$ in the vertical direction:

Downward forces: $(800 + 1000 + 1200)$ N $= 3000$ N

Upward forces: $R_B + R_D = (600 + 2400)$ N $= 3000$ N   (check)

Example Problem 6-4

Compute the reactions at the wall required to support the cantilever beam shown in Figure 6-8.

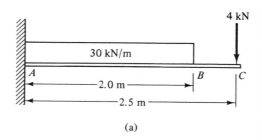

(a)

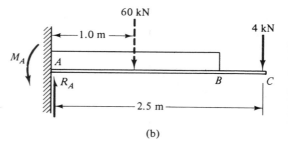

(b)

Figure 6-8   Beam loading.

Solution   In the case of cantilever beams, the reactions at the wall are composed of an upward force R_A which must balance all downward forces on the beam and a reaction moment M_A which must balance the tendency for the applied loads to rotate the beam. These are shown in Figure 6-8(b). Also shown is the resultant, 60 kN, of the distributed load. Then, by summing forces in the vertical direction, we obtain

$$R_A = 60 \text{ kN} + 4 \text{ kN} = 64 \text{ kN}$$

Summing moments about point A yields

$$M_A = 60 \text{ kN} (1.0\text{m}) + 4 \text{ kN} (2.5 \text{ m}) = 70 \text{ kN} \cdot \text{m}$$

6-4 SHEARING FORCES

It will be shown later that the two kinds of stresses developed in a beam are shearing stresses and bending stresses. In order to compute these stresses, it will be necessary to know the magnitude of shearing forces and bending moments at all points in the beam. Therefore, although you may not yet understand the ultimate use of these factors, it is necessary to learn how to determine the variation of shearing forces and bending moments in beams for many types of loading and support combinations.

Shearing forces are internal forces developed in the material of a beam to balance externally applied forces in order to ensure equilibrium in all parts of the beam. The presence of these shearing forces can be visualized by considering a section of the beam and looking at all external forces. For example, Figure 6-9 shows a simple beam carrying a concentrated load at its center. The beam as a whole is in equilibrium under the action of the 1000-N load and the two 500-N reaction forces at the supports. If the whole beam is in equilibrium, then so is any part of it. Consider a part of the beam 0.5 m long, as shown in Figure 6-9(b). For this part of the beam to be in equilibrium,

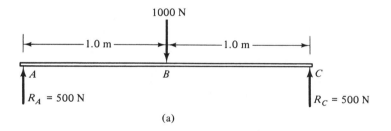

(a)

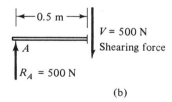

(b)

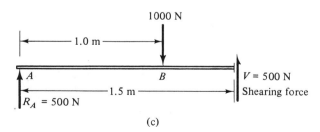

(c)

Figure 6-9 Use of free body diagrams to find shearing forces in beams.

there must be a force of 500 N inside the beam acting downward.* This internal force is called the *shearing force* and will be denoted by the symbol V. Notice that the shearing force would be of the same magnitude if the beam were cut anywhere between A and B.

Now consider a part of the beam 1.5 m long, as shown in Figure 6-9(c). In order for that part to be in equilibrium, an internal shearing force of 500 N upward must exist in the beam. This would be the same if the beam were cut anywhere between B and C.

Shear diagrams. The free-body diagram could be drawn for any part of any beam with any loading to determine the magnitude of the shearing force in the beam. However, it is a time-consuming process, and a simpler method is available to obtain a plot of the variation of shearing force versus position on the beam for the entire beam. The name given to such a plot is the *shear diagram*. The fundamental principle involved in creating a shear diagram is as follows:

> *The magnitude of the shearing force in any part of the beam is equal to the algebraic sum of all external forces acting to the left of the section of interest.*

In drawing the diagram, upward external forces and reactions will be considered positive, and downward external forces will be negative.

Figure 6-10 illustrates the shear diagram for the beam considered earlier. Construction should start at the left end, point A, where the reaction force of 500 N is encountered immediately. Then the shear diagram starts at 500 N. Between points A and B, there are no other forces applied to the beam. Therefore, the shear remains at

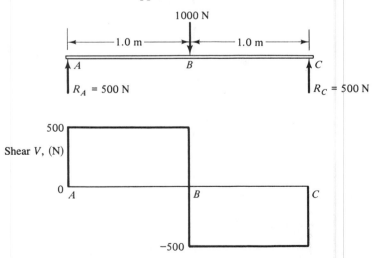

Figure 6-10 Shear diagram.

*Also, a moment would have to exist inside the beam to keep it from rotating. This is called the bending moment and is discussed in the following section.

500 N all the way from A to B. At point B, the 1000-N concentrated load is encountered, causing an abrupt change in shear from positive 500 N to negative 500 N. Typically, the application of a concentraed load will cause an abrupt change in the magnitude of the shear in the beam. Then between B and C, no other loads are applied, resulting in no change in shear. At point C, the 500-N reaction R_C brings the plot abruptly back to zero.

Looking at the complete shear diagram of Figure 6-10, it can be seen that the greatest value of shear is 500 N. Notice that even though there is an applied load of 1000 N, the maximum shearing force in the beam is only 500 N.

Another beam carrying concentrated loads will now be considered in an example problem. The general approach used above will be applicable to any beam carrying concentrated loads.

Example Problem 6-5

Draw the complete shear diagram for the beam shown in Figure 6-11.

Solution   This is the same beam which was analyzed to determine reactions R_A and R_F in Example Problem 6-1. In general, the calculation of the reaction forces would be the first step in the process of drawing the shear diagram.

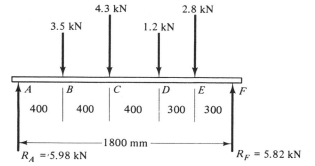

Figure 6-11   Beam loading.

The completed diagram is shown in Figure 6-12. The process used to determine each part of the diagram is described below.

Point A.   Starting at the left end of the beam, the reaction R_A is encountered immediately. Thus, the shear starts abruptly at 5.98 kN.

Point B.   Since no loads are applied between A and B, the shear is constant at 5.98 kN. At B, the 3.5-kN downward load causes an abrupt decrease in shear to 2.48 kN.

Point C.   No change between B and C. At C, shear decreases by 4.3 kN to −1.82 kN.

Point D.   No change between C and D. At D, shear decreases by 1.2 kN to −3.02 kN.

Point E.   No change from D to E. At E, shear decreases by 2.8 kN to −5.82 kN.

Point F.   No change between E and F. At F, the upward reaction $R_F = 5.82$ kN brings the shear back to zero.

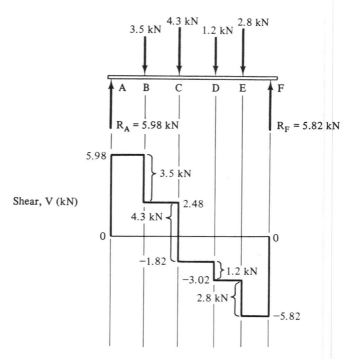

Figure 6-12 Shear diagram.

Notice some general characteristics of shear diagrams for beams having only concentrated loads, as illustrated in this example.

1. The shear diagram starts and ends at zero at the ends of the beam.
2. At each concentrated load or reaction, the value of the shear changes abruptly by an amount equal to the load or reaction force.
3. Between concentrated loads, there is no change in shear, and the shear curve plots as a straight horizontal line.
4. The symbol V is used to identify the shearing force axis, and the units for force are indicated.

Shear diagrams for distributed loads. The variation of shearing force with position on the beam for distributed loads is different from that for concentrated loads. The free-body diagram method is useful to help visualize these variations.

Consider the beam in Figure 6-13, carrying a uniformly distributed load of 1500 N/m over part of its length. It is desired to determine the magnitude of the shear at several points in the beam in order to draw a shear diagram. Let's choose points at intervals of 2 m across the beam, including the two ends of the beam. At the left end of the beam, just to the right of point A, the shearing force in the beam must be 6000 N in order to balance the reaction force R_A. Now, if a segment of the beam 2 m long was considered to be cut from the rest of the beam, the free-body diagram shown in

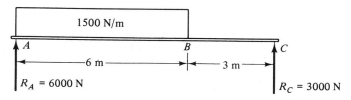

Figure 6-13 Beam loading.

Figure 6-14(a) would be obtained. There must be a shearing force in the beam to balance the external forces in order for the segment to be in equilibrium. The reaction R_A of 6000 N acts upward, while the total distributed load of 3000 N acts downward. The shearing force V must be 3000 N downward.

Analyzing a segment 4 m long in a similar manner would produce the free-body diagram shown in Figure 6-14(b). In this case $V = 0$, since the external loads themselves are balanced.

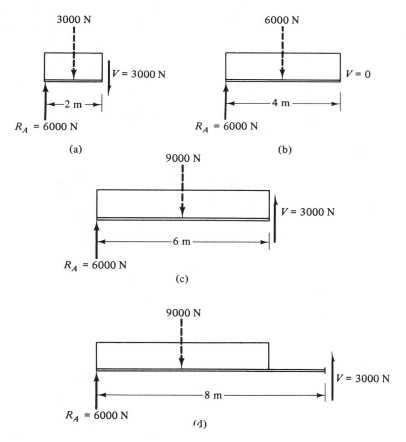

Figure 6-14 Free body diagrams used to find shearing forces.

For a segment 6 m long, Figure 6-14(c) would be the free-body diagram. Now a shearing force V of 3000 N upward must exist. At 8 m, Figure 6-14(d) shows that $V = 3000$ N upward again. This would be true throughout the last 3 m of the beam's length, since there are no external loads applied in this part.

In summary, the shearing forces calculated were:

$$\text{At point } A: \quad V = 6000 \text{ N downward}$$

$$\text{At 2 m}: \quad V = 3000 \text{ N downward}$$

$$\text{At 4 m}: \quad V = 0$$

$$\text{At 6 m}: \quad V = 3000 \text{ N upward}$$

$$\text{Between } B \text{ and } C: \quad V = 3000 \text{ N upward}$$

By convention, downward shearing forces are considered positive, and upward shearing forces are negative. If these values are plotted on a graph of shearing force versus position on the beam, the shear diagram shown in Figure 6-15 would be produced. Notice that for the portion of the beam carrying the uniformly distributed load, the shear curve is a straight line. This is typical for such loads. Also, the following general rules can be derived from this example. For the part of a beam carrying a uniformly distributed load:

1. The *change in shear* between any two points is equal to the area under the load diagram between those points.

2. The slope of the straight-line shear curve is equal to the rate of loading on the beam, that is, the load per unit length.

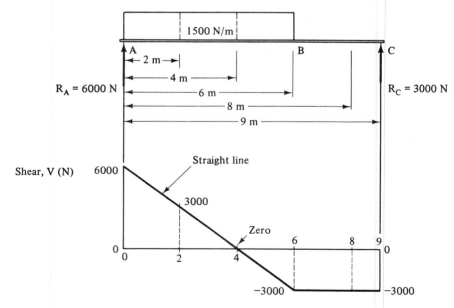

Figure 6-15 Shear diagram.

In the present example, rule 1 is illustrated by noting that between points A and B the shear decreases by 9000 N, from positive 6000 N to negative 3000 N. This is the area under the load curve as computed from

$$(1500 \text{ N/m})(6 \text{ m}) = 9000 \text{ N}$$

Rule 2 states that the shear decreases 1500 N over each meter of length of the beam.

The following example problem shows a beam having a uniformly distributed load plus two concentrated loads. This may occur in the case of a roof beam which carries the roofing material and snow over its entire length, plus a piece of equipment whose legs apply the concentrated loads. The general principles developed for both concentrated loads and distributed loads must be applied in the solution of the problem. It would be of the greatest benefit to you if you would work the problem in steps yourself before looking at the given solution. The following steps should be applied:

1. Solve for the reaction forces at the supports.
2. Make a sketch of the beam. It helps to make the sketch approximately to scale.
3. Draw lines vertically down from key points on the loaded beam to the area below, where the shear diagram will be drawn.
4. Draw the horizontal axis of the shear diagram equal to the length of the beam. Label the vertical shear axis, showing the units for the shearing forces to be plotted.
5. Starting at the left end of the beam, plot the variation in shear across the entire beam. Remember that:
   a. The shear changes abruptly at each point where a concentrated load is applied. The change in shear is equal to the load.
   b. The shear curve is a straight, horizontal line between points where no loads are applied.
   c. The shear curve is a straight, inclined line between points where uniformly distributed loads are applied. The slope of the line is equal to the rate of loading.
   d. The change in shear between points equals the area under the load curve between the points.
6. Show the value of the shear at each point where major changes occur, such as concentrated loads and at the beginning and end of distributed loads.

6-5 BENDING MOMENTS

Bending moments, in addition to shearing forces, are developed in beams as a result of the application of loads acting perpendicular to the beam. It is these bending moments which cause the beam to assume its characteristic curved, or "bent", shape. Pushing on the middle of a thin stick, such as a ruler supported at its ends, illustrates this.

The determination of the magnitude of bending moments in a beam is another application of the principle of static equilibrium. In the preceding section, we analyzed the forces in the vertical direction to determine the shearing forces in the beam which must be developed to maintain all parts of the beam in equilibrium. To do this, it was helpful to consider parts of the beam as free bodies in order to visualize what happens inside the beam. A similar approach helps illustrate bending moments.

Figure 6-16 shows a simply supported beam carrying a concentrated load at its center. The entire beam is in equilibrium, and so is any part of it. Look at the free-body diagrams shown in parts (b), (c), (d), and (e) of Figure 6-16. By summing moments about the point where the beam is cut, the magnitude of the bending moment inside required to keep the segment in equilibrium can be found. The first 0.5-m segment is shown in Figure 6-16(b). Summing moments about point B gives

$$M_B = 500 \text{ N } (0.5 \text{ m}) = 250 \text{ N} \cdot \text{m}$$

The shearing force, determined earlier in Section 6-4, is also shown.

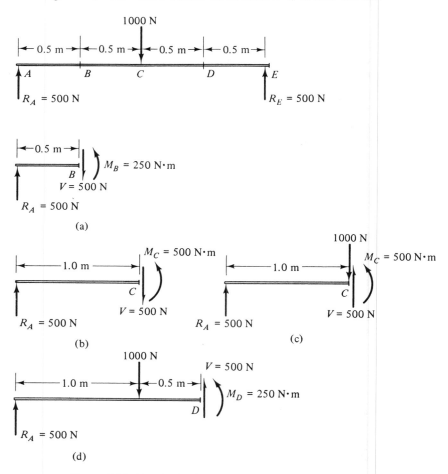

Figure 6-16   Free body diagrams used to find bending moments.

Shearing Forces and Bending Moments in Beams   Chap. 6

In Figure 6-16(c), a segment 1.0 m long, which includes the left half of the beam but *not* the 1000-N load, is drawn as a free body. Summing moments about C gives

$$M_C = 500 \text{ N } (1.0 \text{ m}) = 500 \text{ N} \cdot \text{m}$$

If the 1000-N load had been considered as shown in Figure 6-16(d), the result would be the same, since the load acts right at point C and, therefore, has no moment about that point.

Figure 6-16 shows 1.5 m of the beam isolated as a free body. Summing moments about point D gives

$$M_D = 500 \text{ N}(1.5 \text{ m}) - 1000 \text{ N}(0.5 \text{ m}) = 250 \text{ N} \cdot \text{m}$$

If we consider the entire beam as a free body and sum moments about point E at the right end of the beam, we will get

$$M_E = 500 \text{ N}(2.0 \text{ m}) - 1000 \text{ N}(1.0 \text{ m}) = 0$$

A similar result would be obtained for point A at the left end. In fact, a general rule is:

The bending moments at the ends of a simply supported beam are zero.

In summary, for the beam in Figure 6-16, the bending moments are:

Point A: 0

Point B: 250 N $\cdot$ m

Point C: 500 N $\cdot$ m

Point D: 250 N $\cdot$ m

Point E: 0

Figure 6-17 shows these values plotted on a bending moment diagram below the shear diagram developed earlier for the same beam. Notice that between A and C the bending moment values fall on a straight line. Similarly, between C and E, the points fall on a straight line. This is typical for segments of beams carrying only concentrated loads.

Figure 6-17 also illustrates another general rule:

The change in moment between two points on a beam is equal to the area under the shear curve between the same two points.

This rule can be applied over a segment of any length in a beam to determine the change in bending moment. In general, the rule can be stated that

$$dM = V \, dx \qquad (6\text{-}1)$$

where dM is the change in moment due to a shearing force V acting over a small length segment dx. Figure 6-17 illustrates Equation (6-1).

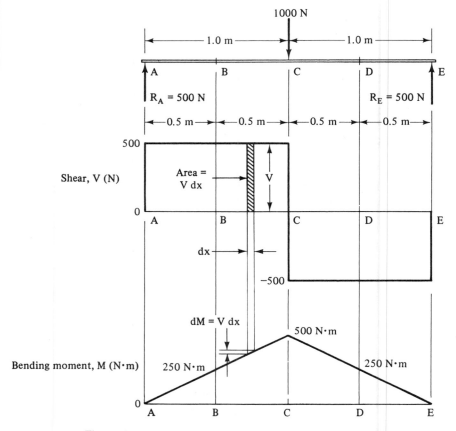

Figure 6-17 Shearing force and bending moment diagrams.

Over a larger length segment, the process of integration can be used to determine the total change in moment over the segment. Between two points, A and B,

$$\int_{M_A}^{M_B} dM = \int_{x_A}^{x_B} V \, dx \tag{6-2}$$

If the shearing force V is constant over the segment, as it is in segment A-B in Figure 6-17, Equation (6-2) becomes

$$\int_{M_A}^{M_B} dM = V \int_{x_A}^{x_B} dx \tag{6-3}$$

Completing the integration gives

$$M_B - M_A = V(x_B - x_A) \tag{6-4}$$

This result is consistent with the rule stated above. Note that $M_B - M_A$ is the *change* in moment between points A and B. The right side of Equation (6-4) is the area under the shear curve between A and B.

Once the principle behind the area rule is understood, it is not necessary to perform the process of integration to solve problems where the areas can be computed by simple geometry. Using the data from Figure 6-17, for example, between A and B,

$$M_B - M_A = V(x_B - x_A) = (500 \text{ N})(0.5 \text{ m} - 0) = 250 \text{ N} \cdot \text{m}$$

That is, the bending moment increased by 250 N · m over the span A-B. But at A the moment $M_A = 0$. Then

$$M_B = M_A + 250 \text{ N} \cdot \text{m} = 0 + 250 \text{ N} \cdot \text{m} = 250 \text{ N} \cdot \text{m}$$

Similarly, between B and C,

$$M_C - M_B = V(x_C - x_B) = (500 \text{ N})(1.0 \text{ m} - 0.5 \text{ m}) = 250 \text{ N} \cdot \text{m}$$

Then

$$M_C = M_B + 250 \text{ N} \cdot \text{m}$$

But $M_B = 250 \text{ N} \cdot \text{m}$. Then

$$M_C = 250 \text{ N} \cdot \text{m} + 250 \text{ N} \cdot \text{m} = 500 \text{ N} \cdot \text{m}$$

Between C and D, note that $V = -500$ N. Then

$$M_D - M_C = V(x_D - x_C) = (-500 \text{ N})(1.5 \text{ m} - 1.0 \text{ m}) = -250 \text{ N} \cdot \text{m}$$

$$M_D = M_C - 250 \text{ N} \cdot \text{m} = 500 \text{ N} \cdot \text{m} - 250 \text{ N} \cdot \text{m} = 250 \text{ N} \cdot \text{m}$$

Between D and E,

$$M_E - M_D = V(x_E - x_D) = (-500 \text{ N})(2.0 \text{ m} - 1.5 \text{ m}) = -250 \text{ N} \cdot \text{m}$$

$$M_E = M_D - 250 \text{ N} \cdot \text{m} = 250 \text{ N} \cdot \text{m} - 250 \text{ N} \cdot \text{m} = 0$$

These results are identical to those found by the free-body diagram method. We will use the area rule for generating the bending moment diagram from the known shear diagram for the remaining problems in this section and whenever area calculations can simply be made.

Example Problem 6-6

Draw the complete bending moment diagram for the beam shown in Figure 6-18. Show the values for the moment at points A, B, C, D, E, and F. The shear diagram shown was developed in Example Problem 6-5.

Solution   It is most convenient to start from the left end of the beam and work toward the right, considering each segment of the beam separately. The segments chosen should be those between points where the shear changes value. We can first note that at the ends of the beam, points A and F, the moment must be zero. Now, considering segment AB, the change in bending moment from A to B is equal to the area under the shear curve betwen A and B. That is,

$$M_B = M_A + 5.98 \text{ kN } (0.4 \text{ m}) = 0 + 2.39 \text{ kN} \cdot \text{m} = 2.39 \text{ kN} \cdot \text{m}$$

This value is plotted on the bending moment diagram, as shown in Figure 6-18, directly

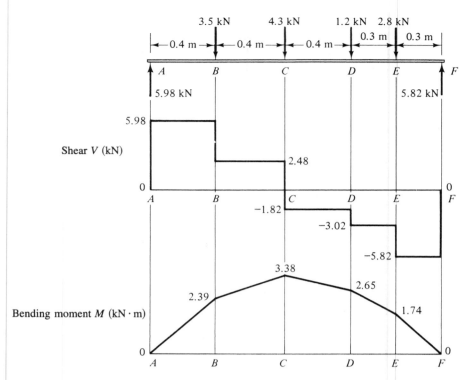

Figure 6-18 Shearing force and bending moment diagrams.

below the shear diagram. The straight line from M_A to M_B shows the variation of moment with position on the beam.

For the segment BC,

$$M_C = M_B + 2.48\,\text{kN}\,(0.4\,\text{m}) = 2.39\,\text{kN}\cdot\text{m} + 0.99\,\text{kN}\cdot\text{m} = 3.38\,\text{kN}\cdot\text{m}$$

For segment CD,

$$M_D = M_C - 1.82\,\text{kN}\,(0.4\,\text{m}) = 3.38\,\text{kN}\cdot\text{m} - 0.73\,\text{kN}\cdot\text{m} = 2.65\,\text{kN}\cdot\text{m}$$

Note that the area between C and D on the shear curve is *negative*.

For segment DE,

$$M_E = M_D - 3.02\,\text{kN}\,(0.3\,\text{m}) = 2.65\,\text{kN}\cdot\text{m} - 0.91\,\text{kN}\cdot\text{m} = 1.74\,\text{kN}\cdot\text{m}$$

For segment EF,

$$M_F = M_E - 5.82\,\text{kN}\,(0.3\,\text{m}) = 1.74\,\text{kN}\cdot\text{m} - 1.74\,\text{kN}\cdot\text{m} = 0$$

The values of the bending moment are plotted in Figure 6-18. The fact that M_F was computed to be zero is a check on the computations, since the moment at the end of a simply supported beam *must* be zero. If this were not the result, some error would have been made, and the computations would have to be checked.

Bending moment diagrams for distributed loads. The preceding examples showed the computation of bending moments and the plotting of bending moment

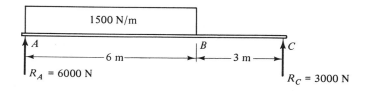

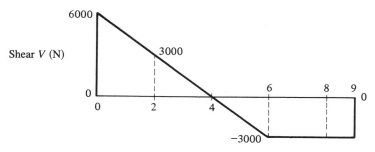

Figure 6-19 Beam loading and shearing force diagram.

diagrams for beams carrying only concentrated loads. Now we will consider distributed loads. The free-body diagram approach will be used again as an aid to visualizing the variation in bending moment as a function of position on the beam.

The beam shown in Figure 6-19 will be used to illustrate the typical results for distributed loads. This is the same beam for which the shearing force was determined, as shown in Figures 6-14 and 6-15. The free-body diagrams for segments of the beam taken in increments of 2 m will be used to compute the bending moments (refer to Figure 6-20).

For a segment at the left end of the beam, 2 m long, we can determine the bending moment in the beam by summing moments about that point due to all external loads to the left of the section, as shown in Figure 6-20(a). Notice that the resultant of the distributed load is shown to be acting at the middle of the 2-m segment. Then because the segment is in equilibrium,

$$M_2 = 6000 \text{ N} (2 \text{ m}) - 3000 \text{ N} (1 \text{ m}) = 9000 \text{ N} \cdot \text{m}$$

The symbol M_2 is being used to indicate the bending moment at the point 2 m out on the beam.

Using a similar method at points 4 m, 6 m, and 8 m out on the beam, as shown in Figure 6-20(b), (c), and (d), we would get

$$M_4 = 6000 \text{ N} (4 \text{ m}) - 6000 \text{ N} (2 \text{ m}) = 12\,000 \text{ N} \cdot \text{m}$$

$$M_6 = 6000 \text{ N} (6 \text{ m}) - 9000 \text{ N} (3 \text{ m}) = 9000 \text{ N} \cdot \text{m}$$

$$M_8 = 6000 \text{ N} (8 \text{ m}) - 9000 \text{ N} (5 \text{ m}) = 3000 \text{ N} \cdot \text{m}$$

Remember that at each end of the beam the bending moment is zero. Now we have several points which can be plotted on a bending moment diagram below the shear diagram, as shown in Figure 6-21. First look at that section of the beam where the distributed load is applied, the first 6 m. By joining the points plotted for bending

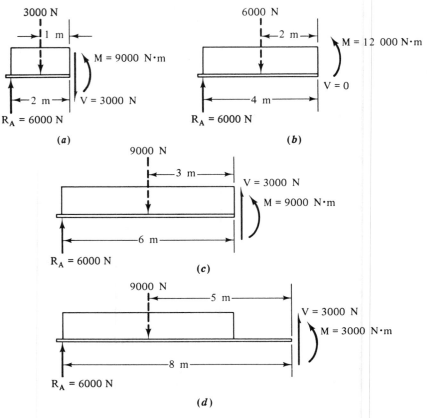

Figure 6-20 Free Body diagrams used to find bending moments.

moment with a smooth curve, the typical shape of a bending moment curve for a distributed load is obtained. For the last 3 m, where no loads are applied, the curve is a straight line, as was the case in earlier examples.

Some important observations can be made from Figure 6-21, which can be generalized into rules for drawing bending moment diagrams.

1. At the ends of a simply supported beam, the bending moment is zero.

2. The *change in bending moment* between two points on a beam is equal to the area under the shear curve between those two points. Thus, when the area under the shear curve is positive (above the axis), the bending moment is increasing, and vice versa.

3. The maximum bending moment occurs at a point where the shear curve crosses its zero axis.

4. On a section of the beam where distributed loads act, the bending moment diagram will be curved.

5. On a section of the beam where no loads are applied, the bending moment diagram will be a straight line.

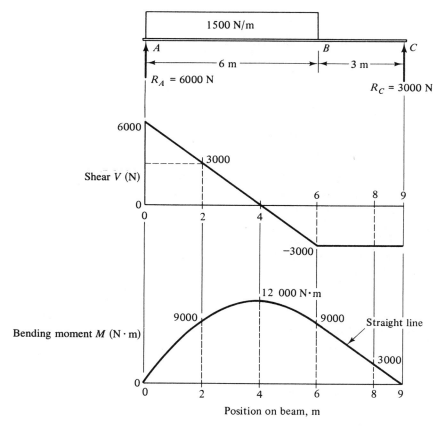

Figure 6-21 Complete beam loading, shearing force and bending moment diagrams.

6. The *slope* of the bending moment curve at any point is equal to the magnitude of the shear at that point.

Consider these rules as they apply to the beam in Figure 6-21. *Rule 1* is obviously satisfied since the moment at each end is zero. *Rule 2* can be used to check the points plotted in the moment diagram at the 2-m intervals. For the first 2 m, the area under the shear curve is composed of a rectangle and a triangle. Then the area is

$$A_{0-2} = 3000 \text{ N}(2 \text{ m}) + \tfrac{1}{2}(3000 \text{ N})(2 \text{ m}) = 9000 \text{ N} \cdot \text{m}$$

This is the change in moment from point 0 to point 2 on the beam. For the segment from 2 to 4, the area under the shear curve is a triangle. Then

$$A_{2-4} = \tfrac{1}{2}(3000 \text{ N})(2 \text{ m}) = 3000 \text{ N} \cdot \text{m}$$

Since this is the change in moment from point 2 to point 4,

$$M_4 = M_2 + A_{2-4} = 9000 \text{ N} \cdot \text{m} + 3000 \text{ N} \cdot \text{m} = 12\,000 \text{ N} \cdot \text{m}$$

Similarly for the remaining segments,

$$A_{4-6} = \tfrac{1}{2}(-3000 \text{ N})(2 \text{ m}) = -3000 \text{ N} \cdot \text{m}$$

$$M_6 = M_4 + A_{4-6} = 12\,000 \text{ N} \cdot \text{m} - 3000 \text{ N} \cdot \text{m} = 9000 \text{ N} \cdot \text{m}$$

$$A_{6-8} = (-3000 \text{ N})(2 \text{ m}) = -6000 \text{ N} \cdot \text{m}$$

$$M_8 = M_6 + A_{6-8} = 9000 \text{ N} \cdot \text{m} - 6000 \text{ N} \cdot \text{m} = 3000 \text{ N} \cdot \text{m}$$

$$A_{8-9} = (-3000 \text{ N})(1 \text{ m}) = (3000 \text{ N} \cdot \text{m})$$

$$M_9 = M_8 + A_{8-9} = 3000 \text{ N} \cdot \text{m} - 3000 \text{ N} \cdot \text{m} = 0$$

Here the fact that $M_9 = 0$ is a check on the process because *rule 1* must be satisfied.

    Rule 3 is illustrated at point 4. Where the maximum bending moment occurs, the shear curve crosses the zero axis.

    Rule 6 will probably take some practice to get familiar with, but it is extremely helpful in the process of sketching moment diagrams. Usually sketching is sufficient. The use of the six rules stated above will allow you to quickly sketch the shape of the diagram and compute key values.

    In applying *rule 6*, remember the basic concepts about the slope of a curve or line, as illustrated in Figure 6-22. Seven different segments are shown, with curves for both the shear diagram and the moment diagram, including those most often encountered in constructing these diagrams. Thus, if you are drawing a portion of a diagram in which a particular shape is observed in the shear curve, the corresponding shape for the moment curve should be as illustrated in Figure 6-22. Applying this

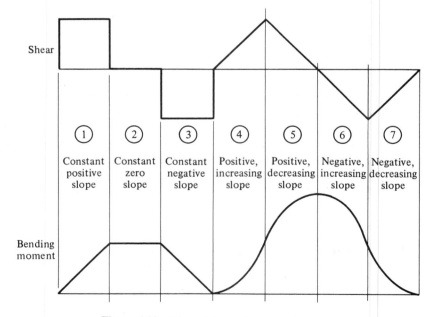

Figure 6-22 General shapes for moment curves.

approach to the moment diagram in Figure 6-21, we can see that the curve from point 0 to point 4 is like that of type 5 in Figure 6-22. Between points 4 and 6, the curve is like type 6. Finally, between points 6 and 9, the straight line with a negative slope, type 3, is used.

6-6 SHEARING FORCES AND BENDING MOMENTS FOR CANTILEVER BEAMS

The manner of support of a cantilever beam causes the analysis of its shearing forces and bending moments to be somewhat different from that for simply supported beams. The most notable difference is that, at the place where the beam is supported, it is fixed and can therefore resist moments. Thus, at the fixed end of the beam, the bending moment is not zero, as it was for simply supported beams. In fact, the bending moment at the fixed end of the beam is usually the *maximum*.

Consider the cantilever beam shown in Figure 6-23. Earlier, in Example Problem 6-4, it was shown that the support reactions at point A are a vertical force $R_A = 64$ kN and a moment $M_A = 70$ kN $\cdot$ m. These are equal to the values of the shearing force and bending moment at the left end of the beam. According to convention, the upward reaction force R_A is positive, and the counterclockwise moment M_A is negative, giving the starting values for the shear diagram and bending moment diagram shown in Figure 6-24. The general rules developed in earlier sections about shear and bending moment diagrams can then be used to complete the diagrams.

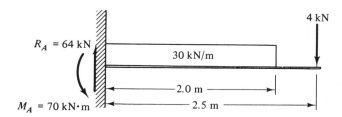

Figure 6-23 Beam loading and reactions.

The shear decreases in a straight-line manner from 64 kN to 4 kN in the interval A to B. Note that the change in shear is equal to the amount of the distributed load. The shear remains constant from B to C, where no loads are applied. The 4-kN load at C returns the curve to zero.

The bending moment diagram starts at -70 kN $\cdot$ m because of the reaction moment M_A. Between points A and B, the curve has a positive but decreasing slope (type 5 in Figure 6-22). The change in moment between A and B is equal to the area under the shear curve between A and B. The area is

$$A_{A-B} = 4\text{kN (2 m)} + \tfrac{1}{2}(60 \text{ kN})(2 \text{ m}) = 68 \text{ kN} \cdot \text{m}$$

Then the moment at B is

$$M_B = M_A + A_{A-B} = -70 \text{ kN} \cdot \text{m} + 68 \text{ kN} \cdot \text{m} = -2\text{kN} \cdot \text{m}$$

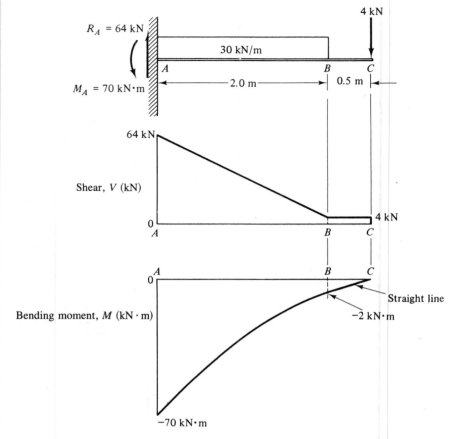

Figure 6-24 Complete load, shear and bending moment diagrams.

Finally, from B to C,

$$M_C = M_B + A_{B-C} = -2\text{kN} \cdot \text{m} + 4 \text{ kN } (0.5 \text{ m}) = 0$$

Since point C is a *free* end of the beam, the moment there must be zero.

6-7 LOADS OF VARYING MAGNITUDE

Some practical cases of beam loading result in varying loads rather than the concentrated loads or uniformly distributed loads considered thus far. For example, a roof on which snow can drift may result in the pattern shown in Figure 6-25.

A platform on which gravel is dumped may result in the pattern shown in Figure 6-26.

The loading on the beams for these cases can be approximated fairly accurately by the free-body diagrams shown in Figures 6-27 and 6-28.

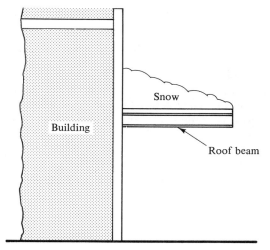

Figure 6-25   Snow load on a roof.

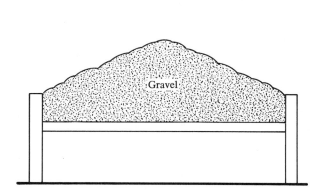

Figure 6-26   Gravel on a platform.

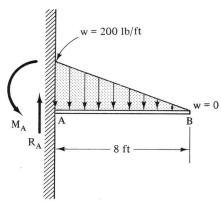

Figure 6-27   Approximate free-body diagram for load on roof beam in Figure 6-25.

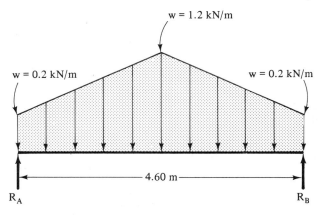

Figure 6-28   Approximate free-body diagram for gravel load on platform in Figure 6-26.

Sec. 6-7   Loads of Varying Magnitude

Each of the loading patterns has a linearly varying rate of loading. Using Figure 6-27 as an example, the loading can be written as

$$w = (-200 + 25x) \text{ lb/ft} \qquad (6\text{-}5)$$

Remember that downward loads are negative.

The complete load, shear, and bending moment diagrams for this loading are shown in Figure 6-29. To find the reaction and the bending moment at the wall, the resultant of the distributed load is needed. The total resultant load is equal to the area under the load curve.

$$R = \tfrac{1}{2}(-200 \text{ lb/ft})(8 \text{ ft}) = -800 \text{ lb}$$

The resultant can be shown to act at the centroid of the load pattern, one-third of the

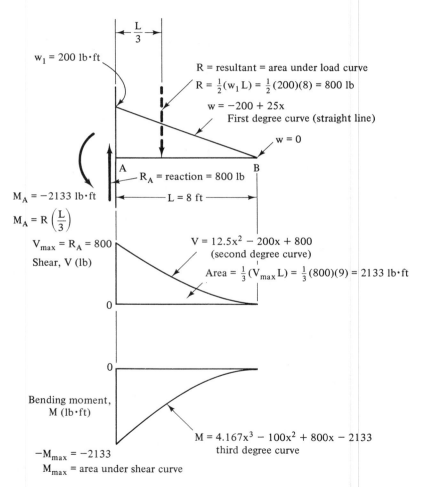

Figure 6-29 Load, shear and bending moment diagrams for beam loading in Figure 6-27.

distance from A. Then the bending moment at the wall is

$$M_A = (R)(8/3 \text{ ft}) = (800 \text{ lb})(2.67 \text{ ft}) = 2133 \text{ lb} \cdot \text{ft}$$

The equation for the shear curve can be found by noting that the shear varies according to the area under the load curve.

$$V = \int w \, dx = \int (-200 + 25x) \, dx = -200x + \frac{25}{2}x^2 + C$$

The constant of integration C can be evaluated by noting the condition:

$$\text{At } x = 0, \ V = 800 \text{ lb}$$

Then

$$V = 800 = -200(0) + 12.5(0)^2 + C$$

$$C = 800$$

Finally, then, after rearranging,

$$V = 12.5x^2 - 200x + 800 \tag{6-6}$$

This second-degree curve is the equation for V as a function of x. The bending moment curve can be defined by

$$M = \int V \, dx = \int (12.5x^2 - 200x + 800) \, dx$$

$$= \frac{12.5}{3}x^3 - \frac{200}{2}x^2 + 800x + C$$

At $x = 0$, $M = -2133$. Then $C = -2133$. The final equation for moment as a function of position can then be written

$$M = 4.167x^3 - 100x^2 + 800x - 2133 \tag{6-7}$$

Equations (6-5), (6-6), and (6-7) define the complete load, shear, and moment diagrams.

Alternative approach using area formulas. The arrangement of diagrams illustrated in Figure 6-29 occurs frequently and often only the maximum shearing force and bending moment along with the general shape of the curves are required. Figure 6-29 also shows the general relationships that can be used instead of the analytical approach.

6-8 FREE-BODY DIAGRAMS OF PARTS OF STRUCTURES

Examples considered thus far have been for generally straight beams with all transverse loads, that is, loads acting perpendicular to the main axis of the beam. Many machine elements and structures are more complex, having parts that extend away from the main beam-like part.

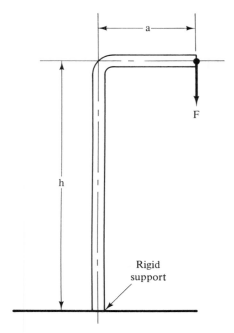

Figure 6-30 Frame for Example Problem 6-7.

For example, consider the simple frame shown in Figure 6-30 composed of vertical and horizontal parts. The vertical post is rigidly secured at its base. At the end of the extended horizontal arm, a downward load is applied. An example of such a loading is a support system for a sign over a highway. Another would be a support post for a basketball basket in which the downward force could be a player hanging from the rim after a slam dunk. A mechanical design application is a bracket supporting machine parts during processing.

Under such conditions, it is convenient to analyze the structure or machine element by considering each part separately and creating a free-body diagram for each part. At the actual joints between parts, one part exerts forces and moments on the other. Using this approach, you will be able to design each part on the basis of how it is loaded using the basic principles of beam analysis in this chapter and those that follow.

Example Problem 6-7

For the support shown in Figure 6-30, draw the complete shearing force and bending moment diagrams for the horizontal and vertical parts.

Solution The support shown in Figure 6-30 can be broken at the right-angle corner, separating off the horizontal arm. Figure 6-31(a) shows the free-body diagram of that part. The applied load acts at the end of the arm, and to balance the applied force there must be an internal force at its left end to hold it in equilibrium. There must also be an internal moment at the break to balance the moment due to the applied force acting at a distance a from the break.

From the free-body diagram, the shear and moment diagrams can be constructed as shown in Figure 6-31(b) and (c). The diagrams are drawn using the principles learned

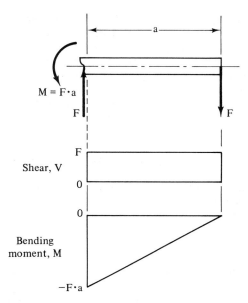

Figure 6-31 Free-body diagram, shearing force, and bending moment diagrams for the horizontal arm.

earlier in this chapter. Practically speaking, the results are similar to those of a cantilever beam.

The free-body diagram for the vertical post is shown in Figure 6-32(a). At the top of the post, a downward force and a clockwise moment are shown, exerted on the vertical

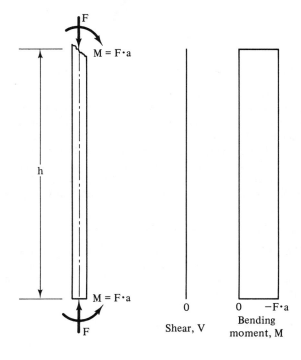

Figure 6-32 Free-body diagram, shearing force and bending moment diagrams for the vertical post.

post by the horizontal arm. Notice the action–reaction pair that exists at joints between parts. Equal but oppositely directed loads act on the two parts. Completing the free-body diagram for the post requires an upward force and a counterclockwise moment at its lower end, provided by the attachment means at its base. Finally, Figure 6-32(b) and (c) show the shear and bending moment diagrams for the post, drawn vertically to relate the values to positions on the post. No shear force exists because there are no transverse forces acting on the post. Where no shear force exists, no change in bending moment occurs and there is a uniform bending moment throughout the post.

Example Problem 6-8

Figure 6-33 shows an L-shaped bracket extending below the main beam carrying an inclined force. The main beam is supported by simple supports at A and C. Support C is designed to react to any unbalanced horizontal force. Draw the complete shearing force and bending moment diagrams for the main beam and the free-body diagrams for all parts of the bracket.

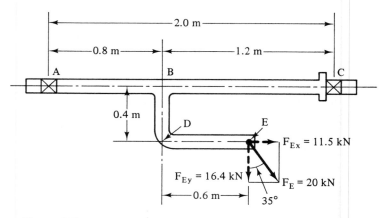

Figure 6-33 Beam with an L-shaped bracket for Example Problem 6-8.

Solution   Three free-body diagrams are convenient to use here: one for the horizontal part of the bracket, one for the vertical part of the bracket, and one for the main beam itself. But first it is helpful to resolve the applied force into its vertical and horizontal components as indicated by the dashed vectors at the end of the bracket.

Figure 6-34 shows the free-body diagrams. Starting with the part DE shown in (a), the applied forces at E must be balanced by the oppositely directed forces at D for equilibrium in the vertical and horizontal directions. But rotational equilibrium must be produced by an internal moment at D. Summing moments with respect to point D shows that

$$M_D = F_{Ey} \cdot d = (16.4 \text{ kN})(0.6 \text{ m}) = 9.84 \text{ kN} \cdot \text{m}$$

In Figure 6-34(b) the forces and moments at D have the same values but opposite directions from those at D in part (a) of the figure. Vertical and horizontal equilibrium conditions show the forces at B to be equal to those at D. The moment at B can be found by summing moments about B as follows.

$$\left(\sum M\right)_B = 0 = M_D - F_{Dx} \cdot c - M_B$$

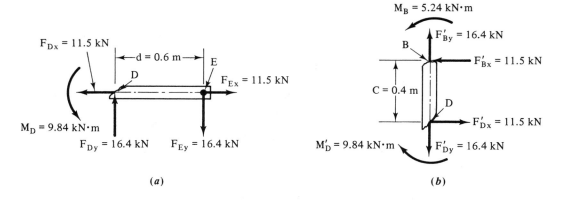

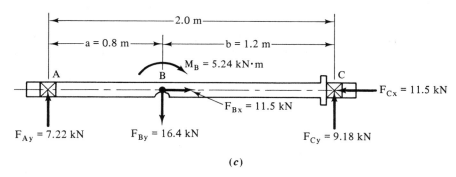

(c)

Figure 6-34 Free-body diagrams. (a) Free-body diagram for part DE (b) Free-body diagram for part BD (c) Free-body diagram for part ABC.

Then

$$M_B = M_D - F_{Dx} \cdot c = 9.84 \text{ kN} \cdot \text{m} - (11.5 \text{ kN})(0.4 \text{ m}) = 5.24 \text{ kN} \cdot \text{m}$$

Now the main beam ABC can be analyzed. The forces and moment are shown applied at B with the values taken from point B on part BD. We now must solve for the reactions at A and C. First summing moments about point C yields

$$(\textstyle\sum M)_C = 0 = F_{By} \cdot b - F_{Ay} \cdot (a + b) - M_B$$

Notice that the moment M_B applied at B must be included. Solving for F_{Ay} gives

$$F_{Ay} = \frac{(F_{By} \cdot b) - M_B}{a + b} = \frac{(16.4 \text{ kN})(1.2 \text{ m}) - 5.24 \text{ kN} \cdot \text{m}}{2.0 \text{ m}} = 7.22 \text{ kN}$$

Similarly, summing moments about point A gives

$$(\textstyle\sum M)_A = 0 = F_{By} \cdot a - F_{Cy} \cdot (a + b) + M_B$$

Notice that the moment M_B applied at B is positive because it acts in the same sense as the moment due to F_{By}. Solving for F_{Cy} gives

$$F_{Cy} = \frac{(F_{By} \cdot a) + M_B}{a + b} = \frac{(16.4 \text{ kN})(0.8 \text{ m}) + 5.24 \text{ kN} \cdot \text{m}}{2.0 \text{ m}} = 9.18 \text{ kN}$$

A check on the calculation for these forces can be made by summing forces in the vertical direction and noting that the sum equals zero.

The completion of the free-body diagram for the main beam requires the inclusion of the horizontal reaction at C equal to the horizontal force at B.

Figure 6-35 shows the shearing force and bending moment diagrams for the main beam ABC. The shear diagram is drawn in the conventional manner with changes in shear occurring at each point of load application. The noticeable difference from previous work is in the moment diagram. The following steps were used to produce it.

1. The moment at A equals zero because A is a simple support.
2. The increase in moment from A to B equals the area under the shear curve between A and B, 5.78 kN $\cdot$ m.
3. At point B the moment M_B is considered to be a *concentrated moment* resulting in an abrupt change in the value of the bending moment by the amount of the applied moment,

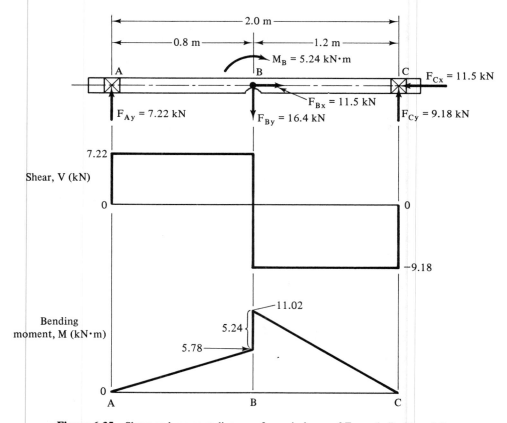

Figure 6-35 Shear and moment diagrams for main beam of Example Problem 6-8.

5.24 kN·m, thus resulting in the peak value of 11.02 kN·m. The convention used here is:
   a. When a concentrated moment is *clockwise*, the moment diagram *rises*.
   b. When a concentrated moment is *counterclockwise*, the moment diagram *drops*.

   4. Between B and C, the moment decreases to zero because of the negative shearing force and the corresponding negative area under the shear curve.

This example problem is concluded.

PROBLEMS

For each of the beams shown in Figures 6-36 to 6-53 perform the following:

   1. Compute the reactions at the supports.
   2. Draw the complete shearing force and bending moment diagrams.
   3. Determine the magnitude and location of the maximum absolute value of the shearing force and bending moment.

Remember that 1 *kip* equals 1000 lb. In the figures, the arrows without labels indicate the locations of the supports.

Each of Figures 6-54 and 6-55 shows a compound beam composed of two or more straight sections. Break the beams apart into sections and show complete free-body diagrams for each section. At the location of the break, show the internal force or moment that is required to maintain the section in equilibrium. Then, for the main horizontal section only,

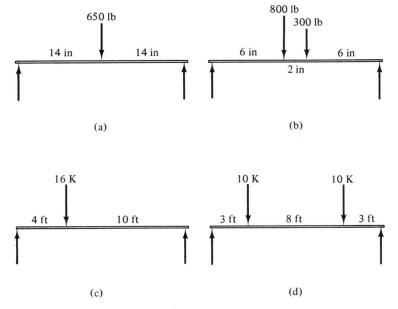

Figure 6-36

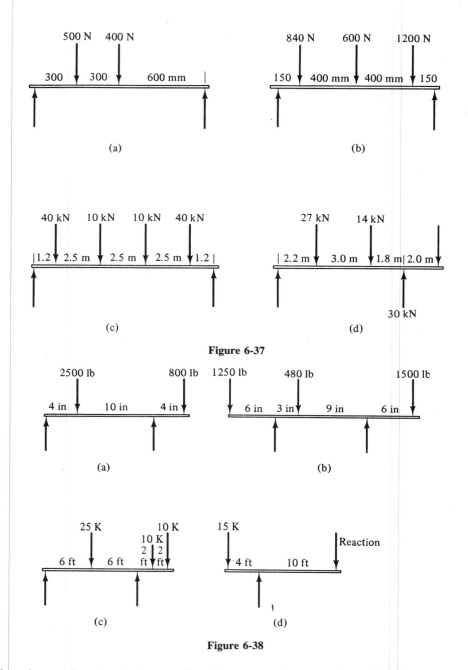

500 N   400 N

300 ↓ 300 ↓   600 mm |

(a)

840 N   600 N   1200 N

150 ↓ 400 mm ↓ 400 mm ↓ 150

(b)

40 kN 10 kN   10 kN 40 kN

|1.2 ↓ 2.5 m ↓ 2.5 m ↓ 2.5 m ↓1.2 |

(c)

27 kN   14 kN

| 2.2 m ↓ 3.0 m ↓1.8 m| 2.0 m ↓

30 kN

(d)

Figure 6-37

2500 lb   800 lb 1250 lb   480 lb   1500 lb

4 in ↓   10 in   4 in ↓

(a)

6 in 3 in↓   9 in   6 in

(b)

25 K   10 K
            10 K
            2 | 2
6 ft ↓ 6 ft   ft ↓ft↓

(c)

15 K   Reaction

↓ 4 ft   10 ft

(d)

Figure 6-38

draw the complete shearing force and bending moment diagrams. Example Problem 6-8 is the model for these problems with its companion illustrations, Figures 6-33, 6-34 and 6-35. Each beam is carried on two supports marked with an "X" which provide only simple support. One support resists unbalanced axial loads.

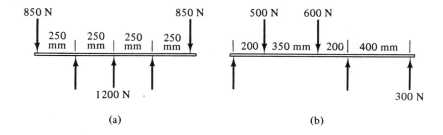

(a)

(b)

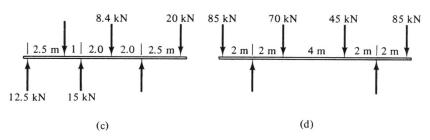

(c)

(d)

Figure 6-39

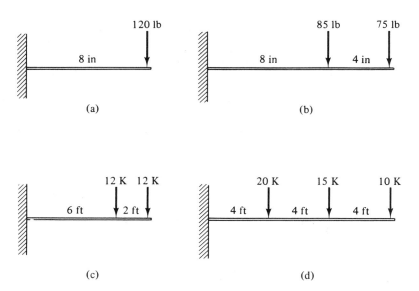

(a)

(b)

(c)

(d)

Figure 6-40

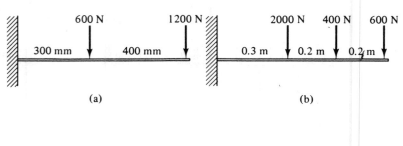

(a)

(b)

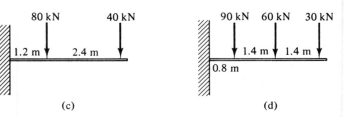

(c)

(d)

Figure 6-41

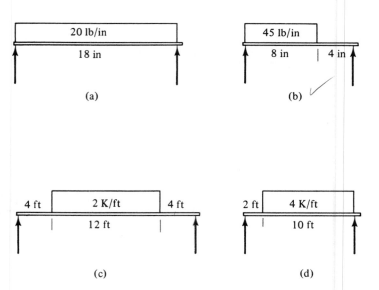

(a)

(b)

(c)

(d)

Figure 6-42

Shearing Forces and Bending Moments in Beams   Chap. 6

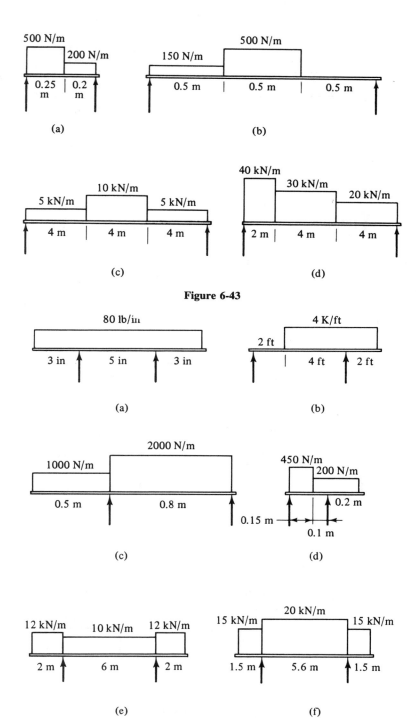

500 N/m

200 N/m

0.25 m | 0.2 m

(a)

150 N/m

500 N/m

0.5 m | 0.5 m | 0.5 m

(b)

5 kN/m

10 kN/m

5 kN/m

4 m | 4 m | 4 m

(c)

40 kN/m

30 kN/m

20 kN/m

2 m | 4 m | 4 m

(d)

Figure 6-43

80 lb/in

3 in | 5 in | 3 in

(a)

4 K/ft

2 ft

4 ft | 2 ft

(b)

1000 N/m

2000 N/m

0.5 m | 0.8 m

(c)

450 N/m

200 N/m

0.2 m

0.15 m

0.1 m

(d)

12 kN/m

10 kN/m

12 kN/m

2 m | 6 m | 2 m

(e)

15 kN/m

20 kN/m

15 kN/m

1.5 m | 5.6 m | 1.5 m

(f)

Figure 6-44

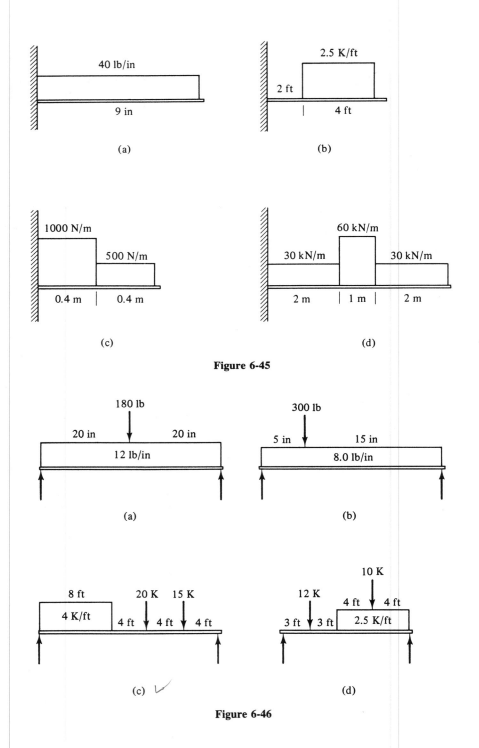

Figure 6-45

Figure 6-46

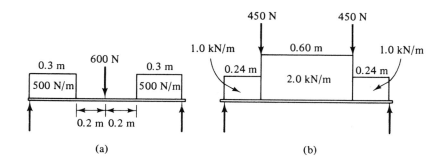

(a)

(b)

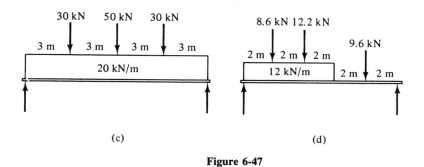

(c)

(d)

Figure 6-47

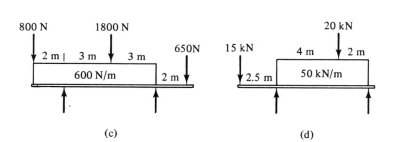

(a)

(b)

(c)

(d)

Figure 6-48

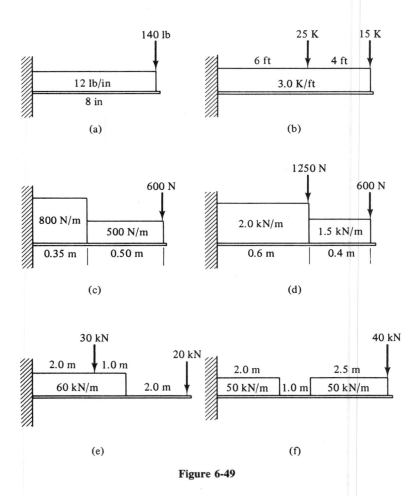

(a)

(b)

(c)

(d)

(e)

(f)

Figure 6-49

Shearing Forces and Bending Moments in Beams   Chap. 6

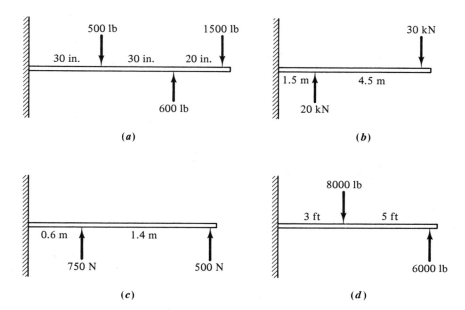

(a)

(b)

(c)

(d)

Figure 6-50

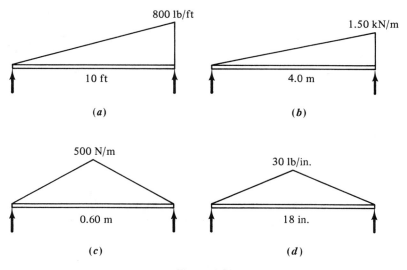

(a)

(b)

(c)

(d)

Figure 6-51

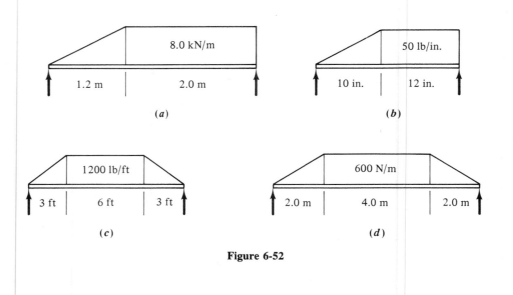

(a)

8.0 kN/m

1.2 m   2.0 m

(b)

50 lb/in.

10 in.   12 in.

(c)

1200 lb/ft

3 ft   6 ft   3 ft

(d)

600 N/m

2.0 m   4.0 m   2.0 m

Figure 6-52

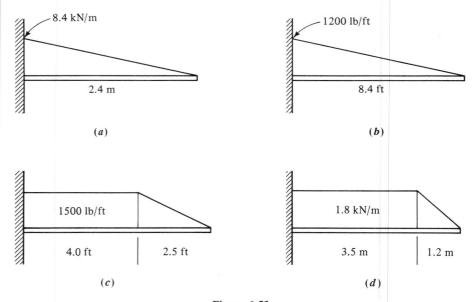

(a)

8.4 kN/m

2.4 m

(b)

1200 lb/ft

8.4 ft

(c)

1500 lb/ft

4.0 ft   2.5 ft

(d)

1.8 kN/m

3.5 m   1.2 m

Figure 6-53

Dimensions in mm

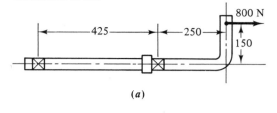

(a)

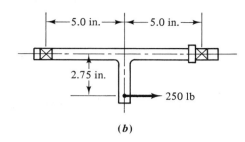

(b)

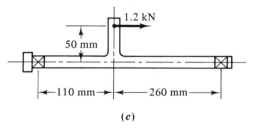

(c)

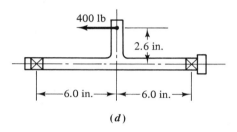

(d)

Figure 6-54

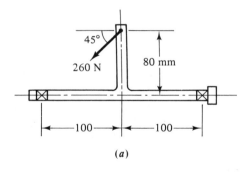

(a)

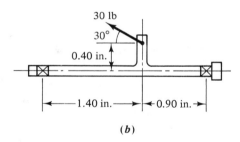

(b)

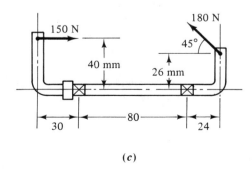

(c)

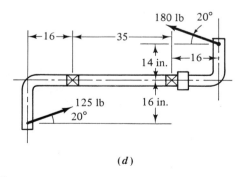

(d)

Figure 6-55

7

Centroids and Moments of Inertia of Areas

7-1 OBJECTIVES OF THIS CHAPTER

In Chapter 6 you learned to determine the value of shearing forces and bending moments in all parts of beams as part of the requirements for computing shearing stresses and bending stresses in later chapters. This chapter continues this pattern by presenting the properties of the shape of the cross section of the beam, also required for complete analysis of the stresses and deformations of beams.

The properties of the cross-sectional area of the beam that are of interest here are the *centroid* and the *moment of inertia with respect to the centroidal axis*. Some readers will already have mastered these topics through a study of *statics*. For them, this chapter should present a worthwhile review and a tailoring of the subject to the applications of interest in strength of materials. For those who have not studied centroids and moments of inertia, the concepts and techniques presented here will enable you to solve the beam analysis problems throughout this book.

After completing this chapter, you should be able to:

1. Define *centroid*.
2. Locate the centroid of simple shapes by inspection.
3. Compute the location of the centroid for complex shapes by treating them as a composite of two or more simple shapes.

4. Define *moment of inertia* as it applies to the cross-sectional area of beams.

5. Use the formulas to compute the moment of inertia for simple shapes with respect to the centroidal axes of the area.

6. Compute the moment of inertia of complex shapes by treating them as a composite of two or more simple shapes.

7. Properly use the *transfer-of-axis theorem* in computing the moment of inertia of complex shapes.

8. Analyze composite beam shapes made from two or more standard structural shapes to determine the resulting centroid location and moment of inertia.

9. Recognize what types of shapes are efficient in terms of providing a large moment of inertia relative to the amount of area of the cross section.

7-2 THE CONCEPT OF CENTROID—SIMPLE SHAPES

The *centroid* of an area is the point about which the area could be balanced if it was supported from that point. The word is derived from the word *center*, and it can be thought of as the geometrical center of an area. For three-dimensional bodies, the term *center of gravity*, or *center of mass*, is used to define a similar point.

For simple areas, such as the circle, the square, the rectangle, and the triangle, the location of the centroid is easy to visualize. Figure 7-1 shows the locations denoted by *C*. If these shapes were carefully made and the location for the centroid carefully found, the shapes could be balanced on a pencil point at the centroid. Of course, a steady hand is required. How's yours?

Appendix A-1 is a more complete source of data for centroids and other properties of areas for a variety of shapes.

7-3 CENTROID OF COMPLEX SHAPES

Most complex shapes can be considered to be made up by combining several simple shapes together. This can be used to facilitate the location of the centroid, as will be demonstrated later.

Another concept which aids in the location of centroids is that if the area has an axis of symmetry, the centroid will be on that axis. Some complex shapes have two axes of symmetry, and therefore the centroid is at the intersection of these two axes. Figure 7-2 shows examples where this occurs.

Where two axes of symmetry do not occur, the *method of composite areas* can be used to locate the centroid. For example, consider the shape shown in Figure 7-3. It has a vertical axis of symmetry but not a horizontal axis of symmetry. Such areas can be considered to be a composite of two or more simple areas for which the centroid can be found by applying the following principle:

> *The product of the total area times the distance to the centroid of the total area is equal to the sum of the products of the area of each component part times the distance to its centroid, with the distances measured from the same reference axis.*

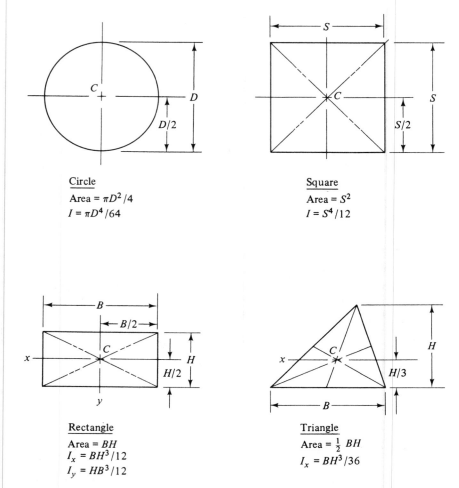

Figure 7-1 Properties of simple areas. The centroid is denoted as C.

This principle uses the concept of the *moment of an area*, that is, the product of the area times the distance from a reference axis to the centroid of the area. The principle states that:

The moment of the total area with respect to a particular axis is equal to the sum of the moments of all the component parts with respect to the same axis.

This can be stated mathematically as

$$A_T\bar{Y} = \Sigma(A_i y_i) \qquad (7\text{-}1)$$

where A_T = total area of the composite shape

$\bar{Y}$ = distance to the centroid of the composite shape measured from some reference axis

Centroids and Moments of Inertia of Areas   Chap. 7

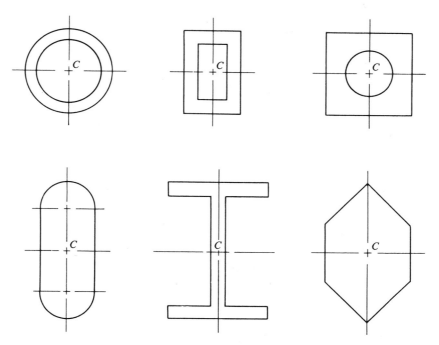

Figure 7-2 Composite shapes having two axes of symmetry. The centroid is denoted as C.

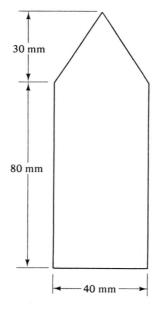

30 mm

80 mm

40 mm

Figure 7-3 Shape for Example Problem 7-1.

A_i = area of one component part of the shape

y_i = distance to the centroid of the component part from the reference axis

The subscript i indicates that there may be several component parts, and the product $A_i y_i$ for each must be formed and then summed together, as called for in Equation (7-1). Since our objective is to compute $\bar{Y}$, Equaton (7-1) can be solved:

$$\bar{Y} = \frac{\Sigma(A_i y_i)}{A_T} \tag{7-2}$$

A tabular form of writing the data helps keep track of the parts of the calculations called for in Equation (7-2). An example problem will illustrate the method.

Example Problem 7-1

Find the location of the centroid of the area shown in Figure 7-3.

Solution   Since the area has a vertical axis of symmetry, we know the centroid lies on that line. Then only the distance up to the horizontal centroidal axis must be found. Consider the total area to be made up of the sum of a triangle and a rectangle. Each of these is a simple area for which the centroid can be easily found (see Figure 7-1). The centroids are shown in Figure 7-4, with their locations measured from the bottom of the section. Now to implement Equation (7-2), the following table is formed.

Part	A_i	y_i	$A_i y_i$
1	3200 mm^2	40 mm	128 000 mm^3
2	600 mm^2	90 mm	54 000 mm^3
A_T = 3800 mm^2			182 000 mm^3 = $\Sigma(A_i y_i)$

Now $\bar{Y}$ can be computed.

$$\bar{Y} = \frac{\Sigma(A_i y_i)}{A_T} = \frac{182\ 000\ \text{mm}^3}{3800\ \text{mm}^2} = 47.9\ \text{mm}$$

This locates the centroid as shown in Figure 7-4.

The composite area method works also for sections where parts are removed as well as added. In this case the removed area is considered negative, illustrated as follows.

Example Problem 7-2

Find the location of the centroid of the area shown in Figure 7-5.

Solution   Again this area has a vertical axis of symmetry. The centroid then falls at the intersection of the horizontal centroidal axis, to be found, and the vertical axis of symmetry. Consider the composite area to be composed of three parts, as labeled in

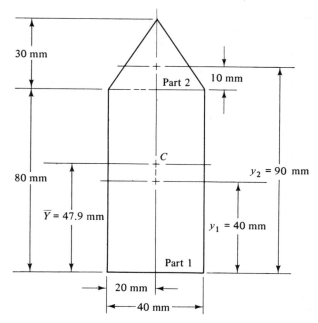

Figure 7-4 Data used in Example Problem 7-1.

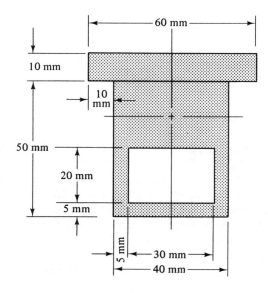

Figure 7-5 Shape for Example Problem 7-2.

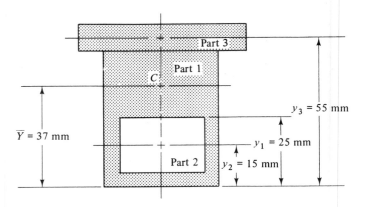

Figure 7-6 Data used in Example Problem 7-2.

Figure 7-6. Notice that part 2 is *removed* from the composite area. The areas and individual centroidal distances are shown in the following table.

Part	A_i	y_i	$A_i y_i$
1	2000 mm²	25 mm	50 000 mm³
2	−600 mm²	15 mm	−9 000 mm³
3	600 mm²	55 mm	33 000 mm³
$A_T = 2000$ mm²			74 000 mm³ $= \Sigma(A_i y_i)$

Then

$$\bar{Y} = \frac{\Sigma(A_i y_i)}{A_T} = \frac{74\ 000\ \text{mm}^3}{2000\ \text{mm}^2} = 37.0\ \text{mm}$$

7-4 THE CONCEPT OF MOMENT OF INERTIA—SIMPLE SHAPES

In the study of strength of materials, the property of moment of inertia is an indication of the stiffness of a particular shape. That is, a shape having a higher moment of inertia would deflect less when subjected to bending moments than one having a lower moment of inertia. The stress developed in a beam is also affected by the moment of inertia of its cross section, as will be shown in Chapter 8. For these reasons, the ability to compute the magnitude of the moment of inertia is very important.

The physical meaning of the moment of inertia is difficult to grasp since it is a derived property. That is, it is defined to be a property because it is important to the calculation of stress and deflection of beams and other processes. But it is not possible to describe it in physical terms, as we can an area or a volume. With practice, you should be able to identify sections which have inherently high moments of inertia because of the effective use of material.

> *The moment of inertia of an area with respect to a particular axis is defined as the sum of the products obtained by multiplying each element of the area by the square of its distance from the axis.*

This definition was applied to mathematically determine the formulas for moment of inertia listed in Figure 7-1 for simple areas. The symbol I is used to denote moment of inertia. If a subscript is shown, it indicates the axis about which the moment of inertia is taken. If no subscript is used, it should be assumed that the axis is the horizontal centroidal axis of the section. That is the one of interest in most strength and deflection calculations. For the simple areas, then, the formulas can be applied directly to compute moment of inertia. The units for I will be *length to the fourth power,* usually mm⁴ in the SI system and in⁴ in the U.S. Customary system.

7-5 MATHEMATICAL DEFINITION OF MOMENT OF INERTIA

As stated in the Section 7-4, the moment of inertia, I, is defined as the sum of the products obtained by multiplying each element of the area by the square of its distance from the reference axis. The mathematical formula for moment of inertia follows from that definition and is given below. Note that the process of summing over an entire area is accomplished by integration.

$$I = \int y^2 \, dA \qquad (7\text{-}3)$$

Refer to Figure 7-7 for an illustration of the terms in this formula for the special case of a rectangle, for which we want to compute the moment of inertia with respect to its centroidal axis. The small element of area is shown as a thin strip parallel to the centroidal axis where the width of the strip is the total width of the rectangle, b, and the thickness of the strip is a small value, dy. Then the area of the element is

$$dA = b \cdot dy$$

The distance, y, is the distance from the centroidal axis to the centroid of the elemental area as shown. Substituting these values into Equation (7-3) allows the derivation of the formula for the moment of inertia of the rectangle with respect to its centroidal axis. Note that to integrate over the entire area requires the limits for the integral to be from $-h/2$ to $+h/2$.

$$I = \int_{-h/2}^{+h/2} y^2 \, dA = \int_{-h/2}^{+h/2} y^2 (b \cdot dy)$$

Because b is a constant, it can be taken outside the integral, giving

$$I = b \int_{-h/2}^{+h/2} y^2 \, dy = b \left[\frac{y^3}{3} \right]_{-h/2}^{+h/2}$$

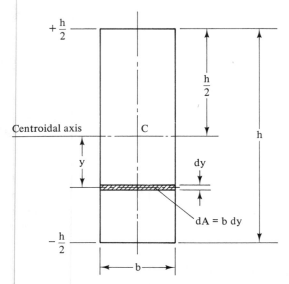

Figure 7-7 Data used in derivation of moment of inertia for a rectangle.

Inserting the limits for the integral gives

$$I = b\left[\frac{h^3}{24} - \frac{(-h)^3}{24}\right] = b\left[\frac{2h^3}{24}\right] = \frac{bh^3}{12}$$

This is the formula reported in the tables. Similar procedures can be used to develop the formulas for other shapes.

7-6 MOMENT OF INERTIA OF COMPLEX SHAPES

The method of composite areas can be used to determine the moment of inertia for a complex shape which is made up of two or more simple areas joined together.

If the component parts of a composite area all have the same centroidal axis, the total moment of inertia can be found by adding or subtracting the moments of inertia of the component parts with respect to the centroidal axis.

The following example problems show the process.

Example Problem 7-3

The cross-shaped section shown in Figure 7-8 has its centroid at the intersection of its axes of symmetry. Compute the moment of inertia of the section with respect to the horizontal axis *x-x*.

Solution The cross can be considered to be made up of the vertical central stem plus the two horizontal arms. All three sections have as their horizontal controidal axes the axis *x-x*. Therefore, the total moment of inertia I_x is the sum of the moments of inertia of each part with respect to the axis *x-x*. That is,

$$I_x = I_1 + I_2 + I_3$$

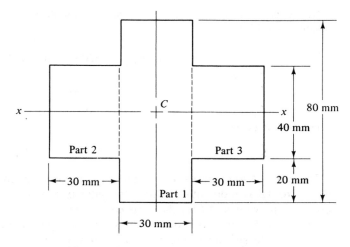

Figure 7-8 Shape for Example Problem 7-3.

Referring to Figure 7-1 gives

$$I_1 = \frac{BH^3}{12} = \frac{30(80)^3}{12} = 1.28 \times 10^6 \text{ mm}^4$$

$$I_2 = I_3 = \frac{30(40)^3}{12} = 0.16 \times 10^6 \text{ mm}^4$$

Then

$$I_x = 1.28 \times 10^6 \text{ mm}^4 + 2(0.16 \times 10^6 \text{ mm}^4) = 1.60 \times 10^6 \text{ mm}^4$$

Example Problem 7-4

Compute the moment of inertia for the section shown in Figure 7-9 with respect to the axis x-x.

Solution   Since the square and the circle both have their centroidal axes coincident with the axis x-x,

$$I_x = I_1 - I_2$$

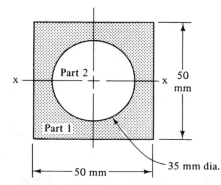

Figure 7-9 Shape for Example Problem 7-4.

Notice that since the circular area is removed from the section, its moment of inertia is subtracted. Now

$$I_1 = \frac{S^4}{12} = \frac{(50)^4}{12} = 520.8 \times 10^3 \text{ mm}^4$$

$$I_2 = \frac{\pi D^4}{64} = \frac{\pi(35)^4}{64} = 73.7 \times 10^3 \text{ mm}^4$$

For the composite section,

$$I_x = I_1 - I_2 = 447.1 \times 10^3 \text{ mm}^4$$

Transfer-of-axis theorem. When a composite section is composed of parts whose centroidal axes do not lie on the centroidal axis of the entire section, the process of summing the values of I for the parts *cannot* be used. It can be said that the effect of material added to or removed from the section at a point far away from the centroid of the composite section is greater than if it were closer to the centroid. This is illustrated in Figure 7-10. Let's start with the simple rectangular section in (a) and add material to it in an attempt to increase its moment of inertia. Two ideas are tried. In (b), an area of 4 in^2 is added at the centroidal axis. In (c), the same area is added at the top of the section. The tee section formed in (c) is stiffer than that in (b). That is, it has a higher moment of inertia. The reason is that the added area was placed far away from the original centroidal axis of the section. To show how much stiffer (c) is, we will compute the moment of inertia of all three sections. Remember, the higher the value of I, the stiffer the section is. The value of I for sections (a) and (b) can be computed using the basic methods already developed since the centroidal axes of all parts are the same. Section (c) will be used as an example to illustrate the transfer-of-axis theorem.

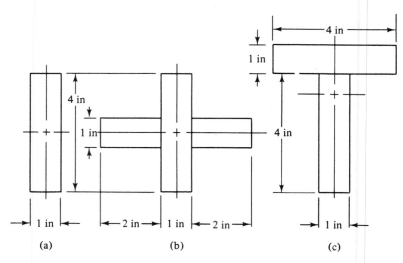

Figure 7-10 Shapes compared in Example Problem 7-5.

Centroids and Moments of Inertia of Areas   Chap. 7

For section (a):

$$I_a = \frac{BH^3}{12} = \frac{1(4)^3}{12} = 5.33 \text{ in}^4$$

For section (b):

$$I_b = I_1 + 2I_2 = \frac{1(4)^3}{12} + 2\left[\frac{2(1)^3}{12}\right] = 5.67 \text{ in}^4$$

Very little increase in I was obtained from (a) to (b). Now I_c will be found.

Example Problem 7-5

Compute the moment of inertia of the tee section shown in Figure 7-10(c), and compare it with the values found for sections (a) and (b).

Solution   The process for finding the moment of inertia of a section whose component parts have centroidal axes different from that for the entire section is outlined as follows:

1. Divide the composite section into component parts which are simple areas.
2. Locate the centroid for the composite section.
3. Compute the moment of inertia of each part with respect to its own centroidal axis. Call these I_1, I_2, etc.
4. Determine the distances d_1, d_2, etc., from the centroid of the composite section to the centroid of each part.
5. Compute the *transfer term* for each part by multiplying the area of the part by the square of the distances d_1, d_2, etc. found in step 4. The transfer terms will then be $A_1 d_1^2$, $A_2 d_2^2$, etc.
6. Compute the total moment of inertia of the composite section from

$$I_T = I_1 + A_1 d_1^2 + I_2 + A_2 d_2^2 + \cdots \tag{7-4}$$

Equation (7-4) is called the transfer-of-axis theorem because it defines how to transfer the moment of inertia of an area from one axis to any parallel axis. For each part of a composite section, the sum $I + Ad^2$ is a measure of its contribution to the total moment of inertia.

Now let's apply the above six-step procedure to compute the value of I for the section in Figure 7-10(c).

Step 1.   Let the vertical stem be part 1 and the flange at the top be part 2. Both are simple rectangles.

Step 2.   The procedure outlined in Section 7-3 can be used to locate the centroid of the composite section.

Part	A_i	y_i	$A_i y_i$
1	4.0 in^2	2.0 in.	8.0 in^3
2	4.0 in^2	4.5 in.	18.0 in^3
$A_T = 8.0$ in^2			26.0 in^3 = $\Sigma(A_i y_i)$

$$\bar{Y} = \frac{\Sigma(A_i y_i)}{A_T} = \frac{26.0 \text{ in}^3}{8.0 \text{ in}^2} = 3.25 \text{ in.}$$

Figure 7-11 shows these dimensions.

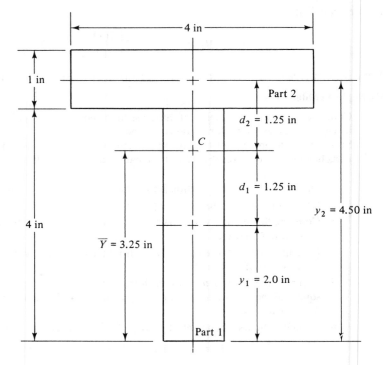

Figure 7-11 Data used in Example Problem 7-5.

Step 3. For part 1,

$$I_1 = \frac{BH^3}{12} = \frac{1(4)^3}{12} = 5.33 \text{ in}^4$$

For part 2,

$$I_2 = \frac{4(1)^3}{12} = 0.33 \text{ in}^4$$

Step 4. Figure 7-11 and the data for step 2 are helpful in finding the required distances.

$$d_1 = \bar{Y} - y_1 = 3.25 \text{ in.} - 2.0 \text{ in.} = 1.25 \text{ in.}$$
$$d_2 = y_2 - \bar{Y} = 4.5 \text{ in.} - 3.25 \text{ in.} = 1.25 \text{ in.}$$

Step 5. For part 1, the transfer term is

$$A_1 d_1^2 = (4.0 \text{ in}^2)(1.25 \text{ in.})^2 = 6.25 \text{ in}^4$$

For part 2,

$$A_2 d_2^2 = (4.0 \text{ in}^2)(1.25 \text{ in.})^2 = 6.25 \text{ in}^4$$

It is just coincidence that the transfer terms for each part are the same in this problem.

Step 6.   The total moment of inertia is

$$I_c = I_T = I_1 + A_1 d_1^2 + I_2 + A_2 d_2^2$$

$$= 5.33 \text{ in}^4 + 6.25 \text{ in}^4 + 0.33 \text{ in}^4 + 6.25 \text{ in}^4$$

$$= 18.16 \text{ in}^4$$

Let's compare the values of I for sections (a), (b), and (c) in Figure 7-10:

$$I_a = 5.33 \text{ in}^4$$

$$I_b = 5.67 \text{ in}^4 \qquad (1.06 \text{ times greater than } I_a)$$

$$I_c = 18.16 \text{ in}^4 \qquad (3.4 \text{ times greater than } I_a \text{ and}$$
$$3.2 \text{ times greater than } I_b)$$

Use of an extended table to aid in computing moment of inertia. Earlier, when introducing the method for computing the location for the centroid of composite sections, it was recommended that a table be created to facilitate keeping track of the data required. Such a table can be extended to aid in the computation of moment of inertia as illustrated below for the data from Example Problem 7-5, just completed. The advantage of using the table is that the new data required, namely the distances d and the areas required to compute the transfer term Ad^2, are taken from other table entries.

After completing the calculation of $\bar{Y}$ to locate the centroidal axis, columns for I_i, d_i, $A_i d_i^2$, and $I_i + A_i d_i^2$ can be added.

Part	A_i	y_i	$A_i y_i$	I_i	d_i	$A_i d_i^2$	$I_i + A_i d_i^2$
1	4.0 in²	2.0 in.	8.0 in³	5.33 in⁴	1.25 in.	6.25 in⁴	11.58 in⁴
2	4.0 in²	4.5 in.	18.0 in³	0.33 in⁴	1.25 in.	6.25 in⁴	6.58 in⁴
	$A_T = 8.0$ in²		26.0 in³ $= \Sigma(A_i y_i)$				18.16 in⁴ $= I_T$

$$\bar{Y} = \frac{\Sigma(A_i y_i)}{A_T} = \frac{26.0 \text{ in}^3}{8.0 \text{ in}^2} = 3.25 \text{ in.}$$

The entries in the last column are the contribution of each component part of the composite section to the total moment of inertia. The sum of all the values in the last column is the total moment of inertia itself.

The benefit from using this type of table becomes greater as the number of component parts gets greater.

7-7 COMPOSITE SECTIONS MADE FROM COMMERCIALLY AVAILABLE SHAPES

In Section 1-14 commercially available structural shapes were described for steel and aluminum. Properties of representative sizes of these shapes are listed in the following appendix tables:

Appendix A-5 for structural steel angles

Appendix A-6 for structural steel channels

Appendix A-7 for structural steel wide-flange shapes

Appendix A-8 for structural steel American Standard beams

Appendix A-9 for structural tubing—square and rectangular

Appendix A-10 for aluminum standard channels

Appendix A-11 for aluminum standard I-beams

Appendix A-12 for standard schedule 40 steel pipe

In addition to being very good by themselves for use as beams, these shapes are often combined to produce special composite shapes with enhanced properties.

When used separately, the properties for designing can be read directly from the tables for area, moment of inertia, and pertinent dimensions. When combined into composite shapes, the area and the moment of inertia of the component shapes with respect to their own centroidal axes are needed and can be read from the tables. Also, the tables give the location of the centroid for the shape which is often needed to determine distances needed to compute the transfer-of-axis term, Ad^2, in the moment of inertia calculation. The following example problems illustrate these processes.

Example Problem 7-6

To increase the stiffness of a standard aluminum I-beam, a 0.50-in.-thick by 6.00-in.-wide plate is welded to both the top and bottom flanges, as shown in Figure 7-12. Compute the moment of inertia of the combined section.

Solution   The centroid of the combined section is the same as the centroid for the beam itself since the added plates are placed symmetrically on the beam. Still, the transfer-of-axis theorem will have to be used to compute the moment of inertia since the centroid of each plate is away from that for the entire section. Calling the beam part 1, the top plate part 2, and the bottom plate part 3, the total moment of inertia will be

$$I_T = I_1 + I_2 + A_2 d_2^2 + I_3 + A_3 d_3^2$$

From Appendix A-11, $I_1 = 155.79$ in^4. Now

$$I_2 = I_3 = \frac{BH^3}{12} = \frac{(6.0)(0.5)^3}{12} = 0.063 \text{ in}^4$$

$$A_1 = A_2 = BH = (6.0)(0.5) = 3.0 \text{ in}^2$$

$$d_1 = d_2 = 5.25 \text{ in.}$$

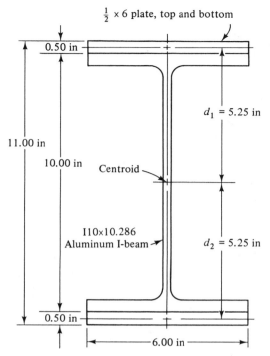

$\frac{1}{2}$ × 6 plate, top and bottom

0.50 in

d_1 = 5.25 in

11.00 in

10.00 in

Centroid

I10×10.286
Aluminum I-beam

d_2 = 5.25 in

0.50 in

6.00 in

Figure 7-12

Then

$$I_T = 155.79 + 0.063 + (3.0)(5.25)^2 + 0.063 + (3.0)(5.25)^2$$
$$= 321.29 \text{ in}^4$$

The moment of inertia was more than doubled by the addition of the plates.

Example Problem 7-7

A built-up beam is made by fastening four L4 × 4 × $\frac{1}{2}$ steel angles to a $\frac{1}{2}$ in. × 16 in. plate, as shown in Figure 7-13. Compute the moment of inertia of the beam.

Solution The beam has two axes of symmetry, so its centroid is at the middle of the plate. The centroid of the angles is away from the centroid of the entire section, so the transfer-of-axis theorem must be used. Notice that all four angles are identical and that all are the same distance from the beam's centroid. Therefore, the total moment of inertia will be

$$I_T = I_1 + 4(I_2 + A_2 d_2^2)$$

where the subscript 1 refers to the plate and the subscript 2 refers to one of the angles. Now

$$I_1 = \frac{BH^3}{12} = \frac{(0.50)(16)^3}{12} = 170.67 \text{ in}^4$$

$$I_2 = 5.56 \text{ in}^4 \qquad \text{(from Appendix A-5)}$$

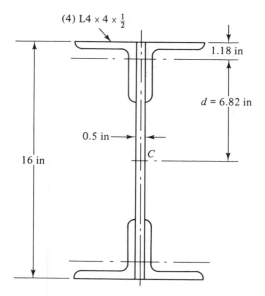

(4) L4 × 4 × $\frac{1}{2}$

1.18 in

$d = 6.82$ in

0.5 in

16 in

C

Figure 7-13

$$A_2 = 3.75 \text{ in}^2$$

$$d_2 = 8.0 \text{ in.} - 1.18 \text{ in.} = 6.82 \text{ in.}$$

The computation of d_2 required the location of the centroidal axis of the angle, which is 1.18 in. down from its top leg, as shown in Table A-5. The calculation can now be completed.

$$I_T = 170.67 + 4[5.56 + 3.75 (6.82)^2]$$
$$= 170.67 + 719.93 = 890.60 \text{ in}^4$$

PROBLEMS

For each of the shapes in Figures 7-14 through 7-22, determine the location of the horizontal centroidal axis and the magnitude of the moment of inertia of the shape with respect to that axis. Figures 7-18 through 7-20 are composed of standard steel or aluminum sections for which the properties are listed in the Appendix. The sections in Figure 7-21 are made of standard wood forms. The sections in Figures 7-16, 7-17 and 7-22 are typical aluminum or plastic extrusions or molded shapes found in business machines, toys, coat hangers, and similar products.

COMPUTER PROGRAMMING ASSIGNMENTS

1. For a generalized I-shape having equal top and bottom flanges similar to that shown in Figure 7-14(c), write a computer program to compute the location of the horizontal centroidal axis, the total area, and the moment of inertia with respect to the horizontal centroidal axis for any set of actual dimensions to be input by the program operator.

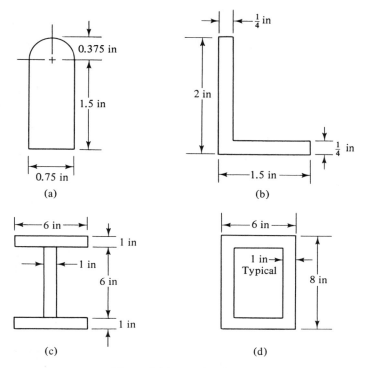

Figure 7-14 Special shapes.

2. For the generalized T-shape similar to that shown in Figure 7-15(a), write a computer program to compute the location of the horizontal centroidal axis, the total area, and the moment of inertia with respect to the horizontal centroidal axis for any set of actual dimensions to be input by the program operator.

3. For the generalized I-shape similar to that shown in Figure 7-15(d), write a computer program to compute the location of the horizontal centroidal axis, the total area, and the moment of inertia with respect to the horizontal centroidal axis for any set of actual dimensions to be input by the program operator.

4. For any generalized shape that can be subdivided into some number of rectangular components with horizontal axes, write a computer program to compute the location of the horizontal centroidal axis, the total area, and the moment of inertia with respect to the horizontal centroidal axis for any set of actual dimensions to be input by the program operator.

5. For the generalized hat-section shape similar to that shown in Figure 7-22(b), write a computer program to compute the location of the horizontal centroidal axis, the total area, and the moment of inertia with respect to the horizontal centroidal axis for any set of actual dimensions to be input by the program operator.

6. Given a set of standard dimension lumber, compute the area and moment of inertia with respect to the horizontal centroidal axis for the generalized box shape similar to that shown in Figure 7-21(b). The data for the lumber should be input by the operator of the program.

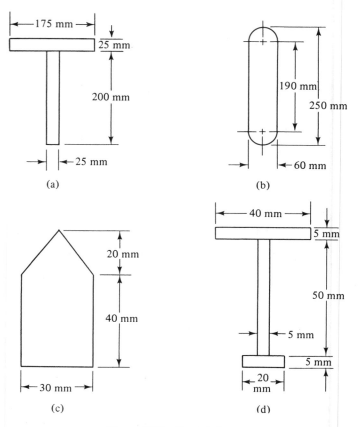

Figure 7-15 Special shapes.

7. Build a data file containing the dimensions of a set of standard dimension lumber. Then permit the operator of the program to select sizes for the top and bottom plates and the two vertical members for the box shape shown in Figure 7-21(b). Then compute the area and moment of inertia with respect to the horizontal centroidal axis for the shape being designed.

8. Write a computer program to compute the area and moment of inertia with respect to the horizontal centroidal axis for a standard W or S beam shape with identical plates attached to both the top and bottom flanges similar to that shown in Figure 7-12. The data for the beam shape and the plates are to be input by the operator of the program.

9. Build a data base of standard W or S beam shapes. Then write a computer program to compute the area and moment of inertia with respect to the horizontal centroidal axis for a selected beam shape with identical plates attached to both the top and bottom flanges as shown in Figure 7-12. The data for the plates are to be input by the operator of the program.

10. Given a standard W or S beam shape and its properties, write a computer program to compute the required thickness for plates to be attached to the top and bottom flanges to produce a specified moment of inertia of the composite section as shown in Figure 7-12.

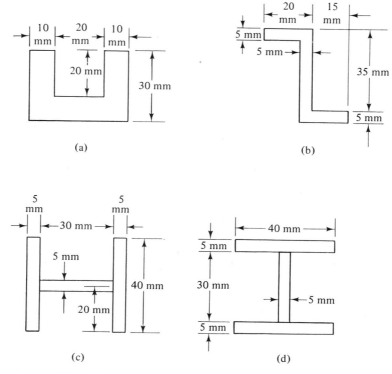

Figure 7-16 Typical extruded plastic or aluminum shapes.

Make the width of the plates equal to the width of the flange. For the resulting section, compute the total area.

11. Using the computer program written for Assignment 1 for the generalized I shape, perform an analysis of the area (A), the moment of inertia (I), and the ratio of I to A, as the thickness of the web is varied over a specified range. Keep all other dimensions for the shape the same. Note that the ratio of I to A is basically the same as the ratio of the stiffness of a beam having this shape to its weight because the deflection of a beam is inversely proportional to the moment of inertia and the weight of the beam is proportional to its cross sectional area.

12. Repeat Assignment 11 but vary the height of the section while keeping all other dimensions the same.

13. Repeat Assignment 11 but vary the thickness of the flange while keeping all other dimensions the same.

14. Repeat Assignment 11 but vary the width of the flange while keeping all other dimensions the same.

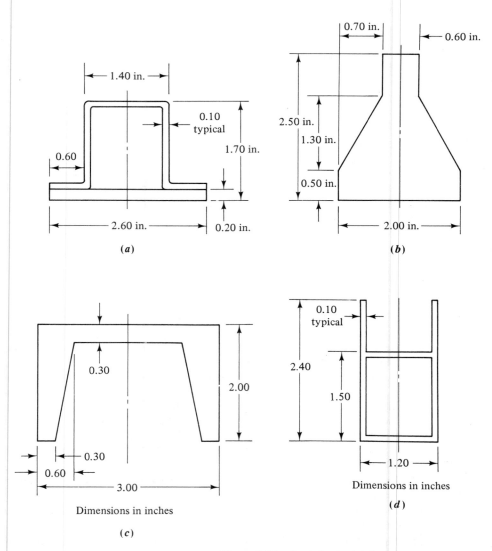

Figure 7-17 Typical extruded shapes.

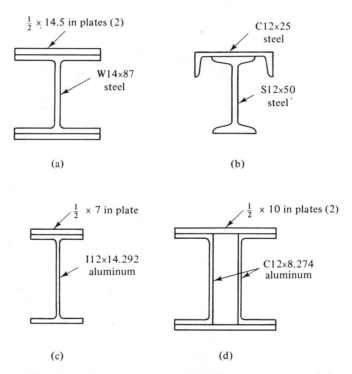

$\frac{1}{2} \times 14.5$ in plates (2)

W14×87 steel

(a)

C12×25 steel

S12×50 steel

(b)

$\frac{1}{2} \times 7$ in plate

I12×14.292 aluminum

(c)

$\frac{1}{2} \times 10$ in plates (2)

C12×8.274 aluminum

(d)

Figure 7-18 Composite shapes made by combining standard structural shapes and plates.

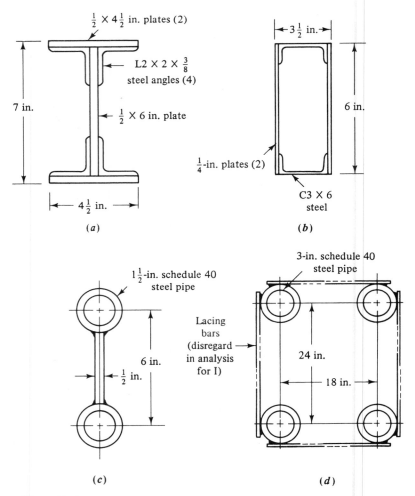

Figure 7-19 Composite shapes made by combining standard structural shapes and plates.

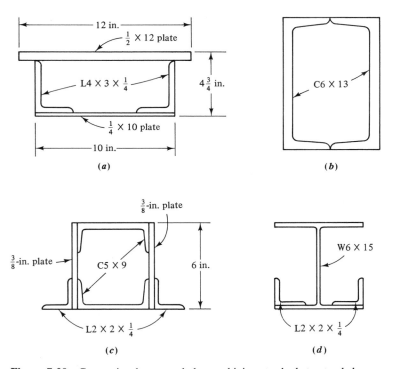

Figure 7-20 Composite shapes made by combining standard structural shapes and plates.

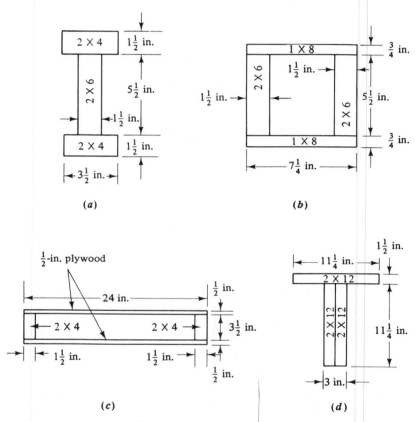

Figure 7-21 Composite shapes made by combining standard wood forms.

Centroids and Moments of Inertia of Areas   Chap. 7

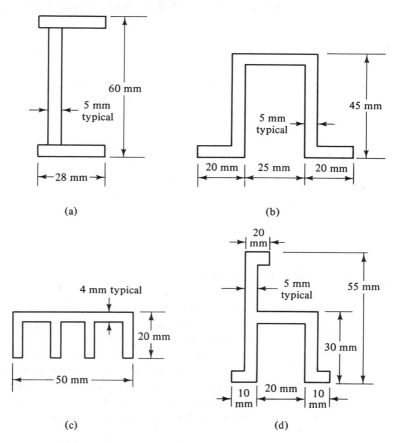

Figure 7-22 Typical extruded plastic or aluminum shapes.

8

Stress Due to Bending

8-1 OBJECTIVES OF THIS CHAPTER

Beams were defined in Chapter 6 to be load-carrying members on which the loads act perpendicular to the long axis of the beam. Methods were presentd there to determine the magnitude of the bending moment at any point on the beam. The bending moment, acting internal to the beam, causes the beam to bend and develops stresses in the fibers of the beam. The magnitude of the stress thus developed is dependent on the moment of inertia of the cross section, found using the methods reviewed in Chapter 7.

This chapter uses the information from previous chapters to enable the calculation of *stress due to bending* in beams. Specific objectives are:

1. To learn the statement of the *flexure formula* and to apply it properly to compute the maximum stress due to bending at the outer fibers of the beam.
2. To be able to compute the stress at any point within the cross section of the beam and to describe the variation of stress with position in the beam.
3. To understand the conditions on the use of the flexure formula.
4. To recognize that it is necessary to ensure that the beam does not twist under the influence of the bending loads.

5. To define the *neutral axis* and to understand that it is coincident with the centroidal axis of the cross section of the beam.
6. To understand the derivation of the flexure formula and the importance of the moment of inertia to bending stress.
7. To determine the appropriate design stress for use in designing beams.
8. To design beams to carry a given loading safely.
9. To define the *section modulus* of the cross section of the beam.
10. To select standard structural shapes for use as beams.
11. To recognize when it is necessary to use stress concentration factors in the analysis of stress due to bending and to apply appropriate factors properly.
12. To define the *flexural center* and describe its proper use in the analysis of stress due to bending.

8-2 THE FLEXURE FORMULA

Beams must be designed to be safe. When loads are applied perpendicular to the long axis of a beam, bending moments are developed inside the beam, causing it to bend. By observing a thin beam, the characteristically curved shape shown in Figure 8-1 is evident. The fibers of the beam near its top surface are shortened and placed in compression. Conversely, the fibers near the bottom surface are stretched and placed in tension.

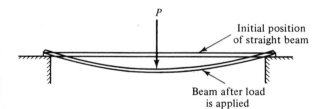

Figure 8-1   Example of a beam.

In designing or analyzing beams, it is usually the objective to determine what the maximum tensile or compressive stress is. The maximums will occur at the outer surfaces of the beam. The flexure formula used to compute the maximum stress due to bending, has the form

$$\sigma_{\max} = \frac{Mc}{I} \tag{8-1}$$

where $\sigma_{\max}$ = maximum stress at the outermost fiber of the beam
M = bending moment at the section of interest
c = distance from the centroidal axis of the beam to the outermost fiber
I = moment of inertia of the cross section with respect to its centroidal axis

The flexure formula is discussed in greater detail later. A simple problem will now be shown to illustrate the application of the formula.

Example Problem 8-1

For the beam shown in Figure 8-1, compute the maximum stress due to bending. The cross section of the beam is a rectangle 100 mm high and 25 mm wide. The load at the middle of the beam is 1500 N, and the length of the beam is 3.40 m.

Solution   The method of solving any beam stress problem involves the following steps:

1. Determine the maximum bending moment on the beam by drawing the shear and bending moment diagrams.
2. Locate the centroid of the cross section of the beam.
3. Compute the moment of inertia of the cross section with respect to its centroidal axis.
4. Compute the distance c from the centroidal axis to the top or bottom of the beam, whichever is greater.
5. Compute the stress from the flexure formula,

$$\sigma_{max} = \frac{Mc}{I}$$

For this problem, refer to Figure 8-2 when completing steps 1–4.

Step 1.   The maximum bending moment is 1275 N·m at the middle of the beam.

Step 2.   The centroid of the rectangular section is at the intersection of its two axes of symmetry.

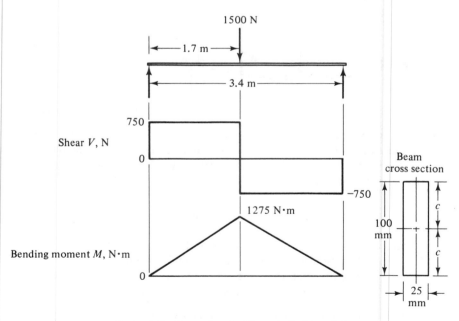

Figure 8-2   Beam data for Example Problem 8-1.

Step 3. For the rectangular section,

$$I = \frac{BH^3}{12} = \frac{25(100)^3}{12} = 2.08 \times 10^6 \text{ mm}^4$$

Step 4.

$$c = 50 \text{ mm}$$

Step 5. The maximum stress due to bending is

$$\sigma_{\max} = \frac{Mc}{I} = \frac{(1275 \text{ N} \cdot \text{m})(50 \text{ mm})}{2.08 \times 10^6 \text{ mm}^4} \times \frac{10^3 \text{ mm}}{\text{m}}$$

$$= 30.6 \text{ N/mm}^2 = 30.6 \text{ MPa}$$

8-3 CONDITIONS ON THE USE OF THE FLEXURE FORMULA

The proper application of the flexure formula requires the understanding of the conditions under which it is valid, listed as follows:

1. The beam must be straight or very nearly so.
2. The cross section of the beam must be uniform.
3. All loads and support reactions must act perpendicular to the axis of the beam.
4. The beam must not twist while the loads are being applied.
5. The beam must be relatively long and narrow in proportion to its depth.
6. The material from which the beam is made must be homogeneous, and it must have an equal modulus of elasticity in tension and compression.
7. The stress resulting from the loading must not exceed the proportional limit of the material.
8. No part of the beam may fail from instability, that is, from the buckling or crippling of thin sections.

Although the list of conditions appears to be long, the flexure formula still applies to a wide variety of real cases. Beams violating some of the conditions can be analyzed by using a modified formula or by using a combined stress approach. For example, for condition 2, a change in cross section will cause stress concentrations which can be handled as described in Section 8-7. The combined bending and axial stress or bending and torsional stress produced by violating condition 3 is discussed in Chapter 11. If the other conditions are violated, special analyses are required which are not covered in this book.

Condition 4 is important, and attention must be paid to the shape of the cross section to ensure that twisting does not occur. In general, if the beam has a vertical axis of symmetry and if the loads are applied through that axis, no twisting will result. Figure 8-3 shows some typical shapes used for beams which satisfy condition 4. Conversely, Figure 8-4 shows several that do not satisfy condition 4; in each of these

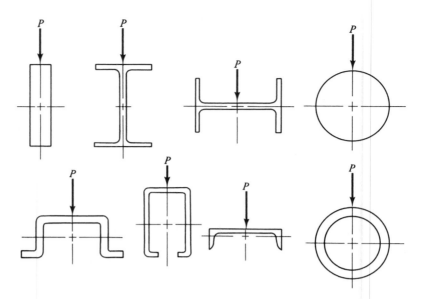

Figure 8-3 Example beam shapes with loads acting through an axis of symmetry.

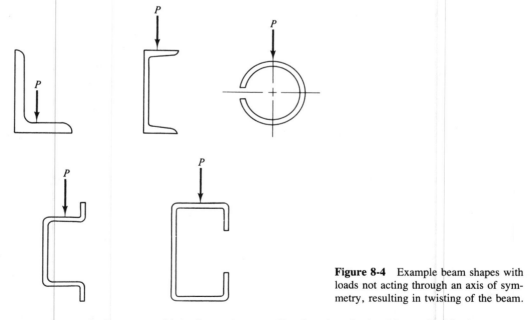

Figure 8-4 Example beam shapes with loads not acting through an axis of symmetry, resulting in twisting of the beam.

cases, the beam would tend to twist as well as bend as the load is applied in the manner shown. Of course, these sections can support some load, but the actual stress condition in them is different from that which would be predicted from the flexure formula. More about these kinds of beams is presented in Section 8-8.

Condition 8 is important because long, thin members and, sometimes, thin sections of members tend to buckle at stress levels well below the yield strength of the material. Such failures are called *instability* and are to be avoided. Frequently, cross braces or local stiffeners are added to beams to relieve the problem of instability. An example can be seen in the wood joist floor construction of many homes and commercial buildings. The relatively slender wood joists are braced near their midpoints to avoid buckling.

One of the basic assumptions made in the derivation of the flexure formula is that a plane section through the beam remains straight as bending occurs. During the action of bending, the plane section will rotate as indicated in Figure 8-5. The result is that fibers near the bottom will be placed in tension and fibers near the top will be in compression, as we have already observed. Furthermore, the plane section will rotate about an axis where neither tension nor compression is developed. This axis is called the *neutral axis*, and it is coincident with the *centroidal axis* of the cross section. We discussed how to locate the centroidal axis in Chapter 7. If no tension or compression exists at the neutral axis, then no stress exists. Another observation can be made from Figure 8-5. There is a linear variation in the strain or deformation of the material as a function of distance away from the neutral axis. Since stress is proportional to strain in a material obeying Hooke's law, stress, too, varies linearly with distance from the neutral axis. The maximum stress occurs at the surface at a distance c from the neutral axis. Then at any other distance y,

$$\sigma = \sigma_{max} \frac{y}{c} \qquad (8-2)$$

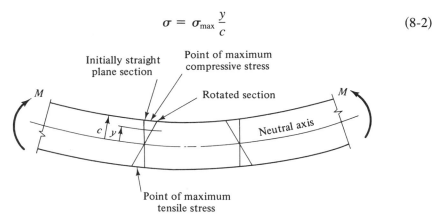

Figure 8-5 Portion of a beam in bending.

This analysis shows that the stress distribution on a section subjected to bending would be as shown in Figure 8-6. The same distribution would occur for any section having a centroidal axis equidistant from the top and bottom surfaces. For this case, the maximum compressive stress would equal the maximum tensile stress.

If the centroidal axis of the section is not the same distance from both the top and bottom surfaces, the stress distribution shown in Figure 8-7 would occur. Still the stress at the neutral axis would be zero. Still the stress would vary linearly with distance from the neutral axis. But now the maximum stress at the bottom of the

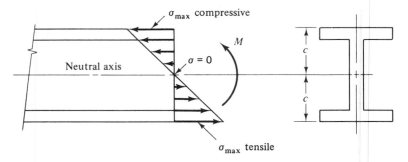

Figure 8-6 Stress distribution on a symmetrical section.

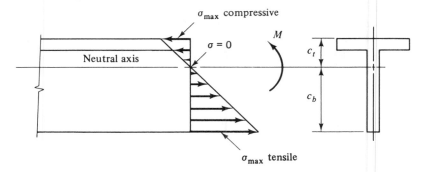

Figure 8-7 Stress distribution on a non-symmetrical section.

section is greater than that at the top because it is farther from the neutral axis. Using the distances c_b and c_t as indicated in Figure 8-7, the stresses would be

$$\sigma_{max} = \frac{Mc_b}{I} \qquad \text{(tension at the bottom)}$$

$$\sigma_{max} = \frac{Mc_t}{I} \qquad \text{(compression at the top)}$$

8-4 DERIVATION OF THE FLEXURE FORMULA

A better understanding of the basis for the flexure formula can be had by following the analysis used to derive it. The principles of static equilibrium are used here to show two concepts that were introduced earlier in this chapter but which were stated without proof. One is that the *neutral axis* is coincident with the *centroidal axis* of the cross section. The second is the flexure formula itself and the significance of the moment of inertia of the cross section.

Refer to Figure 8-6, which shows the distribution of stress over the cross section of a beam. The shape of the cross section is not relevant to the analysis and the I-shape is shown merely for example. The figure shows a portion of a beam, cut at some arbitrary section, with an internal bending moment acting on the section. The stresses, some tensile and some compressive, would tend to produce forces on the cut section in the axial direction. Equilibrium requires that the net sum of these forces must be zero. In general, force equals stress times area. Because the stress varies with position on the cross section, it is necessary to look at the force on any small elemental area and then sum these forces over the entire area using the process of integration. These concepts can be shown analytically as:

Equilibrium condition: $\sum F = 0$

Force on any element of area $dA = \sigma\, dA$

Total force on the cross-sectional area:

$$\sum F = \int_A \sigma\, dA = 0 \tag{8-3}$$

Now we can express the stress σ at any point in terms of the maximum stress by using Equation (8-2):

$$\sigma = \sigma_{max} \frac{y}{c}$$

where y is the distance from the neutral axis to the point where the stress is equal to σ. Substituting this into Equation (8-3) gives

$$\sum F = \int_A \sigma\, dA = \int_A \sigma_{max} \frac{y}{c}\, dA = 0$$

But because σ_{max} and c are constants, they can be taken outside the integral sign.

$$\sum F = \frac{\sigma_{max}}{c} \int_A y\, dA = 0$$

Neither σ_{max} nor c is zero, so the other factor, $\int_A y\, dA$, must be zero. But by definition and as illustrated in Chapter 7,

$$\int_A y\, dA = \bar{Y}(A)$$

where $\bar{Y}$ is the distance to the centroid of the area from the reference axis and A is the total area. Again, A cannot be zero, so, finally, it must be true that $\bar{Y} = 0$. Because the reference axis is the neutral axis, this shows that the neutral axis is coincident with the centroidal axis of the cross section.

The derivation of the flexure formula is based on the principle of equilibrium, which requires that the sum of the moments about any point must be zero. In Figure 8-6 it shows that a bending moment M acts at the cut section. This must be balanced by the net moment created by the stress on the cross section. But moment is the product of force times the distance from the reference axis to the line of action of the force. As used above,

$$\sum F = \int_A \sigma \, dA = \int_A \sigma_{max} \frac{y}{c} \, dA$$

Multiplying this by distance y gives the resultant moment of the force that must be equal to the internal bending moment M. That is,

$$M = \sum F(y) = \int_A \sigma_{max} \frac{y}{c} \, dA \, (y)$$

force

moment arm

area

stress

Simplifying, we obtain

$$M = \frac{\sigma_{max}}{c} \int_A y^2 \, dA$$

By definition, and as illustrated in Chapter 7, the last term in this equation is the moment of inertia I of the cross section with respect to its centroidal axis.

$$I = \int_A y^2 \, dA$$

Then

$$M = \frac{\sigma_{max}}{c} I$$

Solving for σ_{max} yields

$$\sigma_{max} = \frac{Mc}{I}$$

This is the form of the flexure formula shown earlier as Equation (8-1).

8-5 APPLICATIONS—BEAM ANALYSIS

Example Problem 8-2

The tee section shown in Figure 8-8 is from a simply supported beam which carries a bending moment of 100 000 lb · in. due to a load on the top surface. Earlier, this section was analyzed to determine that $I = 18.16$ in^4. The centroid of the section is 3.25 in. up from the bottom of the beam. Compute the stress due to bending in the beam at the six axes a to f indicated in the figure. Then plot a graph of stress versus position in the cross section.

Solution   Let's start at *axis a*. The maximum stress would occur there because it is the

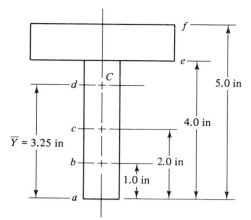

Figure 8-8 Tee section for beam in Example Problem 8-2.

farthest point from the centroidal axis of the beam. Using $c = 3.25$ in. gives

$$\sigma_{max} = \sigma_a = \frac{Mc}{I} = \frac{(100\ 000\ \text{lb} \cdot \text{in.})(3.25\ \text{in.})}{18.16\ \text{in}^4}$$

$$= 17\ 900\ \text{psi} \quad \text{(tension)}$$

The remaining stresses will be computed from Equation (8-2).

At Axis b.

$$\sigma_b = \sigma_{max}\frac{y}{c} = \sigma_a\frac{y}{c}$$

$$y = 3.25\ \text{in.} - 1.0\ \text{in.} = 2.25\ \text{in.}$$

Then

$$\sigma_b = 17\ 900\ \text{psi} \times \frac{2.25}{3.25} = 12\ 400\ \text{psi} \quad \text{(tension)}$$

At Axis c.

$$y = 3.25\ \text{in.} - 2.0\ \text{in.} = 1.25\ \text{in.}$$

$$\sigma_c = 17\ 900\ \text{psi} \times \frac{1.25}{3.25} = 6900\ \text{psi} \quad \text{(tension)}$$

At Axis d.   At the centroid $y = 0$ and $\sigma_d = 0$.

At Axis e.

$$y = 4.0\ \text{in.} - 3.25\ \text{in.} = 0.75\ \text{in.}$$

$$\sigma_e = 17\ 900\ \text{psi} \times \frac{0.75}{3.25} = 4100\ \text{psi} \quad \text{(compression)}$$

At axis f.

$$y = 5.0 \text{ in.} - 3.25 \text{ in.} = 1.75 \text{ in.}$$

$$\sigma_f = 17\ 900 \text{ psi} \times \frac{1.75}{3.25} = 9600 \text{ psi} \quad \text{(compression)}$$

The graph of these data is shown in Figure 8-9. Notice the linear variation of stress with distance from the neutral axis.

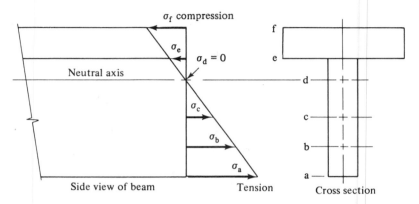

Figure 8-9 Stress distribution on the Tee section for Example Problem 8-3.

Example Problem 8-3

A beam 25 ft long supports loads that cause the bending moments indicated in Figure 8-10. The beam is a W 14 × 43. Compute the maximum stress due to bending in the beam.

Solution Figure 8-10 is the bending moment diagram for the beam, showing the variation in M as a function of position on the beam. The maximum stress would occur

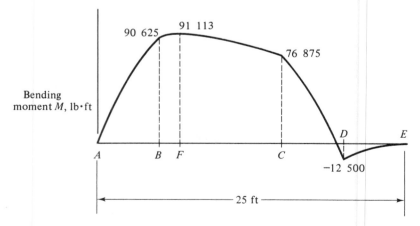

Figure 8-10 Bending moment diagram for beam in Example Problem 8-3.

at point F, where the bending moment is 91 113 lb·ft. Equation (8-1),

$$\sigma_{max} = \frac{Mc}{I}$$

will be used to compute the stress. Appendix A-7, which lists properties of wide-flange shapes, gives $I = 428$ in^4 and the depth of the beam to be 13.66 in. Therefore,

$$c = \frac{13.66 \text{ in.}}{2} = 6.83 \text{ in.}$$

Then

$$\sigma_{max} = 91 \ 113 \text{ lb·ft.} \times \frac{12 \text{ in.}}{\text{ft}} \times \frac{6.83 \text{ in.}}{428 \text{ in}^4} = 17 \ 450 \text{ psi}$$

8-6 APPLICATIONS—BEAM DESIGN

To design a beam, its material, length, placement of loads, placement of supports, and the size and shape of its cross section must be specified. Normally, the length, placement of loads, and placement of supports are given by the requirements of the application. Then the material specification and the size and shape of the cross section are determined by the designer. The primary duty of the designer is to ensure the safety of the design. This requires a stress analysis of the beam and a decision concerning the allowable or design stress which the chosen material may be subjected to. The examples presented here will concentrate on these items. Also of interest to the designer are cost, appearance, physical size, weight, compatibility of the design with other components of the machine or structure, and the availability of the material or beam shape.

Two basic approaches will be shown for beam design. One involves the specification of the *material* from which the beam will be made and the general *shape* of the beam (circular, rectangular, W-beam, etc.), with the subsequent determination of the required dimensions of the cross section of the beam. The second requires specifying the *dimensions* and *shape* of the beam and then computing the required strength of a material from which to make the beam. Then the actual material is specified.

Design stress. In both approaches described above, some representation of *design stresses* must be made. We will use the same approach described in Section 3-3, with design factors selected from Appendix A-20, according to the type of load and the type of material. In general then, the design stress will be computed from

$$\sigma_d = \frac{s_y}{N} \qquad \text{based on yield strength}$$

or

$$\sigma_d = \frac{s_u}{N} \qquad \text{based on ultimate strength}$$

These equations for design stress apply because the kind of stresses developed in a beam are tensile and compressive and the modes of failure are similar to those described in Section 3-3.

Design stress from the AISC specifications. For building applications for which loads are static, the American Institute of Steel Construction (AISC) defines the design stress to be

$$\sigma_d = 0.66 \, s_y$$

Section modulus. The stress analysis will require the use of the flexure formula

$$\sigma_{max} = \frac{Mc}{I}$$

But a modified form is desirable for cases where the determination of the dimensions of a section is to be done. Notice that both the moment of inertia I and the distance c are geometrical properties of the cross-sectional area of the beam. Therefore, the quotient I/c is also a geometrical property. For convenience, we can define a new term, *section modulus*, denoted by the letter S.

$$S = \frac{I}{c} \tag{8-4}$$

The flexure formula then becomes

$$\sigma_{max} = \frac{M}{S} \tag{8-5}$$

This is the most convenient form for use in design. Example problems will demonstrate the use of the section modulus. It should be noted that some designers use the symbol Z in place of S for section modulus.

Example Problem 8-4

A beam is to be designed to carry the load shown in Figure 8-11. A rectangular cross section will be used, with the beam being cut from a $1\frac{1}{4}$-in.-thick plate of A36 structural

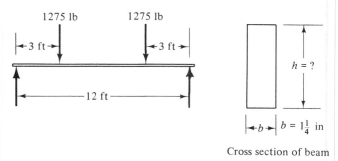

Figure 8-11 Loading and cross section for beam in Example Problem 8-4.

Stress Due to Bending   Chap. 8

steel. (see Appendix A-15.) The required height of the cross section is to be determined. Assume that the forces will be static.

Solution The following steps will be used to determine the required height of the cross section.

1. The shear and bending moment diagrams will be drawn to determine the maximum bending moment M.
2. The design stress σ_d for the A36 steel will be computed.
3. The flexure formula $\sigma_{max} = M/S$ will be solved for S,

$$S = \frac{M}{\sigma_{max}}$$

4. Letting $\sigma_{max} = \sigma_d$, the required section modulus S will be computed.
5. The required height h of the rectangular section will be computed.

   Step 1. Figure 8-12 shows the completed diagrams. The maximum bending moment is 45 900 lb · in.

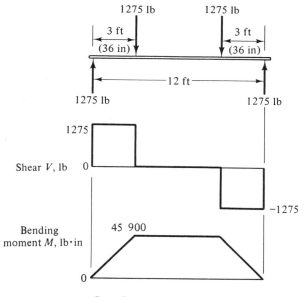

Figure 8-12 Load, shear and bending moment diagrams for Example Problem 8-4.

   Step 2. From Appendix A-15 $s_y = 36\ 000$ psi for A36 steel. For a static load, a design factor of $N = 2$ based on yield strength is reasonable. Then

$$\sigma_d = \frac{s_y}{N} = \frac{36\ 000\ \text{psi}}{2} = 18\ 000\ \text{psi}$$

   Steps 3 and 4. The required S is

$$S = \frac{M}{\sigma_d} = \frac{45\ 900\ \text{lb} \cdot \text{in.}}{18\ 000\ \text{lb/in}^2} = 2.55\ \text{in}^3$$

   Step 5. The formula for the section modulus for a rectangular section with a height h and a thickness b is

$$S = \frac{I}{c} = \frac{bh^3}{12(h/2)} = \frac{bh^2}{6}$$

For the beam in this design problem, b will be $1\frac{1}{4}$ in. Then solving for h gives

$$S = \frac{bh^2}{6}$$

$$h = \sqrt{\frac{6S}{b}} = \sqrt{\frac{6(2.55 \text{ in}^3}{1.25 \text{ in.}}}$$

$$= 3.50 \text{ in.}$$

The beam will then be rectangular in shape, with dimensions of $1\frac{1}{4}$ in. by $3\frac{1}{2}$ in.

Example Problem 8-5

The roof of an industrial building is to be supported by wide-flange beams spaced 4 ft on centers across a 20-ft span, as sketched in Figure 8-13. The roof will be a poured concrete slab, 4 in. thick. The design live load on the roof is 200 lb/ft². Specify a suitable wide-flange beam that will limit the stress in the beam to 22 000 psi.

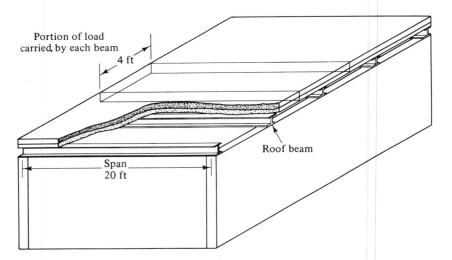

Figure 8-13 Roof structure for building in Example Problem 8-5.

Solution We must first determine the load on each beam of the roof structure. Dividing the load evenly among adjacent beams would result in each beam carrying a 4-ft-wide portion of the roof load. In addition to the 200-lb/ft² live load, the weight of the concrete slab offers a sizeable load. Assuming that the concrete weighs 150 lb/ft³, each square foot of the roof, 4.0 in. thick, would weigh 50 lb. This is called the *dead load*. Then the total loading due to the roof is 250 lb/ft². Now, notice that each foot of length of the beam carries 4 ft² of the roof. Therefore, the load on the beam is a uniformly distributed load of 1000 lb/ft. Figure 8-14 shows the loaded beam and the shear and bending moment diagrams. The maximum bending moment is 50 000 lb·ft.

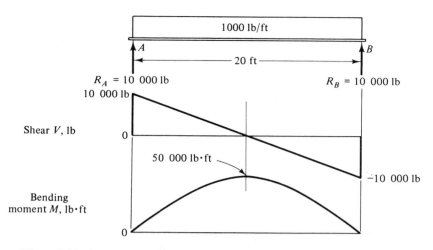

Figure 8-14 Load, shear and bending moment diagrams for Example Problem 8-5.

In order to select a wide-flange beam, the required section modulus must be calculated.

$$\sigma = \frac{M}{S}$$

$$S = \frac{M}{\sigma_d} = \frac{50\ 000\ \text{lb} \cdot \text{ft}}{22\ 000\ \text{lb/in}^2} \times \frac{12\ \text{in.}}{\text{ft}} = 27.3\ \text{in}^3$$

A beam must be found from Appendix A-7 which has a value of S greater than 27.3 in³. In considering alternatives, of which there are many, you should search for the lightest beam which will be safe, since the cost of the beam is based on its weight. Some possible beams are

W14 × 26: $S = 35.3$ in³, 26 lb/ft

W12 × 30: $S = 38.6$ in³, 30 lb/ft

Of these, the W14 × 26 would be preferred because it is the lightest.

8-7 STRESS CONCENTRATIONS

The conditions specified for valid use of the flexure formula in Section 8-3 included the statement that the beam must have a uniform cross section. Changes in cross section result in higher local stresses than would be predicted from the direct application of the flexure formula. Similar observations were made in earlier chapters concerning direct axial stresses and torsional shear stresses. The use of *stress concentration factors* will allow the analysis of beams which do include changes in cross section.

In the design of round shafts for carrying power-transmitting elements, the use of steps in the diameter is encountered frequently. Examples were shown in Chapter

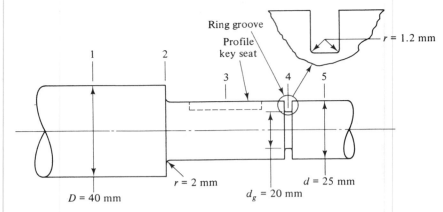

Figure 8-15 Portion of a shaft with several changes in cross section producing stress concentrations.

5, where torsional shear stresses were discussed. Figure 8-15 shows such a shaft. Considering the shaft as a beam subjected to bending moments, there would be stress concentrations at the shoulder (2), the key seat (3), and the groove (4).

At sections where stress concentrations occur, the stress due to bending would be calculated from a modified form of the flexure formula.

$$\sigma_{max} = \frac{McK_t}{I} = \frac{MK_t}{S} \tag{8-6}$$

The stress concentration factor K_t is found experimentally, with the values reported in graphs such as those in Appendix A-21.

Example Problem 8-6

Figure 8-15 shows a portion of a round shaft where a gear is mounted. A bending moment of 30 N · m is applied at this location. Compute the stress due to bending at sections 2, 3, and 4.

Solution   At *section 2*, the step in the shaft causes a stress concentration to occur. Then the stress is

$$\sigma_2 = \frac{MK_t}{S}$$

The smaller of the diameters at section 2 is used to compute S.

$$S = \frac{\pi d^3}{32} = \frac{\pi (25 \text{ mm})^3}{32} = 1534 \text{ mm}^3$$

The value of K_t depends on the ratios r/d and D/d.

$$\frac{r}{d} = \frac{2 \text{ mm}}{25 \text{ mm}} = 0.08$$

$$\frac{D}{d} = \frac{40 \text{ mm}}{25 \text{ mm}} = 1.60$$

From Appendix A-21-7, $K_t = 1.87$. Then

$$\sigma_2 = \frac{MK_t}{S} = \frac{(30 \text{ N} \cdot \text{m})(1.87)}{1534 \text{ mm}^3} \times \frac{10^3 \text{ mm}}{\text{m}} = 36.6 \text{ N/mm}^2 = 36.6 \text{ MPa}$$

At *section 3*, the key seat causes a stress concentration factor of 2.0 as listed in Appendix A-21-6. Then

$$\sigma_3 = \frac{MK_t}{S} = \frac{(30 \text{ N} \cdot \text{m})(2.0)}{1534 \text{ mm}^3} \times \frac{10^3 \text{ mm}}{\text{m}} = 39.1 \text{ N/mm}^2 = 39.1 \text{ MPa}$$

The groove at *section 4* requires the use of Appendix A-21-7 again to find K_t. Note that the nominal stress is based on the root diameter of the groove, d_g. For the groove,

$$\frac{r}{d_g} = \frac{1.2 \text{ mm}}{20 \text{ mm}} = 0.06$$

$$\frac{d}{d_g} = \frac{25 \text{ mm}}{20 \text{ mm}} = 1.25$$

Then $K_t = 1.93$. The section modulus at the root of the groove is

$$S = \frac{\pi d_g^3}{32} = \frac{\pi (20 \text{ mm})^3}{32} = 785 \text{ mm}^3$$

Now the stress at section 4 is

$$\sigma_4 = \frac{MK_t}{S} = \frac{(30 \text{ N} \cdot \text{m})(1.93)}{785 \text{ mm}^3} \times \frac{10^3 \text{ mm}}{\text{m}} = 73.8 \text{ N/mm}^2 = 73.8 \text{ MPa}$$

8-8 FLEXURAL CENTER (SHEAR CENTER)

The flexure formula is valid for computing the stress in a beam provided the applied loads pass through a point called the *flexural center*, or sometimes, the *shear center*. If a section has an axis of symmetry and if the loads pass through that axis, then they also pass through the flexural center. The beam sections shown in Figure 8-3 are of this type. For sections loaded away from an axis of symmetry, the position of the flexural center, indicated by Q, must be found. Such sections were identified in Figure 8-4. In order to result in pure bending, the loads must pass through Q, as shown in Figure 8-16. If they don't, then a condition of *unsymmetrical bending* occurs and other analyses would have to be performed which are not discussed in this book. The sections of the type shown in Figure 8-16 are used frequently in structures. Some lend themselves nicely to production by extrusion and are therefore very economical. But because of the possibility of producing unsymmetrical bending, care must be taken in their application.

Example Problem 8-7

Determine the location of the shear center for the two sections shown in Figure 8-17.

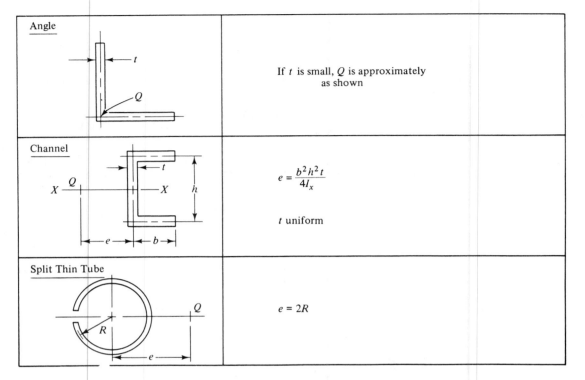

Angle		If t is small, Q is approximately as shown
Channel		$e = \dfrac{b^2 h^2 t}{4I_x}$ t uniform
Split Thin Tube		$e = 2R$

Figure 8-16 Location of flexural center Q.

Solution

      Channel Section (a).   From Figure 8-16, the distance e to the shear center is

$$e = \frac{b^2 h^2 t}{4I_x}$$

Note that the dimensions b and h are measured to the middle of the flange or web. Then $b = 40$ mm and $h = 50$ mm. Because of symmetry about the X axis, I_x can be found by the difference between the value of I for the large outside rectangle (54 mm by 42 mm) and the smaller rectangle removed (46 mm by 38 mm).

$$I_x = \frac{(42)(54)^3}{12} - \frac{(38)(46)^3}{12} = 0.243 \times 10^6 \text{ mm}^4$$

Then

$$e = \frac{(40)^2(50)^2(4)}{4(0.243 \times 10^6)} \text{ mm} = 16.5 \text{ mm}$$

This dimension is drawn to scale in Figure 8-17(a).

      Hat Section (b).   Here the distance e is a function of the ratios c/h and b/h.

$$\frac{c}{h} = \frac{10}{30} = 0.3$$

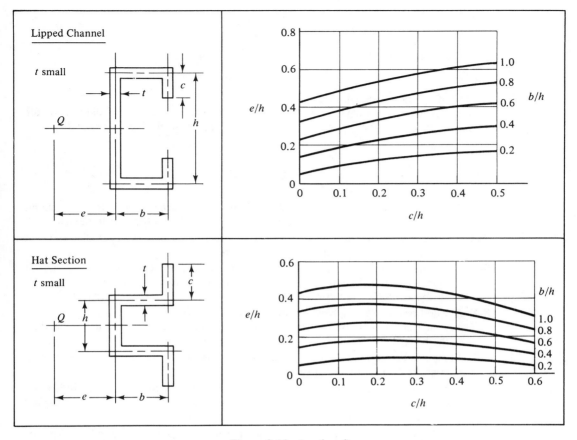

Figure 8-16 (continued)

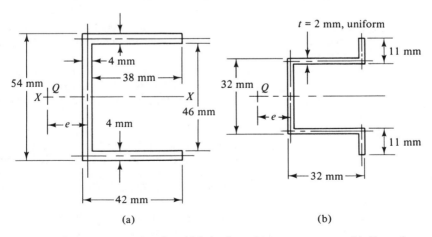

(a)

(b)

Figure 8-17 Beam sections for which the flexural centers are computed in Example Problem 8-7.

$$\frac{b}{h} = \frac{30}{30} = 1.0$$

Then from Figure 8-16, $e/h = 0.45$. Solving for e yields

$$e = 0.45h = 0.45(30 \text{ mm}) = 13.5 \text{ mm}$$

This dimension is drawn to scale in Figure 8-17(b).

Now, can you devise a design for using either section as a beam and provide for the application of the load through the flexural center Q to produce pure bending?

PROBLEMS

8-1. A square bar 30 mm on a side is used as a simply supported beam subjected to a bending moment of 425 N · m. Compute the maximum stress due to bending in the bar.

8-2. Compute the maximum stress due to bending in a round rod 20 mm in diameter if it is subjected to a bending moment of 120 N · m.

8-3. A bending moment of 5800 lb · in. is applied to a rectangular beam made as a rectangle 0.75 in. by 1.50 in. cross section. Compute the maximum bending stress in the beam (a) if the 1.50-in. side is set vertical, and (b) if the 0.75-in. side is vertical.

8-4. A wood beam carries a bending moment of 15 500 lb · in. It has a rectangular cross section 1.50 in. wide by 7.25 in. high. Compute the maximum stress due to bending in the beam.

8-5. Compute the required diameter of a round bar used as a beam to carry a bending moment of 240 N · m with a stress no greater than 125 MPa.

8-6. A rectangular bar is to be used as a beam subjected to a bending moment of 145 N · m. If its height is to be three times its width, compute the required dimensions of the bar to limit the stress to 55 MPa.

8-7. The tee section shown in Figure 7-15(a) is to carry a bending moment of 28.0 kN · m. It is to be made of steel plates welded together. If the load on the beam is a dead load, would AISI 1020 hot-rolled steel be satisfactory for the plates?

8-8. The modified I-section shown in Figure 7-15(d) is to be extruded aluminum. Specify a suitable aluminum alloy if the beam is to carry a repeated load resulting in a bending moment of 275 N · m.

8-9. A standard steel pipe is to be used as a chinning bar for personal exercise. The bar is to be 42 in. long and simply supported at its ends. Specify a suitable size pipe if the bending stress is to be limited to 10 000 psi and a 280-lb man hangs by one hand in the middle.

8-10. A pipeline is to be supported above ground on horizontal beams, 14 ft long. Consider each beam to be simply supported at its ends. Each beam carries the combined weight of 50 ft of 48-in.-diameter pipe and the oil flowing through it, about 42 000 lb. Assuming the load acts at the center of the beam, specify the required section modulus of the beam to limit the bending stress to 20 000 psi. Then specify a suitable wide flange or American Standard beam.

8-11. A wood platform is to be made of standard plywood and finished lumber using the cross section shown in Figure 7-21(c). Would the platform be safe if four men, weighing 250 lb each, were to stand 2 ft apart, as shown in Figure 8-18? Consider only bending stresses (see Chapter 9 for shear stresses).

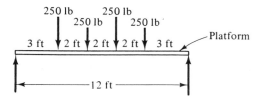

Figure 8-18 Platform loading for Problem 8-7.

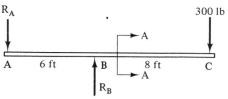

A and B are supports

(a)

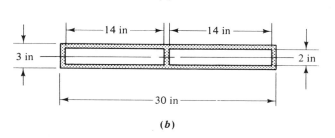

(b)

Figure 8-19 Diving board for Problem 8-12. (a) Loads on diving board (b) Section A-A through board.

8-12. A diving board has a hollow rectangular cross section 30 in. wide and 3.0 in. thick and is supported as shown in Figure 8-19. Compute the maximum stress due to bending in the board if a 300-lb person stands at the end. Would the board be safe if it was made of extruded 6016-T4 aluminum and the person landed at the end of the board with an impact?

8-13. The loading shown in Figure 6-36(d) is to be carried by a W12 × 16 steel beam. Compute the stress due to bending.

8-14. An American Standard beam, S12 × 35, carries the load shown in Figure 6-38(c). Compute the stress due to bending.

8-15. The 24-in. long beam shown in Figure 6-38(b) is an aluminum channel, C4 × 2.331, positioned with the legs down so that the flat 4-in. surface can carry the applied loads. Compute the maximum tensile and maximum compressive stresses in the channel.

8-16. The 650-lb load at the center of the 28-in.-long bar shown in Figure 6-36(a) is carried by a standard steel pipe, $1\frac{1}{2}$-in. schedule 40. Compute the stress in the pipe due to bending.

8-17. The loading shown in Figure 6-37(b) is to be carried by an extruded aluminum hat-section beam having the cross section shown in Figure 7-22(b). Compute the maximum stress due to bending in the beam. If it is made of extruded 6061-T4 aluminum and the loads are dead loads, would the beam be safe?

8-18. The extruded shape shown in Figure 7-22(c) is to be used to carry the loads shown in Figure 6-37(a), which is a part of a business machine frame. The loads are due to a motor mounted on the frame and can be considered dead loads. Specify a suitable aluminum alloy for the beam.

8-19. The loading shown in Figure 6-37(c) is to be carried by the fabricated beam shown in Figure 7-18(d). Compute the stress due to bending in the beam.

8-20. An aluminum I-beam I9 × 8.361, carries the load shown in Figure 6-37(d). Compute the stress due to bending in the beam.

8-21. A beam is being designed to support the loads shown in Figure 8-20. The four shapes proposed are: (a) a round bar, (b) a square bar, (c) a rectangular bar with the height made four times the thickness, and (d) the lightest American Standard beam. Determine the required dimensions of each proposed shape to limit the maximum stress due to bending to 80 MPa. Then compare the magnitude of the cross-sectional areas of the four shapes. Since the weight of the beam is proportional to its area, the one with the smallest area will be the lightest.

8-22. A rack is being designed to support large sections of pipe, as shown in Figure 8-21. Each pipe exerts a force of 2500 lb on the support arm. The height of the arm is to be tapered as suggested in the figure, but the thickness will be a constant 1.50 in. Determine the required height of the arm at sections B and C, considering only bending stress. Use AISI 1040 hot-rolled steel for the arms and a design factor of 4 based on yield strength.

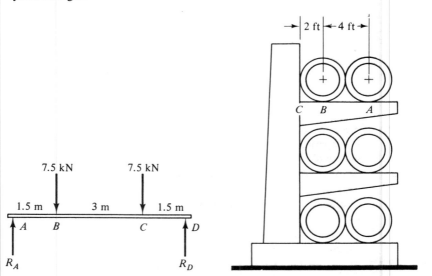

Figure 8-20 Beam for Problem 8-21.

Figure 8-21 Pipe storage rack for Problem 8-22.

8-23. A children's play gym includes a cross beam carrying four swings, as shown in Figure 8-22. Assume that each swing carries 300 lb. It is desired to use a standard steel pipe for the beam, keeping the stress due to bending below 10 000 psi. Specify the suitable size pipe for the beam.

8-24. A part of a truck frame is composed of two channel-shaped members, as shown in Figure 8-23. If the moment at the section is 60 000 lb · ft, compute the bending stress in the frame. Assume that the two channels act as a single beam.

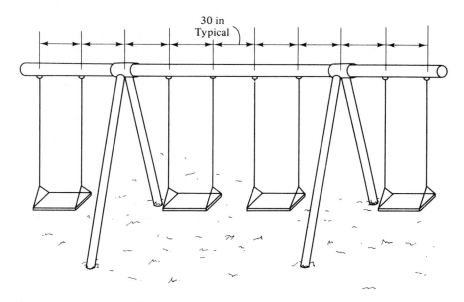

Figure 8-22 Swing set for Problem 8-23.

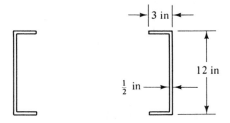

Figure 8-23 Truck frame members for Problem 8-24.

8-25. A 60 in. long beam simply supported at its ends is to carry two 4800 lb loads, each placed 14 in. from an end. Specify the lightest suitable steel tube for the beam, either square or rectangular, to produce a design factor of 4 based on yield strength. The tube is to be cold formed from ASTM A500, grade A steel.

8-26. Repeat Problem 8-25, but specify the lightest standard aluminum I-beam from Appendix A-11. The beam will be extruded using alloy 6061-T6.

8-27. Repeat Problem 8-25, but specify the lightest wide-flange steel shape from Appendix A-7. The beam will be made from ASTM A36 structural steel.

8-28. Repeat Problem 8-25, but specify the lightest structural steel channel from Appendix A-6. The channel is to be installed with the legs down so that the loads can be applied to the flat back of the web of the channel. The channel will be made from ASTM A36 structural steel.

8-29. Repeat Problem 8-25, but specify the lightest standard schedule 40 steel pipe from Appendix A-12. The pipe is to be made from ASTM A501 hot-formed steel.

8-30. Repeat Problem 8-25, but design the beam using any material and shape of your choosing to achieve a safe beam that is lighter than any of the results from Problems 8-25 through 8-29.

8-31. A floor joist for a building is to be made from a standard wooden beam selected from Appendix A-4. If the beam is to be simply supported at its ends and carry a uniformly distributed load of 125 lb/ft over the entire 10-ft length, specify a suitable beam size. The beam will be made from No. 2 grade southern pine. Consider only bending stress.

8-32. A bench for football players is to carry the load shown in Figure 8-24 approximating the case when 10 players, each weighing 300 lb, sit close together, each taking 18 in. of the length of the bench. If the cross section of the bench is made as shown in Figure 8-24, would it be safe for bending stress? The wood is No. 2 grade hemlock.

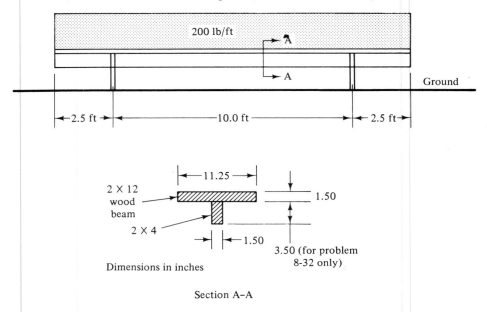

Section A–A

Figure 8-24 Bench and load for Problems 8-32, 8-33, 8-34, and 8-35.

8-33. A bench is to be designed for football players. It is to carry the load shown in Figure 8-24 approximating the case when 10 players, each weighing 300 lb, sit close together, each taking 18 in. of the length of the bench. The bench is to be T-shaped, made from No. 2 grade hemlock, as shown with a 2 × 12 top board. Specify the required vertical member of the tee if the bench is to be safe for bending stress.

8-34. Repeat Problem 8-33, but use the cross-section shape shown in Figure 8-25.

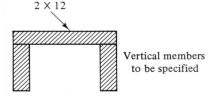

Vertical members to be specified

Figure 8-25 Shape for bench cross section for Problem 8-34.

8-35. Repeat Problem 8-33, but use any cross-section shape of your choosing made from standard wooden beams from Appendix A-4. Try to achieve a lighter design than in either Problem 8-33 or 8-34. Note that a lighter design would have a smaller cross-section area.

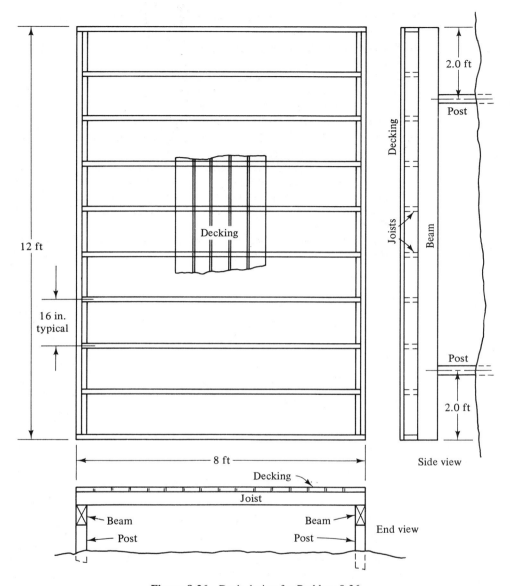

Figure 8-26 Deck design for Problem 8-36.

8-36. A wood deck is being designed to carry a uniformly distributed load over its entire area of 100 lb/ft². Joists are to be used as shown in Figure 8-26, set 16 in. on center. If the deck is to be 8 ft by 12 ft in size, determine the required size for the joists. Use standard wooden beam sections from Appendix A-4 and No. 2 hemlock.

8-37. Repeat Problem 8-36, but run the joists across the 12-ft length rather than the 8-ft width.

8-38. Repeat Problem 8-36, but set the support beams in 18 in. from the ends of the joists instead of at the ends.

8-39. Repeat Problem 8-37, but set the support beams in 18 in. from the ends of the joists instead of at the ends.

8-40. For the deck design shown in Figure 8-26 specify a suitable size for the cross beams that support the joists.

8-41. Design a bridge to span a small stream. Assume that rigid supports are available on each bank, 10.0 ft apart. The bridge is to be 3.0 ft wide and carry a uniformly distributed load of 60 lb/ft² over its entire area. Design only the deck boards and beams. Use two or more beams of any size from Appendix A-4 or others of your own design.

8-42. Would the bridge you designed in Problem 8-41 be safe if a horse and rider weighing 2200 lb walked slowly across it?

8-43. Millwrights in a factory need to suspend a machine weighing 10 500 lb from a beam having a span of 12.0 ft so that a truck can back under it. Assume that the beam is simply supported at its ends. The load is applied by two cables, each 3.0 ft from a support. Design a suitable beam. Consider standard wooden or steel beams or one or your own design.

8-44. In an amateur theater production, a pirate is to "walk the plank." If the pirate weighs 220 lb, would the design shown in Figure 8-27 be safe? If not, design one to be safe.

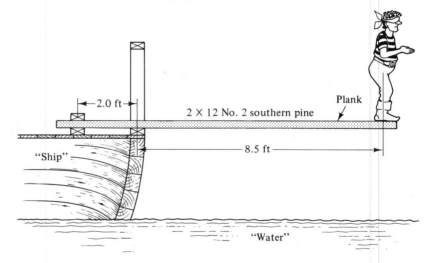

Figure 8-27 Pirate walking the plank in Problem 8-44.

8-45. A branch of a tree has the approximate dimensions shown in Figure 8-28. Assuming the bending strength of the wood to be similar to that of No. 3 grade hemlock, would it be safe for a person having a mass of 135 kg to sit in the swing?

8-46. Would it be safe to use a standard 2 × 4 made from No. 2 grade southern pine as a lever as shown in Figure 8-29 to lift one side of a machine? If not, what would you suggest be used?

8-47. Figure 7-16(a) shows the cross section of an extruded plastic beam made from nylon 6/6. Specify the largest uniformly distributed load in N/mm the beam could carry if it is simply supported with a span of 0.80 m. The maximum stress due to bending is not to exceed one-half of the flexural strength of the nylon.

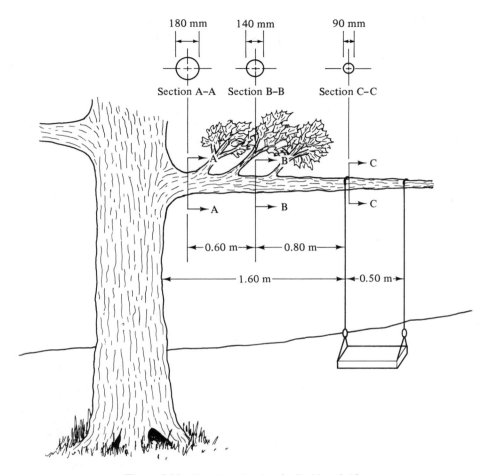

Figure 8-28 Branch and swing for Problem 8-45.

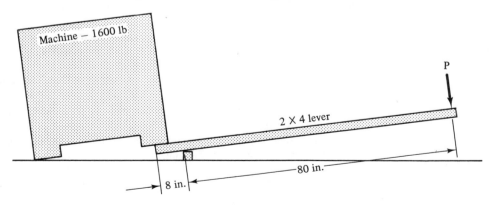

Figure 8-29 2 × 4 used as a lever in Problem 8-46.

8-48. The I-beam shape in Figure 7-16(d) is to carry two identical concentrated loads of 2.25 kN each, symmetrically placed on a simply supported beam with a span of 0.60 m. Each load is 0.2 m from an end. Which of the plastics from Appendix A-19 would carry these loads with a stress due to bending no more than one-third of their flexural strength?

8-49. A bridge for a toy construction set is to be made from polypropylene with the flexural strength listed in Appendix A-19. The cross section of one beam is shown in Figure 7-15(d). What maximum concentrated load could be applied to the middle of the beam if the span was 1.25 m and the ends are simply supported? Do not exceed one-half of the flexural strength of the plastic.

8-50. A structural element in a computer printer is to carry the load shown in Figure 8-30 with the uniformly distributed load representing electronic components mounted on a printed circuit board and the concentrated loads applied from a power supply. It is proposed to make the beam from polycarbonate with the cross-section shape shown in Figure 7-22(c). Compute the stress in the beam and compare it with the flexural strength for polycarbonate from Appendix A-19.

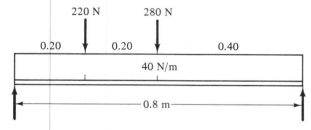

Figure 8-30 Beam in a computer printer for Problem 8-50.

8-51. It is proposed to make the steps for a child's slide from molded polystyrene with the cross section shown in Figure 7-17(c). The steps are to be 14.0 in. wide and will be simply supported at the ends. What maximum weight can be applied at the center of the step if the stress due to bending must not exceed one-third of the flexural strength of the polystyrene listed in Appendix A-19.

8-52. The shape shown in Figure 7-17(d) is used as a beam for a carport. The span of the beam will be 8.0 ft. Compute the maximum uniformly distributed load that can be carried by the beam if it is extruded from 6061-T4 aluminum. Use a design factor of 2 based on yield strength.

8-53. Figure 7-17(a) shows the cross section of an aluminum beam made by fastening a flat plate to the bottom of a roll-formed hat section. If the beam is used as a cantilever, 24 in. long, compute the maximum allowable concentrated load that can be applied at the end if the maximum stress is to be no more than one-eighth of the ultimate strength of 2014-T4 aluminum.

8-54. Repeat Problem 8-53 but use only the hat section without the cover plate.

8-55. Figure 7-17(b) shows the cross section of a beam that is to be cast from aluminum 204.0-T4 casting alloy. If the beam is to be used as a cantilever, 42 in. long, compute the maximum allowable uniformly distributed load it could carry while limiting the stress due to bending to one-sixth of the ultimate strength of the aluminum.

8-56. Repeat Problem 8-55, but use Figure 7-17(c) for the cross section and casting alloy 356.0-T6.

8-57. The beam section shown in Figure 7-17(d) is to be extruded from 6061-T6. aluminum. The allowable tensile strength is 19 ksi. Because of the relatively thin extended legs on

Stress Due to Bending   Chap. 8

the top, the allowable compressive strength is only 14 ksi. The beam is to span 6.5 ft and will be simply supported at its ends. Compute the maximum allowable uniformly distributed load on the beam.

8-58. Repeat Problem 8-57 but turn the section upside down. With the legs pointed downward, they are in tension and can withstand 19 ksi. The part of the section in compression at the top is now well supported and can withstand 21 ksi.

8-59. The shape in Figure 7-16(a) is to carry a single concentrated load at the center of a 1200-mm span. The allowable strength in tension is 100 MPa, while the allowable strength in compression anywhere is 70 MPa. Compute the allowable load.

8-60. Repeat Problem 8-59 with the section turned upside down.

8-61. Repeat Problem 8-59 with the shape shown in Figure 7-16(c).

8-62. Repeat Problem 8-59 with the shape shown in Figure 7-16(d).

8-63. The T-shaped beam cross section shown in Figure 7-15(a) is to be made from gray cast iron, ASTM A48 Grade 40. It is to be loaded with two equal loads P, 1.0 m from the ends of the 2.80 m long beam. Specify the largest static load P that the beam could carry. Use N = 4.

8-64. The modified I-beam shape shown in Figure 7-15(d) is to carry a uniformly distributed static load over its entire 1.20 m length. Specify the maximum allowable load if the beam is made from malleable iron, ASTM A220, class 80002. Use N = 4.

8-65. Repeat Problem 8-64 but turn the beam upside down.

8-66. A wide beam is made as shown in Figure 8-31 from ductile iron, ASTM A536, Grade 120-90-2. Compute the maximum load P that can be carried with a resulting design factor of 10 based on either tensile or compressive ultimate strength.

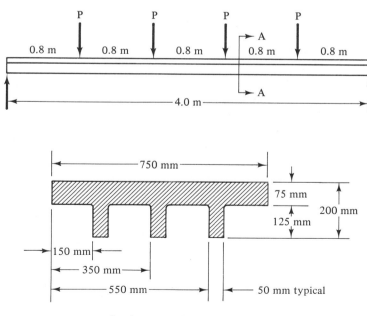

Section A–A — beam cross section

Figure 8-31 Wide beam for Problem 8-66.

8-67. Repeat Problem 8-66 but increase the depth of the vertical ribs by a factor of 2.0.

8-68. Problems 8-63 through 8-67 illustrate that a beam shape made as a modified I-shape more nearly optimizes the use of the available strength of a material having different strengths in tension and compression. Design an I-shape that has a nearly uniform design factor of 6 based on ultimate strength in either tension or compression when made from gray iron, Grade 20, and which carries a uniformly distributed load of 20 kN/m over its 1.20 m length. (Note: You may want to use the computer program written for Assignment 3 at the end of Chapter 7 to facilitate the computations. A trial and error solution may be used.)

For Problems 8-69 through 8-78 using the indicated loading, specify the lightest standard wide-flange beam shape (W shape) which will limit the stress due to bending to the allowable design stress from the AISC specification. All loads are static and the beams are made from ASTM A36 structural steel.

8-69. Use the loading in Figure 6-36(c).

8-70. Use the loading in Figure 6-37(c).

8-71. Use the loading in Figure 6-37(d).

8-72. Use the loading in Figure 6-38(c).

8-73. Use the loading in Figure 6-39(d).

8-74. Use the loading in Figure 6-44(b).

8-75. Use the loading in Figure 6-45(b).

8-76. Use the loading in Figure 6-48(b).

8-77. Use the loading in Figure 6-50(c).

8-78. Use the loading in Figure 6-50(d).

For Problems 8-79 through 8-88, repeat Problems 8-69 through 8-78 but specify the lightest American Standard Beam (S shape).

For Problems 8-89 through 8-98, repeat Problems 8-69 through 8-78 but use ASTM A572 Grade 60 high-strength low-alloy structural steel.

8-99. Specify the lightest wide-flange beam shape (W shape) that can carry a uniformly distributed static load of 2.5 kips/ft over the entire length of a simply supported span, 12.0 ft long. Use the AISC specification and ASTM A36 structural steel.

8-100. A proposal is being evaluated to save weight for the beam application of Problem 8-99. The result for that problem required a W14 × 26 beam that would weigh 312 lb for the 12-ft length. A W12 × 16 beam would only weigh 192 lb but does not have sufficient section modulus S. To increase S, it is proposed to add steel plates, 0.250 in. thick and 3.50 in. wide to both the top and the bottom flange over a part of the middle of the beam. Perform the following analyses.

  (a) Compute the section modulus of the portion of the W12 × 16 beam with the cover plates.

  (b) If the result of part (a) is satisfactory to limit the stress to an acceptable level, compute the required length over which the plates would have to be applied to the nearest 0.5 ft.

  (c) Compute the resulting weight of the composite beam and compare it to the original W14 × 26 beam.

8-101. Figure 7-18(b) shows a composite beam made by adding a channel to an American Standard beam shape. If the beam is simply supported and carries a uniformly distributed load over a span of 15.0 ft, compute the allowable load for the composite beam and for the S shape by itself. The load is static and the AISC specification is to be used for A36 structural steel.

8-102. The shape shown in Figure 7-17(d) is to be made from extruded plastic and used as a simply supported beam, 12 ft long, to carry two electric cables weighing a total of 6.5 lb/ft of length. Specify a suitable plastic for the extrusion to provide a design factor of 4.0 based on flexural strength.

8-103. The loading shown in Figure 6-44(e) represents the load on a floor beam of a commercial building. Determine the maximum bending moment on the beam, and then specify a wide-flange shape that will limit the stress to 150 MPa.

8-104. Figure 6-44(c) represents the loading on a motor shaft; the two supports are bearings in the motor housing. The larger load between the supports is due to the rotor plus dynamic forces. The smaller, overhung load is due to externally applied loads. Using AISI 1141 OQT 1300 steel for the shaft, specify a suitable diameter based on bending stress only. Use a design factor of 8 based on ultimate strength.

8-105. In Figure 8-32 the 4-in. pipe mates smoothly with its support, so that no stress concentration exists at D. At C the $3\frac{1}{2}$-in. pipe is placed inside the 4-in. pipe with a spacer ring to provide a good fit. Then a $\frac{3}{8}$-in. fillet weld is used to secure the section together. Accounting for the stress concentration at the joint, determine how far out point C must be to limit the stress to 20 000 psi. Use Appendix A-21-7 for the stress concentration factor. Is the 4-in. pipe safe at D?

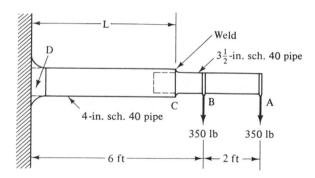

Figure 8-32

8-106. Figure 8-33 shows a round shaft from a gear transmission. Gears are mounted at points $A, C,$ and $E.$ Supporting bearings are at B and $D.$ The forces transmitted from the gears

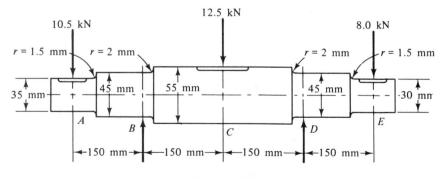

Figure 8-33

to the shaft are shown, all acting downward. Compute the maximum stress due to bending in the shaft, accounting for stress concentrations.

8-107. The forces shown on the shaft in Figure 8-34 are due to gears mounted at B and C. Compute the maximum stress due to bending in the shaft.

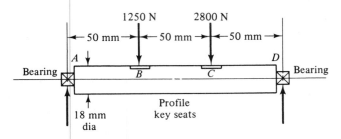

Figure 8-34

8-108. Figure 8-35 shows a machine shaft supported by two bearings at its ends. The two forces are exerted on the shaft by gears. Considering only bending stresses, compute the maximum stress in the shaft and tell where it occurs.

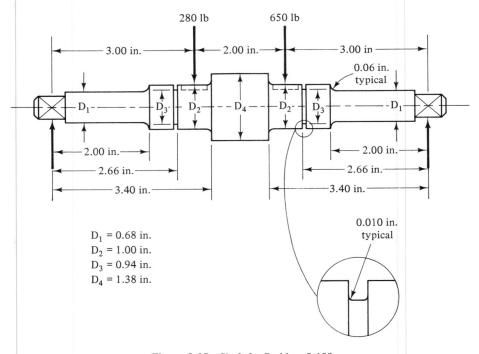

$D_1 = 0.68$ in.
$D_2 = 1.00$ in.
$D_3 = 0.94$ in.
$D_4 = 1.38$ in.

Figure 8-35 Shaft for Problem 8-108.

8-109. Figure 8-36 shows a lever made from a rectangular bar of steel. Compute the stress due to bending at the fulcrum, 20 in. from the pivot and at each of the holes in the bar. The holes are 0.75 in. in diameter.

8-110. Repeat problem 8-109 but use 1.38 in. as the diameter of the holes.

8-111. In Figure 8-36, the holes in the bar are provided to permit the length of the lever to be

Stress Due to Bending   Chap. 8

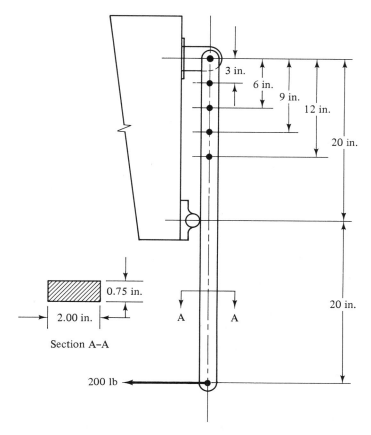

Figure 8-36 Lever for Problems 8-109 to 8-111.

changed relative to the pivot. Compute the maximum bending stress in the lever as the pivot is moved to each hole. Use 1.25 in. for the diameter of the holes.

8-112. The bracket shown in Figure 8-37 carries the opposing forces created by a spring. If the force, F, is 2500 N, compute the bending stress at a section such as A-A, away from the holes.

8-113. If the force, F, in Figure 8-37 is 2500 N, compute the bending stress at a section through the holes, such as B-B. Use $d = 12$ mm for the diameter of the holes.

8-114. Repeat Problem 8-113 but use $d = 15$ mm for the diameter of the holes.

8-115. For the resulting stress computed in Problem 8-114, specify a suitable steel for the bracket if the force is repeated many thousands of times.

8-116. Figure 8-38 shows a stepped flat bar in bending. If the bar is made from AISI 1040 cold-drawn steel, compute the maximum repeated force, F, that can safely be applied to the bar.

8-117. Repeat Problem 8-116, but use $r = 2.0$ mm for the fillet radius.

8-118. For the stepped flat bar shown in Figure 8-38 change the 75-mm dimension that locates the step to a value that makes the bending stress at the step equal to that at the point of application of the load.

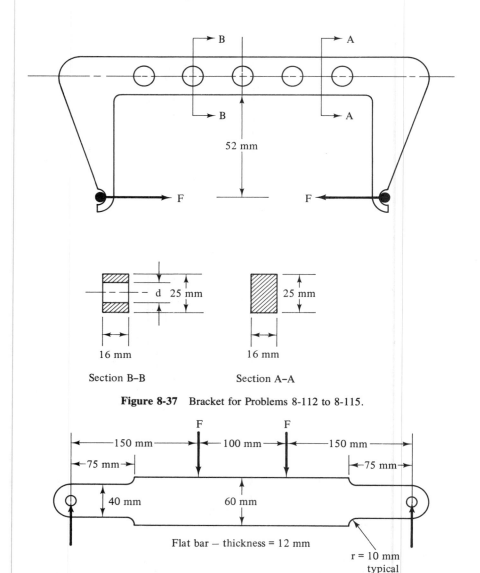

Section B–B   Section A–A

Figure 8-37   Bracket for Problems 8-112 to 8-115.

Figure 8-38   Stepped flat bar for Problems 8-116 to 8-121.

8-119. For the stepped flat bar shown in Figure 8-38 change the size of the fillet radius to make the bending stress at the fillet equal to that at the point of application of the load.

8-120. Repeat Problem 8-116, but change the depth of the bar from 60 mm to 75 mm.

8-121. For the stepped flat bar in Figure 8-38 would it be possible to drill a hole in the middle of the 60-mm depth of the bar between the two forces without increasing the maximum bending stress in the bar? If so, what is the maximum-size hole that can be put in?

8-122. Figure 8-39 shows a stepped flat bar carrying three concentrated loads. Let $P = 200$ N, $L_1 = 180$ mm, $L_2 = 80$ mm, and $L_3 = 40$ mm. Compute the maximum stress due

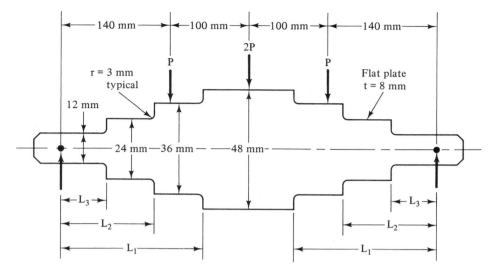

Figure 8-39 Stepped flat bar for Problems 8-122 to 8-126.

to bending and state where it occurs. The bar is braced against lateral bending and twisting. Note that the length dimensions in the figure are not drawn to scale.

8-123. For the data of Problem 8-122, specify a suitable material for the bar to produce a design factor of 8 based on ultimate strength.

8-124. Repeat Problem 8-123 except use $r = 1.50$ mm for the fillet radius.

8-125. For the stepped flat bar shown in Figure 8-39, let $P = 400$ N. The bar is to be made from titanium, Ti-6A1-4V, and a design factor of 8 based on ultimate strength is desired. Specify the maximum permissible lengths, L_1, L_2, and L_3 that would be safe.

8-126. The stepped flat bar in Figure 8-39 is to be made from AISI 4140 OQT 1100 steel. Use $L_1 = 180$ mm, $L_2 = 80$ mm, and $L_3 = 40$ mm. Compute the maximum allowable force P that could be applied to the bar if a design factor of 8 based on ultimate strength is desired.

8-127. Figure 8-40 shows a flat bar that has a uniform thickness of 20 mm. The depth tapers from $h_1 = 40$ mm to $h_2 = 20$ mm in order to save weight. Compute the stress due to

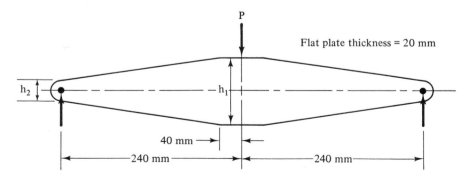

Figure 8-40 Tapered flat bar for Problems 8-127 through 8-129.

bending in the bar at points spaced 40 mm apart from the support to the load. Then create a graph of stress versus distance from the support. The bar is symmetrical with respect to its middle. Let $P = 5.0$ kN.

8-128. For the bar shown in Figure 8-40 let $h_1 = 60$ mm and $h_2 = 20$ mm. The bar is to be made from polycarbonate plastic. Compute the maximum permissible load P that will produce a design factor of 4 based on the flexural strength of the plastic. The bar is symmetrical with respect to its middle.

8-129. In Figure 8-40 the load $P = 1.20$ kN and the bar is to be made from AISI 5160 OQT 1300 steel. Compute the required dimensions h_1 and h_2 that will produce a design factor of 8 based on ultimate strength. The bar is symmetrical with respect to its middle.

8-130. Compute the location of the flexural center of a channel-shaped member shown in Figure 8-41 measured from the left face of the vertical web.

8-131. A company plans to make a series of three channel-shaped beams by roll forming them from flat sheet aluminum. Each channel is to have the same outside dimensions as shown in Figure 8-41 but they will have different material thicknesses, 0.50, 1.60, and 3.00 mm. For each design, compute the moment of inertia with respect to the horizontal centroidal axis and the location of the flexural center, measured from the left face of the vertical web.

8-132. Compute the location of the flexural center for the hat section shown in Figure 8-42 measured from the left face of the vertical web.

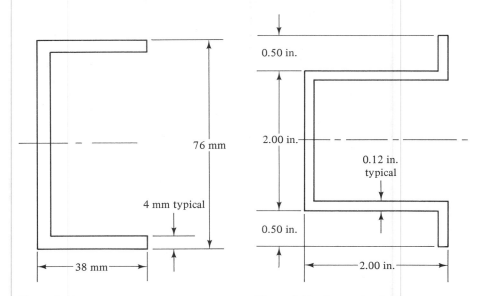

Figure 8-41 Channel shape for Problem 8-130.

Figure 8-42 Hat section for Problem 8-132.

8-133. A company plans to make a series of three hat sections by roll forming them from flat sheet aluminum. Each hat-section is to have the same outside dimensions as shown in Figure 8-42 but they will have different material thicknesses, 0.020, 0.063, and 0.125 in. For each design, compute the location of the flexural center, measured from the left face of the vertical web.

Stress Due to Bending   Chap. 8

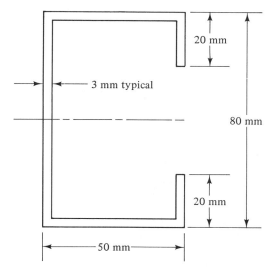

Figure 8-43 Lipped channel for Problem 8-134.

8-134. Compute the location of the flexural center for the lipped channel shown in Figure 8-43, measured from the left face of the vertical web.

8-135. A company plans to make a series of three lipped channels by roll forming them from flat sheet aluminum. Each channel section is to have the same outside dimensions as shown in Figure 8-43, but they will have different material thicknesses, 0.50, 1.60, and 3.00 mm. For each design, compute the location of the flexural center, measured from the left face of the vertical web.

8-136. Compute the location of the flexural center of a split, thin tube if it has an outside diameter of 50 mm and a wall thickness of 4 mm.

8-137. For an aluminum channel C2 × 0.577 with its web oriented vertically, compute the location of its flexural center. Neglect the effect of the fillets between the flanges and the web.

8-138. If the hat section shown in Figure 7-22(b) were turned 90 deg from the position shown, compute the location of its flexural center.

COMPUTER PROGRAMMING ASSIGNMENTS

1. Write a program to compute the maximum bending stress for a simply supported beam carrying a single concentrated load at its center. Allow the operator to input the load, span, and beam section properties. The output should include the maximum bending moment and the maximum bending stress and indicate where the maximum stress occurs.

ENHANCEMENTS

(a) For the computed stress, compute the required strength of the material for the beam to produce a given design factor.

(b) In addition to (a), include a table of properties for a selected material such as the data for steel in Appendix A-13. Then search the table for a suitable steel from which the beam can be made.

2. Repeat Assignment 1 except use a uniformly distributed load.

3. Repeat Assignment 1 except the beam is a cantilever with a single concentrated load at its end.

4. Write a program to compute the maximum bending moment for a simply supported beam carrying a single concentrated load at its center. Allow the operator to input the load and span. Then compute the required section modulus for the cross section of the beam to limit the maximum bending stress to a given level or to achieve a given design factor for a given material. The output should include the maximum bending moment and the required section modulus.

ENHANCEMENTS

(a) After computing the required section modulus, have the program complete the design of the beam cross section for a given general shape, such as rectangular with a given ratio of thickness to depth (see Problem 8-6), or circular.

(b) Include a table of properties for standard beam sections such as any of those in Appendixes A-4 through A-12 and have the program search for a suitable beam section to provide the required section modulus.

5. Repeat Assignment 4 but use a uniformly distributed load.

6. Repeat Assignment 4 but use the load described in Problem 8-21.

7. Repeat Assignment 4 but use any loading pattern assigned by the instructor.

8. Write a computer program to facilitate the solution of Problem 8-68, including the computation of section properties for the modified I-shape using the techniques of Chapter 7. Use the loading pattern of Figure 6-36(c) but allow the user to specify the span of the beam, the magnitude of the load, and the placement of the load.

9. Write a computer program to facilitate the solution of problems of the type given in Problem 8-100. Make the program general, permitting the user to input the loading on the beam, the desired beam section properties, and the dimensions of the plates to be added to the basic beam section.

10. Write a computer program to perform the computations called for in Problem 8-127, but make the program more general, permitting the user to input values for the load, span, beam cross-section dimensions, and the interval for computing the bending stress. If the computer has graphics capability, have the program produce the requested graph of stress versus position on the beam.

11. Write a computer program to compute the location for the flexural center for the generalized channel shape shown in Figure 8-41. Permit the user to input data for all dimensions.

12. Write a computer program to compute the location for the flexural center for the generalized hat section shown in Figure 8-42. Permit the user to input data for all dimensions. Curve-fitting techniques and interpolation may be used to interpret the graph in Figure 8-16.

13. Write a computer program to compute the location for the flexural center for the generalized lipped channel shown in Figure 8-43. Permit the user to input data for all dimensions. Curve-fitting techniques and interpolation may be used to interpret the graph in Figure 8-16.

9

Shearing Stresses in Beams

9-1 OBJECTIVES OF THIS CHAPTER

Continuing the analysis of beams, this chapter is concerned with the stresses created within a beam due to the presence of shearing forces. As shown in Figure 9-1, shearing forces are visualized to act within the beam on its cross section and to be directed transverse, that is perpendicular, to the axis of the beam. Thus they would tend to create *transverse shearing stresses*, sometimes called *vertical shearing stresses*.

But if a small stress element subjected to such shearing stresses is isolated, as shown in Figure 9-2, it can be seen that horizontal shearing stresses must also exist in order to cause the element to be in equilibrium. Thus, both vertical and horizontal shearing stresses, having the same magnitude at a given point, are created by shearing stresses in beams.

After completing this chapter, you should be able to:

1. Describe the conditions under which shearing stresses are created in beams.
2. Compute the magnitude of shearing stresses in beams by using the general shear formula.
3. Define and evaluate the *statical moment* required in the analysis of shearing stresses.

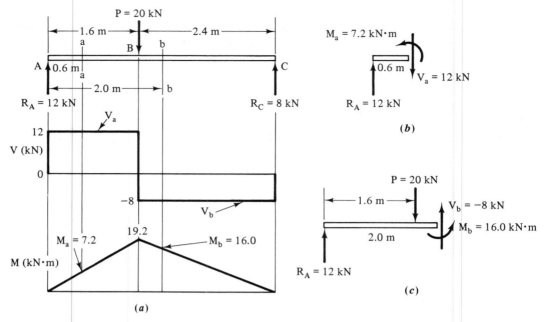

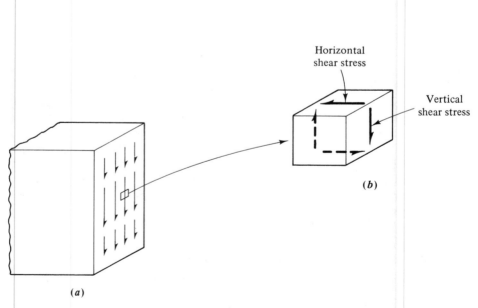

Figure 9-1 Shearing forces in beams. (a) Load, shear and moment diagrams (b) Free body diagram at $a - a$ (c) Free body diagram at $b - b$.

Figure 9-2 Shear stress in a beam. (a) Shear stress on a cut section of a beam (b) Shear stress on a small element.

4. Specify where the maximum shearing stress occurs on the cross section of a beam.

5. Compute the shearing stress at any point within the cross section of a beam.

6. Describe the general distribution of shearing stress as a function of position within the cross section of a beam.

7. Understand the basis for the development of the general shearing stress formula.

8. Describe four design applications where shearing stresses are likely to be critical in beams.

9. Develop and use special shear formulas for computing the maximum shearing stress in beams having rectangular or solid circular cross sections.

10. Understand the development of approximate relationships for estimating the maximum shearing stress in beams having cross sections with tall thin webs or those with thin-walled hollow tubular shapes.

11. Specify a suitable design shearing stress and apply it to evaluate the acceptability of a given beam design.

12. Define *shear flow* and compute its value.

13. Use the shear flow to evaluate the design of fabricated beam sections held together by nails, bolts, rivets, adhesives, welding, or other means of fastening.

9-2 VISUALIZATION OF SHEARING STRESS IN BEAMS

The existence of horizontal shearing stress in beams can also be observed by considering a beam made from several flat strips as illustrated in Figure 9-3. A demonstrator can be made from cardboard, sheet metal, plastic, or other materials.

One thin flat strip would make a very poor beam if it was simply supported near its ends and loaded with a concentrated load at the middle of the span. The beam would deflect a large amount and it would tend to break at a very small load.

Laying several strips on top of each other would produce a beam with increased strength and decreased deflection for a given load, but to only a very small extent. As shown in Figure 9-3(b), the strips would slide over one another at the surfaces of contact and the beam would still be relatively flexible and weak.

A stronger and stiffer beam can be made by fastening the strips together in such a way that the sliding between strips is prevented. This can be done by using an adhesive, welding, brazing, or mechanical fasteners such as rivets, screws, bolts, pins, nails, or even staples. In this way, the tendency for one strip to slide on the next is prevented and the fastening means is subjected to a shearing force directed horizontally, parallel to the neutral axis of the beam. This is a visualization of the *horizontal shearing stress* in a beam.

A similar condition exists in a solid beam. Here the tendency for horizontal sliding of one part of a beam relative to the part above or below it is resisted by *the material of the beam*. Therefore, a shearing stress is developed on any horizontal plane. Again, as shown in Figure 9-2, shearing stresses must exist simultaneously in the vertical plane to maintain equilibrium of any stress element.

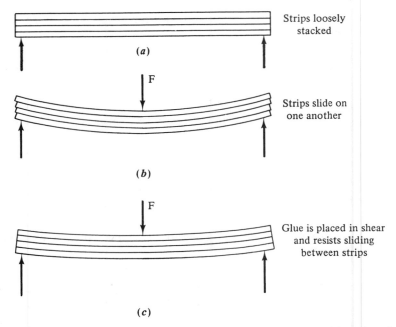

Strips loosely
stacked

(a)

F

Strips slide on
one another

(b)

F

Glue is placed in shear
and resists sliding
between strips

(c)

Figure 9-3 Illustration of the presence of shear stress in a beam. (a) Flat strips of wood, unloaded (b) Flat strips of wood carrying a load (c) Glued strips of wood carrying a load.

9-3 IMPORTANCE OF SHEARING STRESSES IN BEAMS

Several situations exist in practical design in which the mode of failure is likely to be shearing of a part of a beam or of a means of fastening a composite beam together. Five such situations are described here.

Wooden beams. Wood is inherently weak in shear along the planes parallel to the grain of the wood. Consider the beam shown in Figure 9-4, which is similar to the joists used in floor and roof structures for wood-frame construction. The grain runs generally parallel to the long axis in commercially available lumber. When subjected to transverse loads, the initial failure in a wooden beam is likely to be by separation along the grain of the wood, due to excessive horizontal shearing stress. Note in Appendix A-18 that the allowable shearing stress in common species of wood ranges from only 70 to 95 psi (0.48 to 0.66 MPa), very low values.

Thin-webbed beams. An efficient beam cross section would be one with relatively thick horizontal flanges on the top and bottom with a thin vertical web connecting them together. This generally describes the familiar "I-beam," the wide-flange beam, or the American Standard beam, as sketched in Figure 9-5. Actual dimensions for such beam sections are given in Appendixes A-7, A-8, and A-11.

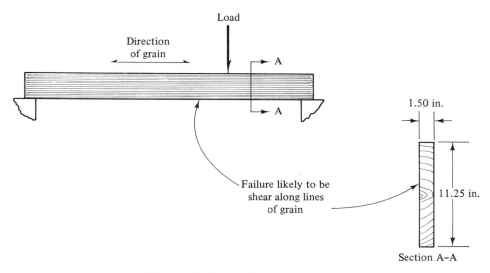

Figure 9-4 Shear failure in a wood beam.

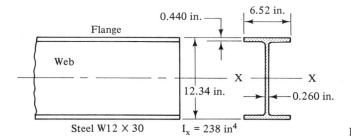

Figure 9-5 Thin-webbed beam shape.

But if the web is excessively thin, it would not have sufficient stiffness and stability to hold its shape and it would fail due to shearing stress in the thin web. When using such sections for beams, it is necessary to compute the value of the shearing stress, to be sure it is safe. The American Institute of Steel Construction (AISC) defines the allowable shear stress in the unstiffened webs of beams to be

$$\tau_d = 0.40s_y$$

when using the *web shear formula*, defined later in this chapter. For very tall, thin-webbed beams and those with stiffeners, special provisions of the AISC are given in Reference 2, Chapter 3.

Short beams. In very short beams, the bending moment, and therefore the bending stress, is likely to be small. In such beams, the shearing stress may be the limiting stress.

Fastening means in fabricated beams. As shown in Figure 9-3, the fasteners in a composite beam section are subjected to shearing stresses. The concept of *shear flow*, developed later, can be used to evaluate the safety of such beams or to specify the required type, number, and spacing of fasteners to use. Also, beams made of composite materials are examples of fabricated beams. Separation of the layers of the composite, called *interlaminar shear*, is a potential mode of failure.

Stressed skin structures. Aircraft and aerospace structures and some ground-based vehicles and industrial equipment are made using a *stressed skin* design. Sometimes called *monocoque* structures, they are designed to carry much of the load in the thin skins of the structure. The method of shear flow is typically used to evaluate such structures, but this application is not developed in this book.

9-4 THE GENERAL SHEAR FORMULA

Presented here is the general shear formula from which you can compute the magnitude of the shearing stress at any point within the cross section of a beam carrying a vertical shearing force. In Section 9-7, the formula itself is developed. It may be desirable to study the development of the formula along with this section.

The general shear formula is stated as follows:

$$\tau = \frac{VQ}{It} \tag{9-1}$$

where

1. *V* is the vertical shearing force at the *section of interest*. The shear diagram developed in Chapter 6 is a plot of the variation of shearing force with position on the beam. If the maximum value of *V* is required, it can be read from the shear diagram. Generally, the maximum absolute value, positive or negative, is used.
2. *I* is the moment of inertia of the *entire* cross section of the beam with respect to its centroidal axis. This is the same value of *I* used in the flexure formula ($\sigma = Mc/I$) to compute bending stresses.
3. The thickness *t* is taken at the point where the shear stress is to be calculated.
4. *Q* is called the *statical moment*, with respect to the centroidal axis, of the area of that part of the cross section that lies away from the axis where the shear stress is to be calculated. By definition,

$$Q = A_p \overline{y} \tag{9-2}$$

where A_p = area of that *part* of the cross section that lies away from the axis where the shear stress is to be calculated

$\overline{y}$ = distance to the centroid of A_p from the neutral axis of the cross section, that is, from the centroid of the entire area

Note that the statical moment is the *moment of an area*, that is, area times distance. Therefore, it will have the units of length cubed, such as in^3, m^3, or mm^3.

Careful evaluation of the statical moment *Q* is critical to proper use of the general

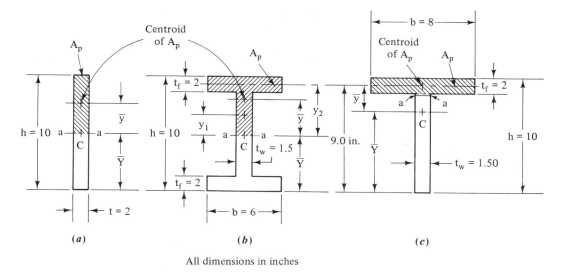

Figure 9-6 Examples of A_p and $\bar{y}$ for use in computing Q.

shear formula. It is helpful to draw a sketch of the beam cross section and then to highlight the partial area, A_p. Then show the location of the centroid of the partial area on the sketch. Figure 9-6 shows three examples for which this has been done. In each example, the objective is to calculate the shear stress at the axis labeled *a-a*. The crosshatched area is A_p, shown as that part away from the axis *a-a*.

The following three example problems illustrate the method of computing Q. In each, this is the procedure used.

1. Locate the neutral axis for the entire cross section by computing the location of the centroid.
2. Draw in the axis where the shear stress is to be calculated.
3. Identify the partial area A_p away from the axis of interest and shade it for emphasis.

If the partial area A_p is a simple area for which the centroid is readily found by simple calculations, use steps 4–7 to compute Q. Otherwise, use steps 8–12.

4. Compute the magnitude of A_p.
5. Locate the centroid of the partial area.
6. Compute the distance $\bar{y}$ from the neutral axis of the full section to the centroid of the partial area.
7. Compute $Q = A_p\bar{y}$.

For cases in which the partial area is itself a composite area made up of several component parts, steps 8–11 are used.

Sec. 9-4   The General Shear Formula

8. Divide A_p into component parts which are simple areas and label them A_1, A_2, A_3, and so on. Compute their values.

9. Locate the centroid of each component area.

10. Determine the distances from the neutral axis to the centroid of each component area, calling them y_1, y_2, y_3, and so on.

11. Note that, by the definition of the centroid,

$$A_p\bar{y} = A_1 y_1 + A_2 y_2 + A_3 y_3 + \cdots$$

Now, because $Q = A_p\bar{y}$, the most convenient way to calculate Q is

$$Q = A_1 y_1 + A_2 y_2 + A_3 y_3 + \cdots \tag{9-3}$$

Example Problem 9-1

For each of the sections shown in Figure 9-6, compute the statical moment Q as it would be used to determine the vertical shear stress at the section marked a-a.

Solution (a) For the rectangular shape in Figure 9-6(a):

Step 1. The neutral axis for the rectangular section is at its midheight, $h/2$ from the bottom; for this problem $h/2 = 5.00$ in.

Step 2. The axis of interest is a-a, coincident with the neutral axis for this example.

Step 3. The partial area A_p is shown crosshatched in the figure to be the upper half of the rectangle. Because the partial area is itself a simple rectangle, steps 4–7 are used to compute Q.

Step 4. The partial area is

$$A_p = \frac{h}{2}t = (5.0 \text{ in.})(2.0 \text{ in.}) = 10.0 \text{ in}^2$$

Step 5. The centroid of the partial area is at its midheight, 2.5 in. above a-a.

Step 6. Because the neutral axis is coincident with the axis a-a, $\bar{y} = 2.5$ in.

Step 7. Now Q can be computed.

$$Q = A_p\bar{y} = (10.0 \text{ in}^2)(2.5 \text{ in.}) = 25.0 \text{ in}^3$$

(b) For the I-shape in Figure 9-6(b):

Step 1. The I-shape is symmetrical and, therefore, the neutral axis lies at half the height from its base, 5.0 in.

Step 2. The axis of interest is a-a, coincident with the neutral axis for this example.

Step 3. The partial area A_p is shown crosshatched in the figure to be the upper half of the I-shape. Because the partial area is in the form of a "T," steps 8–11 are used to compute Q.

Step 8. The T-shape is divided into two parts: the upper half of the vertical web is part 1 and the entire top flange is part 2. The magnitudes of these areas are

$$A_1 = \left(\frac{h}{2} - t_f\right)(t_w) = (5.0 \text{ in.} - 2.0 \text{ in.})(1.5 \text{ in.}) = 4.5 \text{ in}^2$$

$$A_2 = bt_f = (6.0 \text{ in.})(2.0 \text{ in.}) = 12.0 \text{ in}^2$$

Step 9.   Each part is a rectangle for which the centroid is at its midheight, as shown in the figure.

Step 10.   The required distances are

$$y_1 = \tfrac{1}{2}\left(\frac{h}{2} - t_f\right) = \tfrac{1}{2}(5.0 \text{ in.} - 2.0 \text{ in.}) = 1.5 \text{ in.}$$

$$y_2 = \left(\frac{h}{2} - \frac{t_f}{2}\right) = (5.0 \text{ in.} - 1.0 \text{ in.}) = 4.0 \text{ in.}$$

Step 11.   Using Equation (9-3) gives us

$$Q = A_1y_1 + A_2y_2$$
$$= (4.5 \text{ in}^2)(1.5 \text{ in.}) + (12.0 \text{ in}^2)(4.0 \text{ in.})$$
$$= 54.75 \text{ in}^3$$

(c) For the T-shape in Figure 9-6(c):

Step 1.   Locate the centroid.

$$\bar{Y} = \frac{A_wy_w + A_fy_f}{A_w + A_f}$$

where the subscript w refers to the vertical web and the subscript f refers to the top flange. Then

$$\bar{Y} = \frac{(12)(4) + (16)(9)}{12 + 16} = 6.86 \text{ in.}$$

Step 2.   The axis of interest, a-a, is at the very top of the web, just below the flange.

Step 3.   The partial area above a-a is the entire flange.

Step 4. $A_p = (8 \text{ in.})(2 \text{ in.}) = 16 \text{ in}^2$

Step 5.   The centroid of A_p is 1.0 in. down from the top of the flange, which is 9.0 in. above the base of the tee.

Step 6. $\bar{y} = 9.0 \text{ in.} - \bar{Y} = 9.0 \text{ in.} - 6.86 \text{ in.} = 2.14 \text{ in.}$

Step 7. $Q = A_p\bar{y} = (16 \text{ in}^2)(2.14 \text{ in.}) = 34.2 \text{ in}^3$

This completes the example problem.

9-5 USE OF THE GENERAL SHEAR FORMULA

Example problems are presented here to illustrate the use of the general shear formula [Equation (9-1)] to compute the vertical shear stress in a beam. The following procedure is typical of that used in solving such problems.

The overall objective is to compute the shear stress at any specified position on the beam at any specified axis within the cross section using the general shear formula,

$$\tau = \frac{VQ}{It} \tag{9-1}$$

1. Determine the vertical shearing force V at the section of interest. This may require preparation of the complete shearing force diagram using the procedures of Chapter 6.
2. Locate the centroid of the entire cross section and draw the neutral axis through the centroid.
3. Compute the moment of inertia of the section with respect to the neutral axis.
4. Identify the axis for which the shear stress is to be computed and determine the thickness t at that axis. Include all parts of the section that are cut by the axis of interest when computing t.
5. Compute the statical moment of the partial area away from the axis of interest with respect to the neutral axis. Use the procedure developed in Section 9-4.
6. Compute the shear stress using Equation (9-1).

Example Problem 9-2

A wood joist, used as a part of a floor system, carries a uniformly distributed load of 100 lb/ft over its entire 14 ft length. It is proposed that a standard 2 × 8 wood beam made from No. 2 grade southern pine be used for the joist. Compute the maximum shear stress in the beam and compare it with the allowable stress for the wood as listed in Appendix A-18. Appendix A-4 gives the dimensions for the beam section.

Solution The procedure outlined above will be used to solve for the shear stress. But it is not yet known where the maximum shear stress occurs on the cross section of the beam. To verify where this occurs and to help develop an important concept, the shear stress will be computed at several axes within the cross section. For the first calculation, assume that the axis of interest is coincident with the neutral axis of the beam. Now proceed with the shear stress analysis.

The remainder of the problem solution is done in the programmed format in which each step is evaluated within a panel enclosed by bars running across the page. You should try to develop the solution to each step yourself before reading the given solution in the following panel. To do step 1, we have to decide at what section along the beam the maximum shearing force will occur. The complete shearing force diagram can be used to determine this. Draw the diagram now.

Step 1. The complete shearing force diagram is shown in Figure 9-7. The total load on the beam is 1400 lb and because of the symmetrical load pattern, each reaction force at the supports is 700 lb. This is the maximum shearing force on the beam. Now do step 2.

Step 2. The cross section of the beam is a simple rectangle and the neutral axis is at its midheight, as shown in Figure 9-6(a). Now do step 3 using the $h = 7.25$ in. and $t = 1.50$ in.

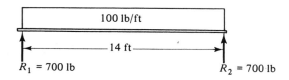

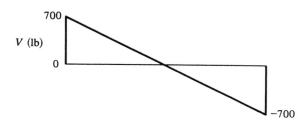

Figure 9-7 Reactions and shearing force diagram.

Step 3. The moment of inertia with respect to the neutral axis is 47.6 in⁴. It could be read from Appendix A-4 or computed from

$$I = \frac{th^3}{12} = \frac{(1.5)(7.25)^3}{12} = 47.6 \text{ in}^4$$

Now complete step 4.

Step 4. The thickness at the axis of interest, and at any axis, is 1.5 in. Now complete step 5.

Step 5. The partial area used for the calculation of the statical moment, Q, is the upper half of the rectangular cross section.

$$A_p = \left(\frac{h}{2}\right)(t) = (3.625)(1.5) = 5.438 \text{ in}^2$$

$$\bar{y} = \frac{3.625}{2} = 1.813 \text{ in.}$$

Then

$$Q = A_p\bar{y} = (5.438)(1.813) = 9.86 \text{ in}^3$$

We can now complete step 6, the final calculation of the shear stress.

Step 6. You should have $\tau = 96.7$ psi, found from Equation (9-1).

$$\tau = \frac{VQ}{It} = \frac{(700 \text{ lb})(9.86 \text{ in}^3)}{(47.6 \text{ in}^4)(1.5 \text{ in.})} = 96.7 \text{ psi}$$

In order to decide what the actual maximum shear stress is on this cross section, let's select a series of axes above the neutral axis that was just used and compute the shear stress at those axes. Then we can plot a graph of shear stress versus position on the cross section. The choice of axes is arbitrary and let's use three axes 1.0 in. apart and a fourth axis at the top of the section, as shown in Figure 9-8.

Sec. 9-5   Use of the General Shear Formula

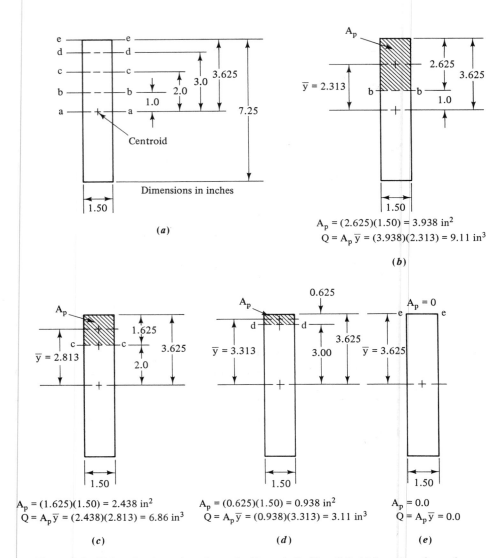

Figure 9-8 Values for Q at selected axes for Example Problem 9-2. (a) Location of axes for computing shear stress (b) Q for axis $b - b$ (c) Q for axis $c - c$ (d) Q for axis $d - d$ (e) Q for axis $e - e$.

Note that when we compute the shear stress at each axis using Equation (9-1), the values of V, I, and t are all the same as those used for the neutral axis. The only quantity that will change for the other axes is the statical moment Q. Compute the value of Q for each of the four additional axes now.

You should have,

$$Q_b = 9.11 \text{ in}^3$$

$$Q_c = 6.86 \text{ in}^3$$

$$Q_d = 3.11 \text{ in}^3$$

$$Q_e = 0.00 \text{ in}^3$$

The data are shown in Figure 9-8. Using these values for Q, now compute the shear stress at each axis.

Including axis a-a done before, you should have these values for shear stress, all computed using Equation (9-1):

$$\tau_a = 96.7 \text{ psi}$$

$$\tau_b = 89.3 \text{ psi}$$

$$\tau_c = 67.3 \text{ psi}$$

$$\tau_d = 30.5 \text{ psi}$$

$$\tau_e = 0.0 \text{ psi}$$

Figure 9-9 is a plot of shear stress versus position on the cross section. The points below the neutral axis are identical to those above the axis because of the symmetry of the rectangular section.

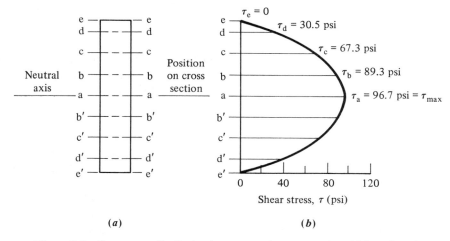

Figure 9-9 Shear stress distribution in a rectangular cross section. (a) Location of axes (b) Plot of τ vs position.

We can see from Figure 9-9 that the maximum shear stress occurs at the neutral axis, axis a-a, and has the value of 96.7 psi. Refer to Appendix A-18 to see if this is an acceptable level of stress for the No. 2 southern pine wood.

It is not. Appendix A-18 lists the allowable horizontal shear stress as 70 psi, lower than the actual maximum shear stress. It should be noted that the beam sizes listed in Appendix A-18 are somewhat larger than the dimensions of the 2 × 8 beam section used in this problem and that the allowable stress may be different from 70 psi. However, for the problems in this book, we rely on the data in the tables.

This example problem is completed. We will refer back to this problem in Section 9-6, where we discuss how to determine where the maximum shear stress occurs.

9-6 DISTRIBUTION OF SHEARING STRESS IN BEAMS

Most applications require that the maximum shearing stress be determined to evaluate the acceptability of the stress relative to some criterion of design. Example Problem 9-2 in the preceding section accomplished this by the rather time-consuming method of computing the stress at several sections and observing which was the greatest. Note that the greatest shearing stress occurred at the neutral axis.

It would have been desirable to know at the beginning where the maximum shearing stress occurs. For most sections used for beams, the maximum shearing stress occurs at the neutral axis, coincident with the centroidal axis, about which bending occurs. The following rule can be used to decide when to apply this observation.

Provided that the thickness at the centroidal axis is not greater than at some other axis, the maximum shearing stress in the cross section of a beam occurs at the centroidal axis.

The result for Example Problem 9-2, shown in Figure 9-9, illustrates this principle. The thickness of the rectangular cross section is the same at any axis. Thus the computation of the shearing stress only at the centroidal axis would have produced the maximum shearing stress in the section, making the computations at other axes unnecessary.

The logic behind this rule can be seen by examining Equation (9-1), the general shear formula. To compute the shearing stress at any axis, the values of the shearing force V and the moment of inertia I are the same. The smallest thickness t would tend to produce the largest shearing stress as implied in the statement of the rule. But the value of the statical moment Q also varies at different axes, decreasing as the axis of interest moves toward the outside of the section. Recall that Q is the product of the partial area A_p and the distance $\bar{y}$ to the centroid of A_p. For axes away from the centroidal axis, the area decreases at a faster rate than $\bar{y}$ increases, resulting in the value of Q decreasing. Thus the maximum value of Q will be that for stress computed at the centroidal axis. It follows that the maximum shearing stress will always occur at the centroidal axis *unless the thickness at some other axis is smaller than that at the centroidal axis*.

Figure 9-10(a) shows a triangular cross section for which the centroidal axis, marked *c-c*, is one-third of the height from the bottom. Obviously, the thickness there is not the smallest. Any axis above *c-c* would have a smaller thickness. Thus the rule stated above *does not apply* and the maximum shearing stress *may* occur at some other axis. Example Problem 9-3 demonstrates this using the method of computing the stress

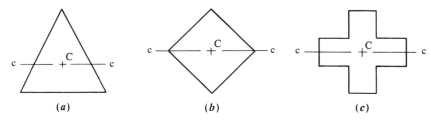

(a)   (b)   (c)

Figure 9-10 Beam cross sections for which the maximum shearing stress may not occur at the centroidal axis.

at several axes and drawing a graph of stress versus position within the section. Figure 9-10(b) and (c) shows two other shapes for which the rule would *not apply*.

Example Problem 9-3

For the triangular beam cross section shown in Figure 9-11, compute the shear stress which occurs at the axes *a* through *g*, each 50 mm apart. Plot the variation of stress with position on the section. The shearing force is 50 kN.

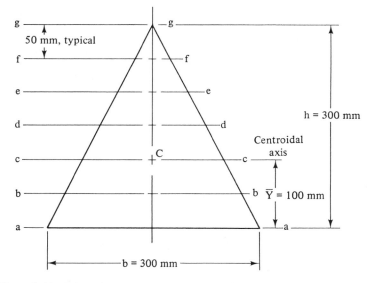

Figure 9-11 Triangular cross section for a beam for which the maximum shearing stress does not occur at the centroidal axis.

Solution   In the general shear formula, the values of V and I will be the same for all computations. V is given to be 50 kN and

$$I = \frac{bh^3}{36} = \frac{(300)(300)^3}{36} = 225 \times 10^6 \text{ mm}^4$$

Table 9-1 shows the remaining computations. Obviously, the value for Q for axes *a-a* and *g-g* is zero because the area outside each axis is zero. Note that because of the unique shape of the given triangle, the thickness t at any axis is equal to the height of the triangle above the axis.

TABLE 9-1

Axis	A_p (mm^2)	$\bar{y}$ (mm)	$Q = A_p\bar{y}$ (mm^3)	t (mm)	τ (MPa)
a-a	0	100	0	300	0
b-b	13 750	75.8	1.042×10^6	250	0.92
c-c	20 000	66.7	1.333×10^6	200	1.48
d-d	11 250	100.0	1.125×10^6	150	1.67
e-e	5 000	133.3	0.667×10^6	100	1.48
f-f	1 250	166.7	0.208×10^6	50	0.92
g-g	0	200	0	0	0

Figure 9-12 shows a plot of these stresses. The maximum shear stress occurs at half the height of the section, and the stress at the centroid (at $h/3$) is lower. This illustrates the general statement made earlier that for sections whose minimum thickness does not occur at the centroidal axis, the maximum shear stress may occur at some axis other than the centroidal axis.

One further note can be made about the computations shown for the triangular section. For the axis b-b, the partial area A_p was taken as that area *below* b-b. The resulting section is the trapezoid between b-b and the bottom of the beam. For all other axes, the partial area A_p was taken as the triangular area *above* the axis. The area below the axis could have been used, but the computations would have been more difficult. When computing Q, it does not matter whether the area above or below the axis of interest is used for computing A_p and $\bar{y}$. This completes the example problem.

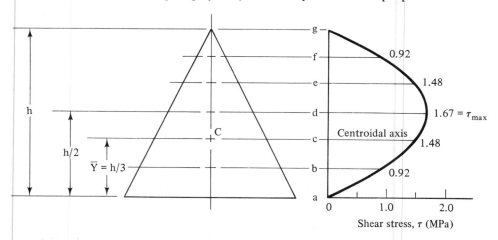

Figure 9-12 Shearing stress distribution in the triangular cross section for Problem 9-3.

Summary of observations about the distribution of shearing stress in the cross section of a beam. By reviewing the results of the several example problems worked thus far in this chapter, the following conclusions can be drawn:

1. The shearing stress at the outside of the section away from the centroidal axis is zero.

Shearing Stresses in Beams   Chap. 9

2. The maximum shearing stress in the cross section occurs at the centroidal axis provided that the thickness there is no greater than at some other axis.

3. Within a part of the cross section where the thickness is constant, the shearing stress varies in a curved fashion, decreasing as the distance from the centroidal axis increases. The curve is actually a part of a parabola.

4. At an axis where the thickness changes abruptly, as where the web of a tee or an I-shape joins the flange, the shearing stress also changes abruptly, being much smaller in the flange than in the thinner web.

9-7 DEVELOPMENT OF THE GENERAL SHEAR FORMULA

This section presents the background information on the general shear formula. Figure 9-13 shows a beam carrying two transverse loads and the corresponding shearing force and bending moment diagrams.

The *moment-area* principle of beam diagrams states that *the change in bending moment between two points on a beam is equal to the area under the shear curve between those two points*. For example, consider two points in segment *A-B* of the beam in Figure 9-13, marked x_1 and x_2, a small distance dx apart. The moment at x_1

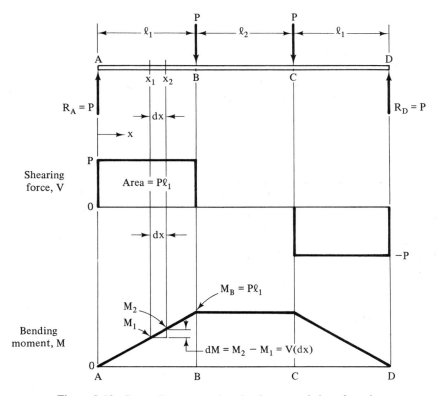

Figure 9-13 Beam diagrams used to develop general shear formula.

is M_1 and the moment at x_2 is M_2. Then the moment-area rule states that

$$M_2 - M_1 = V(dx) = dM$$

This can also be stated,

$$V = \frac{dM}{dx} \tag{9-4}$$

That is, the differential change in bending moment for a differential change in position on the beam is equal to the shearing force occurring at that position.

Equation (9-4) can also be developed by looking at a free-body diagram of the small segment of the beam between x_1 and x_2 as shown in Figure 9-14(a). As this is a cut section from the beam, the internal shearing forces and bending moments are shown acting on the cut faces. Because the beam itself is in equilibrium, this segment is also. Then the sum of moments about a point in the left face at O must be zero. This gives

$$\sum M_O = 0 = M_1 - M_2 + V(dx) = -dM + V(dx)$$

Or, as shown before,

$$V = \frac{dM}{dx}$$

Any *part* of the beam segment in Fig. 9-14(a) must also be in equilibrium. The shaded portion isolated in Figure 9-14(b) is acted on by forces parallel to the axis of the beam. On the left side, F_1 is due to the bending stress acting at that section on the area. On the right side, F_2 is due to the bending stress acting at that section on the area.

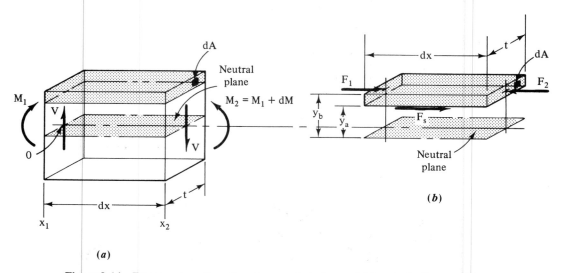

Figure 9-14 Forces on a portion of a cut segment of a beam. (a) Free body diagram of beam segment (b) Isolated portion of segment.

In general, the values of F_1 and F_2 will be different and there must be a third force acting on the bottom face of the shaded portion of the segment to maintain equilibrium. This is the shearing force, F_s, which causes the shearing stress in the beam. Figure 9-14(b) shows F_s acting on the area $t(dx)$. Then the shearing stress is

$$\tau = \frac{F_s}{t(dx)} \tag{9-5}$$

By summing forces in the horizontal direction, we find

$$F_s = F_2 - F_1 \tag{9-6}$$

We will now develop the equations for the forces F_1 and F_2. Each force is the product of the bending stress times the area over which it acts. But the bending stress varies with position in the cross section. From the flexure formula, the bending stress at any position y relative to the neutral axis is

$$\sigma = \frac{My}{I}$$

Then the total force acting on the shaded area of the left face of the beam segment is

$$F_1 = \int_A \sigma \, dA = \int_{y_a}^{y_b} \frac{M_1 y}{I} \, dA \tag{9-7}$$

where dA is a small area within the shaded area. The values of M_1 and I are constant and can be taken outside the integral sign. Equation (9-7) then becomes

$$F_1 = \frac{M_1}{I} \int_{y_a}^{y_b} y \, dA \tag{9-8}$$

Now the last part of Equation (9-8) corresponds to the definition of the centroid of the shaded area. That is,

$$\int_{y_a}^{y_b} y \, dA = \overline{y} A_p \tag{9-9}$$

where A_p is the area of the shaded portion of the left face of the segment and $\overline{y}$ is the distance from the neutral axis to the centroid of A_p. This product of $\overline{y} A_p$ is called the *statical moment* Q in the general shear formula. Making this substitution in Equation (9-8) gives

$$F_1 = \frac{M_1}{I} \int_{y_a}^{y_b} y \, dA = \frac{M_1}{I} \overline{y} A_p = \frac{M_1 Q}{I} \tag{9-10}$$

Similar reasoning can be used to develop the relationship for the force F_2 on the right face of the segment.

$$F_2 = \frac{M_2 Q}{I} \tag{9-11}$$

Substitutions can now be made for F_1 and F_2 in Equation (9-6) to complete the development of the shearing force.

$$F_s = F_2 - F_1 = \frac{M_2 Q}{I} - \frac{M_1 Q}{I} = \frac{Q}{I}(M_2 - M_1) \qquad (9\text{-}12)$$

Earlier we defined $(M_2 - M_1) = dM$. Then

$$F_s = \frac{Q(dM)}{I} \qquad (9\text{-}13)$$

Then, in Equation (9-5),

$$\tau = \frac{F_s}{t(dx)} = \frac{Q(dM)}{It(dx)}$$

But, from Equation (9-4), $V = dM/dx$. Then

$$\tau = \frac{VQ}{It}$$

This is the form of the general shear formula [Equation (9-1)] used in this chapter.

9-8 SPECIAL SHEAR FORMULAS

As demonstrated in several example problems, the general shear formula can be used to compute the shearing stress at any axis on any cross section of the beam. However, frequently it is desired to know only the *maximum shearing stress*. For many common shapes used for beams, it is possible to develop special simplified formulas that will give the maximum shearing stress quickly. The rectangle, circle, thin-walled hollow tube, and thin-webbed shapes can be analyzed this way. The formulas are developed in this section.

For all of these section shapes, the maximum shearing stress occurs at the neutral axis. The rectangle and the thin-webbed shapes conform to the rule stated in Section 9-6 because the thickness at the neutral axis is no greater than at other axes in the section. The circle and the thin-walled tube do not conform to the rule. However, it can be shown that the ratio Q/t in the general shear formula decreases continuously as the axis of interest moves away from the neutral axis, resulting in the decrease in the shearing stress.

Rectangular shape. Figure 9-15 shows a typical rectangular cross section having a thickness t and a height h. The three geometrical terms in the general shear formula can be expressed in terms of t and h.

$$I = \frac{th^3}{12}$$

$$t = t$$

$$Q = A_p \overline{y} \qquad \text{(for area above centroidal axis)}$$

$$Q = \frac{th}{2} \cdot \frac{h}{4} = \frac{th^2}{8}$$

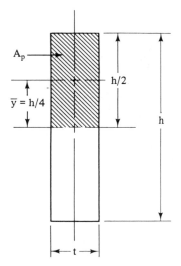

Figure 9-15 Rectangular shape.

Putting these terms in the general shear formula gives

$$\tau = \frac{VQ}{It} = V \cdot \frac{th^2}{8} \cdot \frac{12}{th^3} \cdot \frac{1}{t} = \frac{3}{2} \frac{V}{th}$$

But since *th* is the total area of the section,

$$\tau = \frac{3V}{2A} \qquad (9\text{-}14)$$

Equation (9-14) can be used to compute exactly the maximum shear stress in a rectangular beam at its centroidal axis.

Note that $\tau = V/A$ represents the *average* shearing stress on the section. Thus the maximum shearing stress on a rectangular cross section is 1.5 times higher than the average.

Example Problem 9-4

Compute the maximum shear stress which would occur in the rectangular cross section of a beam like that shown in Figure 9-15. The shearing force is 1000 lb, $t = 2.0$ in., and $h = 8.0$ in.

Solution   Using Equation (9-14) yields

$$\tau = \frac{3V}{2A} = \frac{3(1000 \text{ lb})}{2(2 \text{ in.})(8 \text{ in.})} = 93.8 \text{ psi}$$

Circular shape.   The special shear formula for the circular shape is developed in a similar manner to that used for the rectangular shape. Equations for Q, I, and t are written in terms of the primary size variable for the circular shape, its diameter. Then the general shear formula is simplified (refer to Figure 9-16).

$$t = D$$

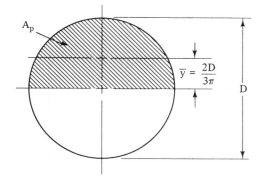

Figure 9-16 Circular shape.

$$I = \frac{\pi D^4}{64}$$

$$Q = A_p \bar{y} \quad \text{(for the semicircle above the centroid)}$$

$$Q = \frac{\pi D^2}{8} \cdot \frac{2D}{3\pi} = \frac{D^3}{12}$$

Then the maximum shear stress is

$$\tau = \frac{VQ}{It} = V \cdot \frac{D^3}{12} \cdot \frac{64}{\pi D^4} \cdot \frac{1}{D} = \frac{64V}{12\pi D^2}$$

To refine the equation, factor out a 4 from the numerator and then note that the total area of the circular section is $A = \pi D^2/4$.

$$\tau = \frac{16(4)V}{12\pi D^2} = \frac{16V}{12A}$$

$$\tau = \frac{4V}{3A} \tag{9-15}$$

This shows that the maximum shearing stress is 1.33 times higher than the average on the circular section.

Example Problem 9-5

Compute the maximum shear stress that would occur in a circular shaft, 50 mm in diameter if it is subjected to a vertical shearing force of 110 kN.

Solution Equation (9-15) will give the maximum shear stress at the horizontal diameter of the shaft.

$$\tau = \frac{4V}{3A}$$

But

$$A = \frac{\pi D^2}{4} = \frac{\pi (50 \text{ mm})^2}{4} = 1963 \text{ mm}^2$$

Then

$$\tau = \frac{4(110 \times 10^3 \text{ N})}{3(1963 \text{ mm}^2)} = 74.7 \text{ MPa}$$

Hollow thin-walled tubular shape. Removing material from the center of a circular cross section tends to increase the local value of the shearing stress, especially near the diameter where the maximum shearing stress occurs. Although not giving a complete development here, it is observed that the maximum shearing stress in a thin-walled tube is approximately twice the average. That is,

$$\tau_{\text{max}} \approx 2\frac{V}{A} \qquad (9\text{-}16)$$

where A is the total cross-sectional area of the tube.

Example Problem 9-6

Compute the approximate maximum shearing stress that would occur in a 3-in. schedule 40 steel pipe if it is used as a beam and subjected to a shearing force of 6200 lb.

Solution Equation (9-16) should be used. From Appendix A-12 we find that the cross-sectional area of the 3-in. schedule 40 steel pipe is 2.228 in². Then an estimate of the maximum shearing stress in the pipe, occurring near the horizontal diameter, is

$$\tau_{\text{max}} \approx 2\frac{V}{A} = \frac{2(6200 \text{ lb})}{2.228 \text{ in}^2} = 5566 \text{ psi}$$

Thin-webbed shapes. Structural shapes such as W- and S-beams have relatively thin webs. The distribution of shear stress in such beams is typically like that shown in Figure 9-17. The maximum shear stress is at the centroidal axis. It decreases slightly in the rest of the web and then drastically in the flanges. Thus most of the resistance to the vertical shearing force is provided by the web. Also, the average shear stress in the web would be just slightly smaller than the maximum shear stress. For these reasons, the *web shear formula* is often used to get a quick estimate of the shear stress in thin-webbed shapes.

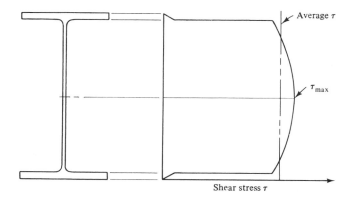

Average τ

τ_{max}

Shear stress τ

Figure 9-17 Distribution of shearing stress in a thin-webbed shape.

$$\tau \approx \frac{V}{A_{web}} = \frac{V}{th} \tag{9-17}$$

The thickness of the web is t. The simplest approach would be to use the full height of the beam for h. This would result in a shear stress approximately 15% lower than the actual maximum shear stress at the centroidal axis for typical beam shapes. Using just the web height between the flanges would produce a closer approximation of the maximum shear stress, probably less than 10% lower than the actual value. In problems using the web shear formula, we use the full height of the cross section unless otherwise stated.

In summary, for thin-webbed shapes, compute the shearing stress from the web shear formula using the full height of the beam for h and the actual thickness of the web for t. Then, to obtain a more accurate estimate of the maximum shearing stress, increase this value by about 15%.

Example Problem 9-7

Using the web shear formula, compute the shear stress in a W12 × 16 beam if it is subjected to a shearing force of 25 000 lb.

Solution In Appendix A-7 for W-beams, it is found that the web thickness is 0.220 in. and the overall depth (height) of the beam is 11.99 in. Then, using Equation (9-17), we have

$$\tau = \frac{V}{th} = \frac{25\ 000\ \text{lb}}{(0.220\ \text{in.})(11.99\ \text{in.})} = 9478\ \text{psi}$$

9-9 DESIGN SHEAR STRESS

As a general guide, the same design shear stresses that were used in Chapter 3 can be used for shearing stresses in beams. Usually, the design stress is based conservatively on the yield strength of the material in shear. That is,

$$\tau_d = \frac{s_{ys}}{N} = \frac{0.5s_y}{N} \tag{9-18}$$

The design factor can be selected from Appendix A-20.

For shear stress in the webs of rolled steel beam shapes, the AISC recommends

$$\tau_d = 0.40s_y \tag{9-19}$$

but exceptions are cited in Reference 2, Chapter 3.

9-10 SHEAR FLOW

Built-up sections used for beams, such as those shown in Figures 9-18 and 9-19, must be analyzed to determine the proper size and spacing of fasteners. The discussion in preceding sections showed that horizontal shearing forces exist at the planes joined by the nails, bolts, and rivets. Thus the fasteners are subjected to shear. Usually, the size

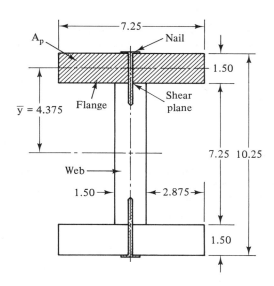

Dimensions in inches

Dimensions in mm

Figure 9-18 Beam shape for Example Problem 9-8.

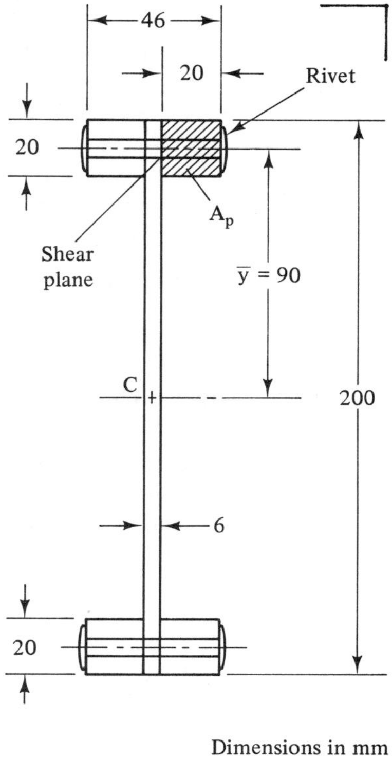

Figure 9-19 Built-up beam shape for Example Problem 9-9.

and material of the fastener will permit the specification of an allowable shearing force on each. Then the beam must be analyzed to determine a suitable spacing for the fasteners which will ensure that all parts of the beam act together.

The term *shear flow* is useful for analyzing built-up sections. Called q, the shear flow is found by multiplying the shear stress at a section by the thickness at that section. That is,

$$q = \tau t \tag{9-20}$$

But from the general shear formula,

$$\tau = \frac{VQ}{It}$$

Then

$$q = \tau t = \frac{VQ}{I} \tag{9-21}$$

The units for q are *force per unit length,* such as N/m, N/mm, or lb/in. The shear flow is a measure of how much shearing force must be resisted at a particular section per unit length. Knowing the shearing force capacity of a fastener then allows the determination of a safe spacing for the fasteners.

For example, if a particular style of nail can safely withstand 150 lb of shearing force, we will define

$$F_{sd} = 150 \text{ lb}$$

Then, if in a particular place on a beam fabricated by nailing boards together, the shear flow is computed to be $q = 28.5$ lb/in., the required minimum spacing, s, of the nail is

$$s = \frac{F_{sd}}{q} = \frac{150 \text{ lb}}{28.5 \text{ lb/in.}} = 5.26 \text{ in.} \tag{9-22}$$

Example Problem 9-8

Determine the proper spacing of the nails used to secure the flange boards to the web of the built-up I-beam shown in Figure 9-18. All boards are standard 2 × 8 wood shapes. The nails to be used can safely resist 250 lb of shearing force each. The load on the beam is shown in Figure 9-13 with $P = 500$ lb.

Solution   The shear flow must be calculated.

$$q = \frac{VQ}{I}$$

At the place where the nails join the boards, Q is evaluated for the area of the top (or bottom) flange board.

$$Q = A_p \bar{y} = (1.5 \text{ in.})(7.25 \text{ in.})(4.375 \text{ in.}) = 47.6 \text{ in}^3$$

The moment of inertia can be computed by subtracting the two open-space rectangles at the sides of the web from the full rectangle surrounding the I-shape.

$$I = \frac{7.25(10.25)^3}{12} - \frac{2(2.875)(7.25)^3}{12} = 468.0 \text{ in}^4$$

The maximum shearing force on the beam is 500 lb, occurring between each support and the applied loads. Then the shear flow is

$$q = \frac{VQ}{I} = \frac{(500 \text{ lb})(47.6 \text{ in}^3)}{468 \text{ in}^4} = 50.9 \text{ lb/in.}$$

This means that 50.9 lb of force must be resisted along each inch of length of the beam at the point between the flange and the web boards. Since each nail can withstand 250 lb, the required spacing is

$$s = \frac{F_{sd}}{q} = \frac{250 \text{ lb}}{50.9 \text{ lb/in.}} = 4.92 \text{ in.}$$

A spacing of 4.5 in. would be reasonable.

The principle of shear flow also applies for sections like that shown in Figure 9-19, in which a beam section is fabricated by riveting square bars to a vertical web plate to form an I-shape. The shear flow occurs from the web plate to the flange bars.

Thus, when evaluating the statical moment Q, the partial area, A_p, is taken to be the area of one of the flange bars.

Example Problem 9-9

A fabricated beam is made by riveting square aluminum bars to a vertical plate, as shown in Figure 9-19. The bars are 20 mm square. The plate is 6 mm thick and 200 mm high. The rivets can withstand 800 N of shear force across one cross section. Determine the required spacing of the rivets if a shearing force of 5 kN is applied.

Solution In the shear flow equation,

$$q = \frac{VQ}{I}$$

I is the moment of inertia of the entire cross section, and Q is the product of $A_p\bar{y}$ for the area *outside* of the section where the shear is to be calculated. In this case, the partial area A_p is the 20-mm square area *to the side* of the web. For the beam in Figure 9-19,

$$I = \frac{6(200)^3}{12} + 4\left[\frac{20^4}{12} + (20)(20)(90)^2\right]$$

$$= 17.0 \times 10^6 \text{ mm}^4$$

$$Q = A_p\bar{y} \qquad \text{(for one square bar)}$$

$$= (20)(20)(90) \text{ mm}^3 = 36\ 000 \text{ mm}^3$$

Then for $V = 5$ kN,

$$q = \frac{VQ}{I} = \frac{(5 \times 10^3 \text{ N})(36 \times 10^3 \text{ mm}^3)}{17.0 \times 10^6 \text{ mm}^4} = 10.6 \text{ N/mm}$$

Thus a shear force of 10.6 N is to be resisted for each millimeter of length of the beam. Since each rivet can withstand 800 N of shear force, the required spacing is

$$s = \frac{F_{sd}}{q} = \frac{800 \text{ N}}{10.6 \text{ N/mm}} = 75.5 \text{ mm}$$

PROBLEMS

For Problems 9-1 through 9-20, compute the shear stress at the horizontal neutral axis for a beam having the cross-sectional shape shown in the given figure and for the given shearing force.

9-1. Use a rectangular shape having a width of 50 mm and a height of 200 mm. $V = 7500$ N.

9-2. Use a rectangular shape having a width of 38 mm and a height of 180 mm. $V = 5000$ N.

9-3. Use a rectangular shape having a width of 1.5 in. and a height of 7.25 in. $V = 12\ 500$ lb.

9-4. Use a rectangular shape having a width of 3.5 in. and a height of 11.25 in. $V = 20\ 000$ lb.

9-5. Use a circular shape having a diameter of 50 mm. $V = 4500$ N.

9-6. Use a circular shape having a diameter of 38 mm. $V = 2500$ N.

9-7. Use a circular shape having a diameter of 2.00 in. $V = 7500$ lb.

9-8. Use a circular shape having a diameter of 0.63 in. $V = 850$ lb.

9-9. Use the shape shown in Figure 7-14(a). $V = 1500$ lb.

9-10. Use the shape shown in Figure 7-14(c). $V = 850$ lb.

9-11. Use the shape shown in Figure 7-14(d). $V = 850$ lb.

9-12. Use the shape shown in Figure 7-15(a). $V = 112$ kN.

9-13. Use the shape shown in Figure 7-15(b). $V = 71.2$ kN.

9-14. Use the shape shown in Figure 7-15(c). $V = 1780$ N.

9-15. Use the shape shown in Figure 7-15(d). $V = 675$ N.

9-16. Use the shape shown in Figure 7-16(a). $V = 2.5$ kN.

9-17. Use the shape shown in Figure 7-16(c). $V = 10.5$ kN.

9-18. Use the shape shown in Figure 7-17(a). $V = 1200$ lb.

9-19. Use the shape shown in Figure 7-17(d). $V = 775$ lb.

9-20. Use the shape shown in Figure 7-20(a). $V = 2500$ lb.

For Problems 9-21 through 9-30, assume that the indicated shape is the cross section of a beam made from wood having an allowable shearing stress of 70 psi, which is that of No. 2 grade southern pine listed in Appendix A-18. Compute the maximum allowable shearing force for each shape.

9-21. Use a standard 2 × 4 wooden beam with the long dimension vertical.

9-22. Use a standard 2 × 4 wooden beam with the long dimension horizontal.

9-23. Use a standard 2 × 12 wooden beam with the long dimension vertical.

9-24. Use a standard 2 × 12 wooden beam with the long dimension horizontal.

9-25. Use a standard 10 × 12 wooden beam with the long dimension vertical.

9-26. Use a standard 10 × 12 wooden beam with the long dimension horizontal.

9-27. Use the shape shown in Figure 7-21(a).

9-28. Use the shape shown in Figure 7-21(b).

9-29. Use the shape shown in Figure 7-21(c).

9-30. Use the shape shown in Figure 7-21(d).

9-31. For a beam having the I-shape cross section shown in Figure 7-14(c), compute the shearing stress on horizontal axes 0.50 in. apart from the bottom to the top. At the ends of the web where it joins the flanges, compute the stress in both the web and the flange. Use a shearing force of 500 lb. Then plot the results.

9-32. For a beam having the box-shape cross section shown in Figure 7-14(d), compute the shearing stress on horizontal axes 0.50 in. apart from the bottom to the top. At the ends of the vertical sides where they join the flanges, compute the stress in both the web and the flange. Use a shearing force of 500 lb. Then plot the results.

9-33. For a standard W14 × 43 steel beam, compute the shearing stress at the neutral axis when subjected to a shearing force of 33 500 lb. Use the general shear formula. Neglect the fillets at the intersection of the web with the flanges.

9-34. For the same conditions listed in Problem 9-33, compute the shearing stress at several axes and plot the variation of stress with position in the beam.

9-35. For a standard W14 × 43 steel beam, compute the shearing stress from the web shear formula when it carries a shearing force of 33 500 lb. Compare this value with that computed in Problem 9-33 and plot it on the graph produced for Problem 9-34.

9-36. For an Aluminum Association Standard I8 × 6.181 beam, compute the shearing stress at the neutral axis when subjected to a shearing force of 13 500 lb. Use the general shear formula. Neglect the fillets at the intersection of the web with the flanges.

9-37. For the same conditions listed in Problem 9-36, compute the shearing stress at several axes and plot the variation of stress with position in the beam.

9-38. For an aluminum I8 × 6.181 beam, compute the shearing stress from the web shear formula when the beam carries a shearing force of 13 500 lb. Compare this value with that computed in Problem 9-36 and plot it on the graph produced for Problem 9-37.

Note: In problems calling for design stresses, use the following:

For structural steel:

$$\text{In bending:} \qquad \sigma_d = 0.66s_y$$

$$\text{In shear:} \qquad \tau_d = 0.4s_y$$

For any other metal:

$$\text{In bending:} \qquad \sigma_d = \frac{s_y}{N}$$

$$\text{In shear:} \qquad \tau_d = 0.5\frac{s_y}{N}$$

9-39. The loading shown in Figure 6-36(d) is to be carried by a W10 × 15 steel beam. Compute the shearing stress using the web shear formula. Also compute the maximum bending stress. Then compare these stresses to the design stresses for ASTM A36 structural steel.

9-40. Specify a suitable wide-flange beam to be made from ASTM A36 structural steel to carry the load shown in Figure 6-36(d) based on the design stress in bending. Then, for the beam selected, compute the shearing stress from the web shear formula and compare it with the design shear stress.

9-41. Specify a suitable wide-flange beam to be made from ASTM A36 structural steel to carry the load shown in Figure 6-48(b) based on the design stress in bending. Then, for the beam selected, compute the shearing stress from the web shear formula and compare it with the design shear stress.

9-42. Specify a suitable wide-flange beam to be made from ASTM A36 structural steel to carry the load shown in Figure 6-48(d) based on the design stress in bending. Then, for the beam selected, compute the shearing stress from the web shear formula and compare it with the design shear stress.

9-43. Specify a suitable standard steel pipe from Appendix A-12 to be made from AISI 1020 hot-rolled steel to carry the load shown in Figure 6-48(a) based on the design stress in bending with a design factor of 3. Then, for the pipe selected, compute the shearing stress from the special shear formula for hollow tubes and compute the resulting design factor from the design shear stress formula.

9-44. An Aluminum Association standard channel (Appendix A-10) is to be specified to carry the load shown in Figure 6-38(a) to produce a design factor of 4 in bending. The legs of the channel are to point down. The channel is made from 6061-T6 aluminum. For the channel selected, compute the maximum shearing stress.

9-45. A wood joist in the floor of a building is to carry a uniformly distributed load of 200 lb/ft over a length of 12.0 ft. Specify a suitable standard wood beam shape for the joist, made from No. 2 grade hemlock, to be safe in both bending and shear (see Appendixes A-4 and A-18).

9-46. A wooden beam in an outdoor structure is to carry the load shown in Figure 6-48(c). If it is to be made from No. 3 grade Douglas fir, specify a suitable standard wood beam to be safe in both bending and shear (see Appendixes A-4 and A-18).

9-47. The box beam shown in Figure 7-21(b) is to be made from No. 1 grade southern pine. It is to be 14 ft. long and carry two equal concentrated loads, each 3 ft. from an end. The beam is simply supported at its ends. Specify the maximum allowable load for the beam to be safe in both bending and shear.

9-48. An aluminum I-beam, I9 × 8.361, carries the load shown in Figure 6-37(d). Compute the shear stress in the beam using the web shear formula.

9-49. Compute the bending stress for the beam in Problem 9-48.

9-50. A 2 × 8 wooden floor joist in a home is 12 ft. long and carries a uniformly distributed load of 80 lb/ft. Compute the shear stress in the joist. Would it be safe if it is made from No. 2 southern pine wood?

9-51. A steel beam is made as a rectangle, 0.50 in. wide by 4.00 in. high.
   (a) Compute the shear stress in the beam if it carries the load shown in Figure 6-38(b).
   (b) Compute the stress due to bending.
   (c) Specify a suitable steel for the beam to produce a design factor of 3 for either bending or shear.

9-52. An aluminum beam is made as a rectangle, 16 mm wide by 60 mm high.
   (a) Compute the shear stress in the beam if it carries the load shown in Figure 6-37(b).
   (b) Compute the stress due to bending.
   (c) Specify a suitable aluminum for the beam to produce a design factor of 3 for either bending or shear.

9-53. It is planned to use a rectangular bar to carry the load shown in Figure 6-47(a). Its thickness is to be 12 mm and it is to be made from aluminum 6061-T6. Determine the required height of the rectangle to produce a design factor of 4 in bending based on yield strength. Then compute the shear stress in the bar and the resulting design factor for shear.

9-54. A round shaft, 40 mm in diameter, carries the load shown in Figure 6-47(b).
   (a) Compute the maximum shear stress in the shaft.
   (b) Compute the maximum stress due to bending.
   (c) Specify a suitable steel for the shaft to produce a design factor of 4 based on yield strength in either shear or bending.

9-55. Compute the required diameter of a round bar to carry the load shown in Figure 6-47(a) while limiting the stress due to bending to 120 MPa. Then compute the resulting shear stress in the bar and compare it to the bending stress.

9-56. Compute the maximum allowable vertical shearing force on a wood dowel, 1.50 in. in diameter, if the maximum allowable shearing stress is 70 psi.

9-57. A standard steel pipe is to be selected from Appendix A-12 to be used as a chinning bar in a gym. It is to be simply supported at the ends of its 36-in. length. Men weighing up to 400 lb are expected to hang on the bar with either one or two hands at any place along its length. The pipe is to be made from AISI 1020 hot-rolled steel. Specify a suitable pipe to provide a design factor of 6 based on yield strength in either bending or shear.

9-58. A standard steel pipe is to be simply supported at its ends and carry a single concentrated load of 2800 lb at its center. The pipe is to be made from AISI 1020 hot-rolled steel. The minimum design factor is to be 4 based on yield strength for either bending or shear.

Specify a suitable pipe size from Appendix A-12 if the length of the pipe is (a) 1.5 in.; (b) 3.0 in.; (c) 4.5 in.; (d) 6.0 in.

9-59. The shape in Figure 7-17(a) is to be made by gluing the flat plate to the hat section. If the beam made from this section is subjected to a shearing force of 1200 lb, compute the shear flow at the joint. What must be the shearing strength of the glue in psi?

9-60. The shape in Figure 7-18(b) is to be made by using a metal-bonding glue between the S-beam and the web of the channel. Compute the shear flow at the joint and the required shear strength of the glue for a shearing force of 2500 lb.

9-61. The shape in Figure 7-20(a) is to be fabricated by riveting the bottom plate to the angles and then welding the top plate to the angles. When used as a beam, there are four potential failure modes: bending stress, shearing stress in the angles, shear in the welds, and shear of the rivets. The shape is to be used as the seat of a bench carrying a uniformly distributed load over a span of 10.0 ft. Compute the maximum allowable distributed load for the following design limits.

    (a) The material of all components is aluminum 6061-T4 and a design factor of 4 is required for either bending or shear.

    (b) The allowable shear flow on each weld is 1800 lb/in.

    (c) The rivets are spaced 4.0 in. apart along the entire length of the beam. Each rivet can withstand 600 lb of shear.

9-62. An alternative design for the bench described in problem 9-61 is to use the built-up wood T-shape shown in Figure 7-21(d). The wood is to be No. 3 grade southern pine. One nail is to be driven into each vertical 2 × 12. Each nail can withstand 160 lb in shear and the nails are spaced 6.0 in. apart along the length of the beam. Compute the maximum allowable distributed load on the beam.

9-63. The shape shown in Figure 7-21(a) is glued together and the allowable shearing strength of the glue is 800 psi. The components are No. 2 grade Douglas fir. If the beam is to be simply supported and carry a single concentrated load at its center, compute the maximum allowable load. The length is 10 ft.

9-64. The I-section shown in Figure 7-21(a) is fabricated from three wooden boards by nailing through the top and bottom flanges into the web. Each nail can withstand 180 lb of shearing force. If the beam having this section carries a vertical shearing force of 300 lb, what spacing would be required between nails?

9-65. The built-up section shown in Figure 7-21(b) is nailed together by driving one nail through each side of the top and bottom boards into the $1\frac{1}{2}$-in.-thick sides. If each nail can withstand 150 lb of shearing force, determine the required nail spacing when the beam carries a vertical shearing force of 600 lb.

9-66. The platform whose cross section is shown in Figure 7-21(c) is glued together. How much force per unit length of the platform must the glue withstand if it carries a vertical shearing force of 500 lb?

9-67. The built-up section shown in Figure 7-18(a) is fastened together by passing two $\frac{3}{8}$-in. rivets through the top and bottom plates into the flanges of the beam. Each rivet will withstand 2650 lb in shear. Determine the required spacing of the rivets along the length of the beam if it carries a shearing force of 175 kN.

9-68. A fabricated beam having the cross section shown in Figure 7-18(b) carries a shearing force of 50 kN. The channel is riveted to the S-beam with two $\frac{1}{4}$-in.-diameter rivets which can withstand 1750 lb each in shear. Determine the required rivet spacing.

10

The General Case of Combined
Stress and Mohr's Circle

10-1 OBJECTIVES OF THIS CHAPTER

In the earlier chapters of this book we focused on the computation of simple stresses, those cases in which only one type of stress was of interest. We studied direct stresses due to tension, compression, bearing, and shear; torsional shear stress; stress due to bending; and shearing stresses in beams. Also presented were many practical problems for which the computation of the simple stress was the appropriate method of analysis.

But a much wider array of real, practical problems involve *combined stresses*, situations in which two or more different components of stress act at the same point in a load-carrying member. In this chapter we develop the general approaches used to properly combine stresses. In Chapter 11 we develop several practical special cases involving combined stress.

After completing this chapter, you should be able to:

1. Recognize cases for which combined stresses occur.
2. Represent the stress condition on a stress element.
3. Understand the development of the equations for combined stresses, from which you can compute the following:
    a. The maximum and minimum principal stresses
    b. The orientation of the principal stress element

c. The maximum shear stress on an element

d. The orientation of the maximum shear stress element

e. The normal stress that acts along with the maximum shear stress

f. The normal and shear stress that occurs on the element oriented in any direction

4. Construct Mohr's circle for biaxial stress.

5. Interpret the information available from Mohr's circle for the stress condition at a point in any orientation.

6. Use the data from the Mohr's circle to draw the principal stress element and the maximum shear stress element.

10-2 THE STRESS ELEMENT

In general, combined stress refers to cases in which two or more types of stress are exerted on a given point at the same time. The component stresses can be either *normal* (i.e., tension or compression) or *shear* stresses.

When a load-carrying member is subjected to two or more different kinds of stresses, the first task is to compute the stress due to each component. Then a decision is made about which point within the member has the highest *combination* of stresses and the combined stress analysis is completed for that point. For some special cases, it is desired to know the stress condition at a given point regardless of whether or not it is the point of maximum stress. Examples would be near welds in a fabricated structure, along the grain of a wood member, or near a point of attachment of one member to another.

After the point of interest is identified, the stress condition at that point is determined from the classical stress analysis relationships presented in this book if possible. At times, because of the complexity of the geometry of the member or the loading pattern, it is not possible to complete a reliable stress analysis entirely by computations. In such cases experimental stress analysis can be employed in which strain gauges, photoelastic models, or strain-sensitive coatings give data experimentally. Also, with the aid of finite element stress analysis techniques, the stress condition can be determined using computer-based analysis.

Then, after using one of these methods, you should know the information required to construct the *initial stress element*, as shown in Figure 10-1. The element is assumed to be infinitesimally small and aligned with known directions on the member being analyzed. The complete element, as shown, could have a normal stress (either tensile or compressive) acting on each pair of faces in mutually perpendicular directions, usually named the x and y axes. As the name *normal stress* implies, these stresses act normal (perpendicular) to the faces. As shown, σ_x is aligned with the x-axis and is a tensile stress tending to pull the element apart. Recall that tensile stresses are considered positive. Then σ_y is shown to be compressive, tending to crush the element. Compressive stresses are considered negative.

In addition, there may be shear stresses acting along the sides of the element as if each side were being cut away from the adjacent material. Recall from earlier discussion of shearing stresses that a set of four shears, all equal in magnitude, exist

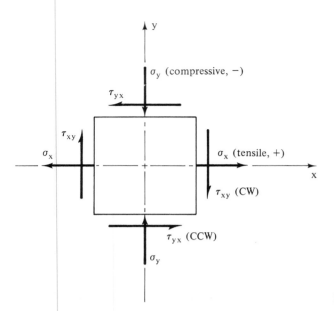

Figure 10-1 Complete stress element.

on any element in equilibrium. On any two opposite faces the shear stresses will act in opposite directions, thus creating a *couple* that tends to rotate the element. Then there must exist a pair of shear stresses on the adjacent faces producing an oppositely directed couple for the element to be in equilibrium. We will refer to each *pair* of shears using a double subscript notation. For example, τ_{xy} refers to the shear stress acting perpendicular to the x-axis and parallel to the y-axis. Conversely, τ_{yx} acts perpendicular to the y-axis and parallel to the x-axis. Rather than establish a convention for signs of the shear stresses, we will refer to them as *clockwise*, (CW) or *counterclockwise* (CCW), according to how they tend to rotate the stress element.

10-3 STRESS DISTRIBUTION CREATED BY BASIC STRESSES

Figures 10-2, 10-3, and 10-4 show direct tension, direct compression, and bending stresses in which normal stresses were developed. It is important to know the distribution of stress within the member as shown in the figures. Also shown are stress elements subjected to these kinds of stresses. Listed below are the principal formulas for computing the value of these stresses.

Direct tension: $\qquad \sigma = \dfrac{P}{A} \qquad$ (see Figure 10-2)

Direct compression: $\qquad \sigma = \dfrac{-P}{A} \qquad$ (see Figure 10-3)

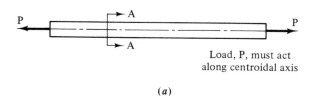

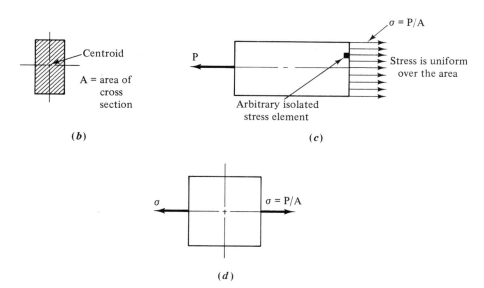

Figure 10-2 Distribution of normal stress for direct tension. (a) Load condition (b) Section *A-A* (shape is arbitrary) (c) Internal stress distribution on an isolated segment (d) Stress element showing normal tensile stress.

Stress due to bending: $\quad \sigma = \pm \dfrac{Mc}{I} \quad$ (maximum stress)
(see Figure 10-4)

$$\sigma = \pm \dfrac{My}{I} \quad \text{(stress at any point } y\text{)}$$

Figures 10-5, 10-6, and 10-7 show three cases in which shear stresses are produced along with the stress distributions and the stress elements subjected to these types of stress. Listed below are the principal formulas used to compute shearing stresses.

Direct shear: $\qquad\qquad \tau = \dfrac{P}{A_s} \quad$ (see Figure 10-5)

Torsional shear: $\qquad\qquad \tau = \dfrac{Tc}{J} \quad$ (maximum)
(see Figure 10-6)

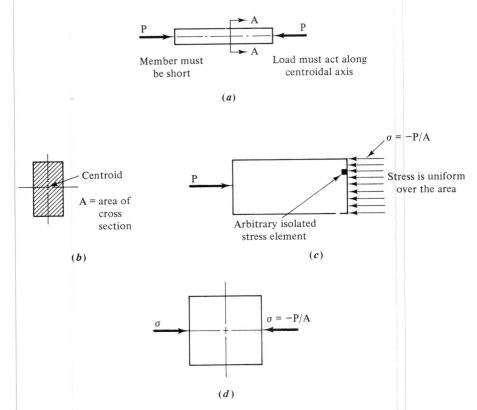

Figure 10-3 Distribution of normal stress for direct compression. (a) Load condition (b) Section A-A (shape is arbitrary) (c) Internal stress distribution on an isolated segment (d) Stress element showing normal compressive stress.

$$\tau = \frac{Tr}{J} \qquad \text{(at any radius)}$$

Shearing stresses in beams: $\qquad \tau = \dfrac{VQ}{It} \qquad$ (see Figure 10-7)

10-4 CREATING THE INITIAL STRESS ELEMENT

One major objective of this chapter is to develop relationships from which the *maximum principal stresses* and the *maximum shear stress* can be determined. Before this can be done, it is necessary to know the state of stress at a point of interest in some orientation. In this section we demonstrate the determination of the initial stress condition by direct calculation from the formulas for basic stresses.

Figure 10-8 shows an L-shaped lever attached to a rigid surface with a downward load *P* applied to its end. The shorter segment at the front of the lever is loaded as a

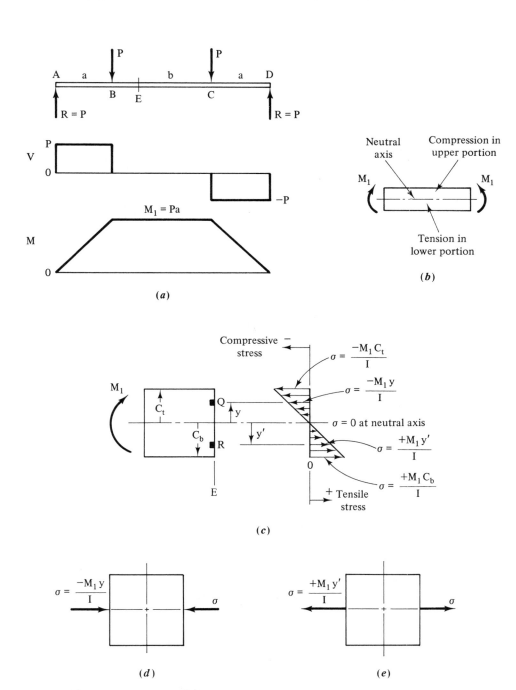

Figure 10-4 Distribution of normal stress for bending. (a) Example of a beam in positive bending (b) Isolated segment near E (c) Stress distribution at isolated section E (d) Stress element Q (e) Stress element R.

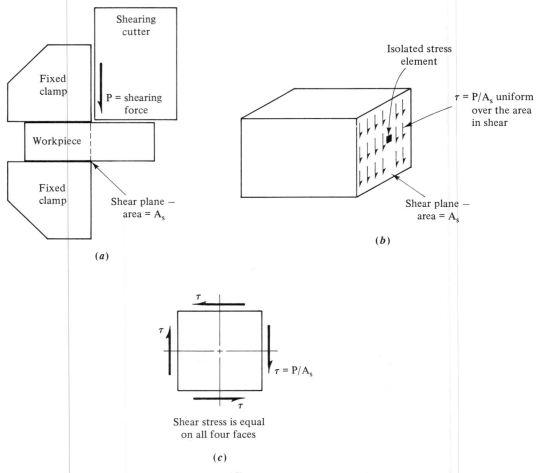

Figure 10-5 Distribution of shear stress for direct shear. (a) Example of direct shear (b) Shear stress distribution (c) Stress element showing shear stress.

cantilever beam as shown in Figure 10-9(a), with the moment at its left end resisted by the other segment of the lever.

Figure 10-9(b) shows the longer segment as a free-body diagram. At the front, the force P and the torque T are the reactions to the force and moment at the left end of the shorter segment in (a). Then at the rear, there must be reactions M, P, and T to maintain equilibrium. The bar is subjected to a combined stress condition with the following kinds of stresses:

Bending stress due to the bending moment
Torsional shear stress due to the torque
Shearing stress due to the vertical shearing force

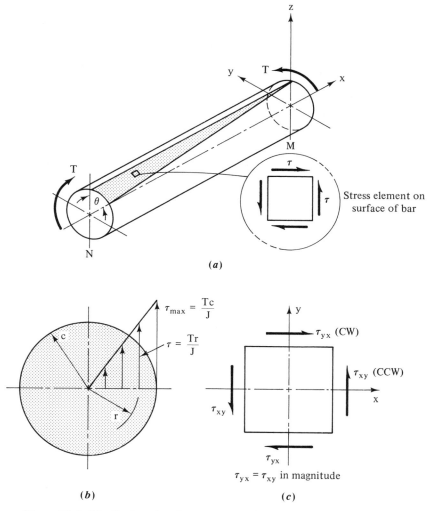

Figure 10-6 Distribution of torsional shear stress in a solid round shaft. (a) Circular bar loaded in torsion (b) Distribution of shear stress on a cross section of the bar (c) Stress element for pure torsional shear stress.

Reviewing Figures 10-4, 10-6, and 10-7, we can conclude that one of the points where the stress is likely to be the highest is on the top of the long segment near the support. Called element K in Figure 10-9(b), it would see the maximum tensile bending stress and the maximum torsional shear stress. But the vertical shear stress would be zero because it is at the outer surface away from the neutral axis.

Element K is then subjected to a combined stress, as shown in Figure 10-10. The tensile normal stress σ_x, is directed parallel to the x-axis along the top surface of the bar. The applied torque tends to produce a shear stress τ_{xy} in the negative y-direction

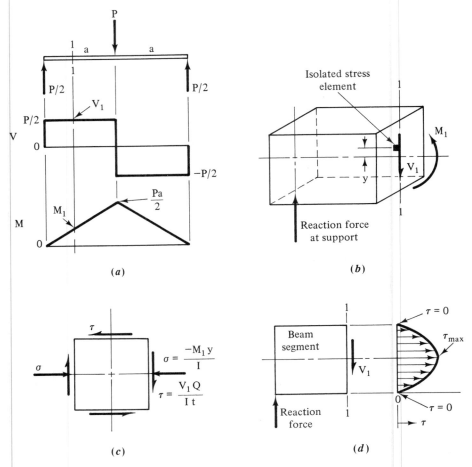

Figure 10-7 Distribution of shearing stress in a beam. (a) Example of shear force on a beam in bending (b) Isolated segment 1-1 showing shearing force V and bending moment M internal to beam (c) Stress element subjected to combined shearing stress and normal stress due to bending (d) Distribution of shearing stress only within the beam.

on the face toward the front of the bar and in the positive y-direction on the face toward the rear. Together they create a counterclockwise couple on the element. The element is completed by showing the shear stresses τ_{yx} producing a clockwise couple on the other faces.

The example problem shown next includes illustrative calculations for the values of the stresses that are shown in Figure 10-10.

Example Problem 10-1

Referring to Figure 10-8, let $P = 1500$ N, $a = 150$ mm, and $b = 300$ mm. The lever has a diameter of 30 mm. Compute the stress condition that exists on a point on top of the lever near the support.

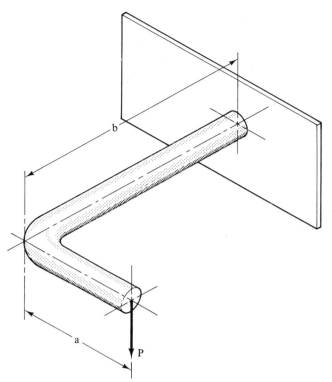

b

a

P

Figure 10-8 L-shaped lever illustrating combined stresses.

Solution   Using the free-body diagrams in Figure 10-9, we can show that

$$M_1 = T = Pa = (1500 \text{ N})(150 \text{ mm}) = 225\ 000 \text{ N} \cdot \text{mm}$$

$$M_2 = Pb = (1500 \text{ N})(300 \text{ mm}) = 450\ 000 \text{ N} \cdot \text{mm}$$

Then the bending stress on the top of the bar, shown as element K in Figure 10-9(b), is

$$\sigma_x = \frac{M_2 c}{I} = \frac{M_2}{S}$$

But the section modulus S is

$$S = \frac{\pi D^3}{32} = \frac{\pi (30 \text{ mm})^3}{32} = 2651 \text{ mm}^3$$

Then

$$\sigma_x = \frac{450\ 000 \text{ N} \cdot \text{mm}}{2651 \text{ mm}^3} = 169.8 \text{ N/mm}^2 = 169.8 \text{ MPa}$$

The torsional shear stress is at its maximum value all around the outer surface of the bar, with the value of

$$\tau = \frac{T}{Z_p}$$

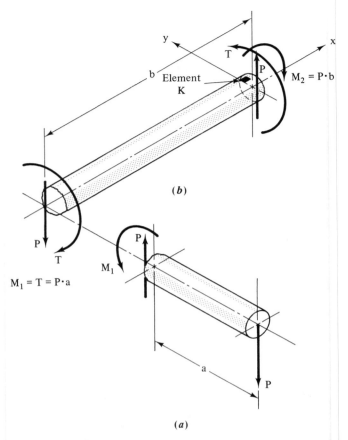

Figure 10-9 Free body diagrams of segments of an L-shaped lever. (a) Shorter segment of lever (b) Longer segment of lever.

But the polar section modulus Z_p is

$$Z_p = \frac{\pi D^3}{16} = \frac{\pi(30 \text{ mm})^3}{16} = 5301 \text{ mm}^3$$

Then

$$\tau = \frac{225\,000 \text{ N} \cdot \text{mm}}{5301 \text{ mm}^3} = 42.4 \text{ N/mm}^2 = 42.4 \text{ MPa}$$

The results are shown on the stress element in Figure 10-10.

10-5 EQUATIONS FOR STRESSES IN ANY DIRECTION

The *initial stress element* discussed in Section 10-4 was oriented in a convenient direction related to the member being analyzed. The methods of this section allow you to compute the stresses in any direction and to compute the maximum normal stresses and the maximum shear stress directly.

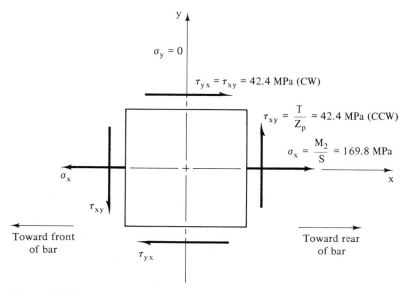

$\sigma_y = 0$

$\tau_{yx} = \tau_{xy} = 42.4 \text{ MPa (CW)}$

$\tau_{xy} = \dfrac{T}{Z_p} = 42.4 \text{ MPa (CCW)}$

$\sigma_x = \dfrac{M_2}{S} = 169.8 \text{ MPa}$

σ_x

τ_{xy}

τ_{yx}

x

y

Toward front
of bar

Toward rear
of bar

Figure 10-10   Stress on element K from Figure 10-9 with data from Example
Problem 10-1.

Figure 10-11 shows a stress element with orthogonal axes u and v superimposed
on the initial element such that the u-axis is at an angle ϕ relative to the given x-axis.
In general, there will be a normal stress σ_u and a shear stress τ_{uv} acting on the inclined
surface AC. The following development will result in equations for those stresses.

Before going on, note that Figure 10-11(a) shows only two dimensions of an
element that is really a three-dimensional cube. Part (b) of the figure shows the entire
cube with each side having the dimension h.

Visualize a wedge-shaped part of the initial element as shown in Figure 10-12.
Note how the angle ϕ is located. The side AB is the original left side of the initial
element having the height h. The bottom of the wedge, side BC, is only a part of the
bottom of the initial element where the length is determined by the angle ϕ itself.

$$BC = h \cdot \tan \phi$$

Also, the length of the sloped side of the wedge, AC, is

$$AC = \frac{h}{\cos \phi}$$

These lengths are important because now we are going to consider all the forces
that act on the wedge. Because force is the product of stress times area, we need to
know the area on which each stress acts. Starting with σ_x, it acts on the entire left face
of the wedge having an area equal to h^2. Remember that the depth of the wedge
perpendicular to the paper is also h. Then,

$$\text{force due to } \sigma_x = \sigma_x h^2$$

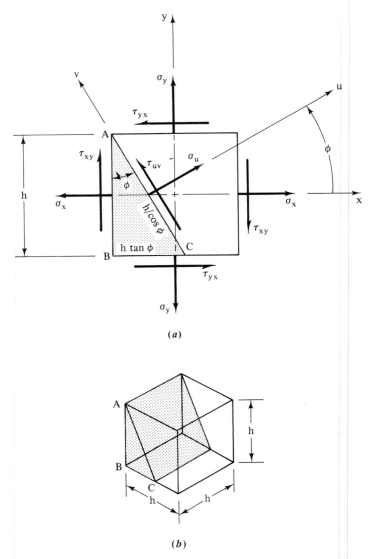

(a)

(b)

Figure 10-11 Initial stress element with u and v axes included. (a) Stress element with inclined surface (b) 3-dimensional element showing wedge.

Using similar logic,

$$\text{force due to } \sigma_y = \sigma_y h^2 \tan \phi$$

$$\text{force due to } \tau_{xy} = \tau_{xy} h^2$$

$$\text{force due to } \tau_{yx} = \tau_{yx} h^2 \tan \phi$$

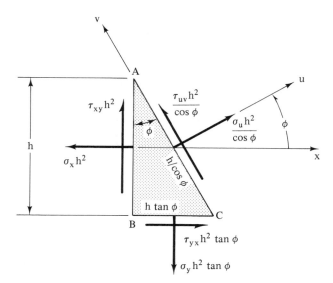

Figure 10-12 Free body diagram of wedge showing forces acting on each face.

The stresses acting on the inclined face of the wedge must also be considered:

$$\text{force due to } \sigma_u = \frac{\sigma_u h^2}{\cos \phi}$$

$$\text{force due to } \tau_{uv} = \frac{\tau_{uv} h^2}{\cos \phi}$$

Now using the principle of equilibrium, we can sum forces in the u-direction. From the resulting equation, we can solve for σ_u. The process is facilitated by resolving all forces into components perpendicular and parallel to the inclined face of the wedge. Figure 10-13 shows this for each force except for those due to σ_u and τ_{uv} which are already aligned with the u and v axes. Then,

$$\sum F_u = 0 = \frac{\sigma_u h^2}{\cos \phi} - \sigma_x h^2 \cos \phi - \sigma_y h^2 \tan \phi \sin \phi$$
$$+ \tau_{xy} h^2 \sin \phi + \tau_{yx} h^2 \tan \phi \cos \phi$$

To begin solving for σ_u, all terms include h^2, which can be canceled. Also, we have noted that $\tau_{xy} = \tau_{yx}$ and we can let $\tan \phi = \sin \phi/\cos \phi$. The equilibrium equation then becomes

$$0 = \frac{\sigma_u}{\cos \phi} - \sigma_x \cos \phi - \frac{\sigma_y \sin \phi \sin \phi}{\cos \phi} + \tau_{xy} \sin \phi + \frac{\tau_{xy} \sin \phi \cos \phi}{\cos \phi}$$

Now multiply by $\cos \phi$ to obtain

$$0 = \sigma_u - \sigma_x \cos^2 \phi - \sigma_y \sin^2 \phi + \tau_{xy} \sin \phi \cos \phi + \tau_{xy} \sin \phi \cos \phi$$

Combine the last two terms and solve for σ_u.

$$\sigma_u = \sigma_x \cos^2 \phi + \sigma_y \sin^2 \phi - 2\tau_{xy} \sin \phi \cos \phi$$

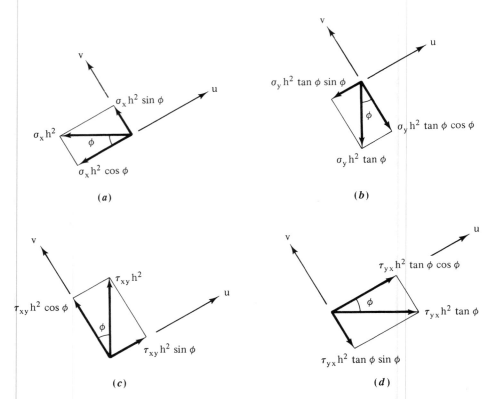

Figure 10-13 Resolution of forces along u and v directions (a) Components of force due to σ_x (b) Components of force due to σ_y (c) Components of force due to τ_{xy} (d) Components of force due to τ_{yx}.

This is a usable formula for computing σ_u, but a more convenient form can be obtained by using the following trigonometric identities:

$$\cos^2 \phi = \tfrac{1}{2} + \tfrac{1}{2} \cos 2\phi$$

$$\sin^2 \phi = \tfrac{1}{2} - \tfrac{1}{2} \cos 2\phi$$

$$\sin \phi \cos \phi = \tfrac{1}{2} \sin 2\phi$$

Making the substitutions gives

$$\sigma_u = \tfrac{1}{2}\sigma_x + \tfrac{1}{2}\sigma_x \cos 2\phi + \tfrac{1}{2}\sigma_y - \tfrac{1}{2}\sigma_y \cos 2\phi - \tau_{xy} \sin 2\phi$$

Combining terms, we obtain

$$\sigma_u = \tfrac{1}{2}(\sigma_x + \sigma_y) + \tfrac{1}{2}(\sigma_x - \sigma_y) \cos 2\phi - \tau_{xy} \sin 2\phi \qquad (10\text{-}1)$$

Equation (10-1) can be used to compute the normal stress in any direction provided that the stress condition in some direction, indicated by the x and y axes, is known.

Now the equation for the shearing stress, τ_{uv}, acting parallel to the cut plane and perpendicular to σ_u will be developed. Again referring to Figures 10-12 and 10-13, we can sum the forces on the wedge-shaped element acting in the v-direction.

$$\sum F_v = 0 = \frac{\tau_{uv} h^2}{\cos \phi} + \sigma_x h^2 \sin \phi - \sigma_y h^2 \tan \phi \cos \phi$$
$$+ \tau_{xy} h^2 \cos \phi - \tau_{yx} h^2 \tan \phi \sin \phi$$

Using the same techniques as before, this equation can be simplified and solved for τ_{uv}, resulting in

$$\tau_{uv} = -\tfrac{1}{2}(\sigma_x - \sigma_y) \sin 2\phi - \tau_{xy} \cos 2\phi \qquad (10\text{-}2)$$

Equation (10-2) can be used to compute the shearing stress that acts on the face of the element at any angular orientation.

10-6 PRINCIPAL STRESSES

In design and stress analysis, the maximum stresses are frequently desired in order to ensure the safety of the load-carrying member. Equation (10-1) can be used to compute the maximum normal stress if we know at what angle ϕ it occurred.

From the study of calculus, we know that the value of the angle ϕ at which the maximum or minimum normal stress occurs can be found by differentiating the function and setting the result equal to zero, then solving for ϕ. Differentiating equation (10-1) gives

$$\frac{d\sigma_u}{d\phi} = 0 = 0 + \tfrac{1}{2}(\sigma_x - \sigma_y)(-\sin 2\phi)(2) - \tau_{xy} \cos 2\phi (2)$$

Dividing by $\cos 2\phi$ and simplifying gives

$$0 = -(\sigma_x - \sigma_y) \tan 2\phi - 2\tau_{xy}$$

Solving for $\tan 2\phi$ gives

$$\tan 2\phi = \frac{-2\tau_{xy}}{\sigma_x - \sigma_y} = \frac{-\tau_{xy}}{\tfrac{1}{2}(\sigma_x - \sigma_y)} \qquad (10\text{-}3)$$

The angle ϕ is then

$$\phi = \tfrac{1}{2} \tan^{-1} \frac{-\tau_{xy}}{\tfrac{1}{2}(\sigma_x - \sigma_y)} \qquad (10\text{-}4)$$

If we substitute the value of ϕ defined by Equations (10-3) and (10-4) into Equation (10-1), we can develop an equation for the maximum normal stress on the element. In addition, we can develop the equation for the minimum normal stress. These two stresses are called the *principal stress* with σ_1 used to denote the *maximum principal stress* and σ_2 denoting the *minimum principal stress*.

Note from Equation (10-1) that we need values for $\sin 2\phi$ and $\cos 2\phi$. Figure 10-14 is a graphical aid to obtaining expressions for these functions. The right triangle has the opposite and adjacent sides defined by the terms of the tangent function in Equation (10-3).

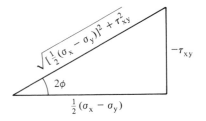

The triangle is defined by Equation (10-3):

$$\tan 2\phi = \frac{-\tau_{xy}}{\frac{1}{2}(\sigma_x - \sigma_y)}$$

$$\sin 2\phi = \frac{-\tau_{xy}}{\sqrt{[\frac{1}{2}(\sigma_x - \sigma_y)]^2 + \tau_{xy}^2}}$$

$$\cos 2\phi = \frac{\frac{1}{2}(\sigma_x - \sigma_y)}{\sqrt{[\frac{1}{2}(\sigma_x - \sigma_y)]^2 + \tau_{xy}^2}}$$

Figure 10-14 Development of $\sin 2\phi$ and $\cos 2\phi$ for principal stress formulas.

Substituting into Equation (10-1) and simplifying gives

$$\sigma_{max} = \sigma_1 = \frac{1}{2}(\sigma_x + \sigma_y) + \sqrt{[\frac{1}{2}(\sigma_x - \sigma_y)]^2 + \tau_{xy}^2} \qquad (10\text{-}5)$$

Because the square root has two possible values, $+$ and $-$, we can also find the expression for the minimum principal stress, σ_2.

$$\sigma_{min} = \sigma_2 = \frac{1}{2}(\sigma_x + \sigma_y) - \sqrt{[\frac{1}{2}(\sigma_x - \sigma_y)]^2 + \tau_{xy}^2} \qquad (10\text{-}6)$$

Conceivably, there would be a shear stress existing along with these normal stresses. But it can be shown that by substituting the value of ϕ from Equation (10-4) into the shear stress equation (10-2) that the result will be zero. In conclusion, *on the element on which the principal stresses act, the shear stress is zero.*

10-7 MAXIMUM SHEAR STRESS

The same technique can be used to find the maximum shear stress by working with Equation (10-2). Differentiating with respect to ϕ and setting the result equal to zero gives

$$\frac{d\tau_{uv}}{d\phi} = 0 = -\frac{1}{2}(\sigma_x - \sigma_y)(\cos 2\phi)(2) - \tau_{xy}(-\sin 2\phi)(2)$$

Dividing by $\cos 2\phi$ and simplifying gives

$$\tan 2\phi = \frac{(\sigma_x - \sigma_y)}{2\tau_{xy}} = \frac{\frac{1}{2}(\sigma_x - \sigma_y)}{\tau_{xy}} \qquad (10\text{-}7)$$

Solving for ϕ yields

$$\phi = \tfrac{1}{2} \tan^{-1} \frac{\tfrac{1}{2}(\sigma_x - \sigma_y)}{\tau_{xy}} \qquad (10\text{-}8)$$

Obviously, the value for ϕ from Equation (10-8) is different from that of Equation (10-4). In fact, we will see that the two values are *always 45 deg apart*.

Figure 10-15 shows the right triangle, from which we can find $\sin 2\phi$ and $\cos 2\phi$ as we did in Figure 10-14. Substituting these values into Equation (10-2) gives the maximum shear stress.

$$\tau_{max} = \pm\sqrt{[\tfrac{1}{2}(\sigma_x - \sigma_y)]^2 + \tau_{xy}^2} \qquad (10\text{-}9)$$

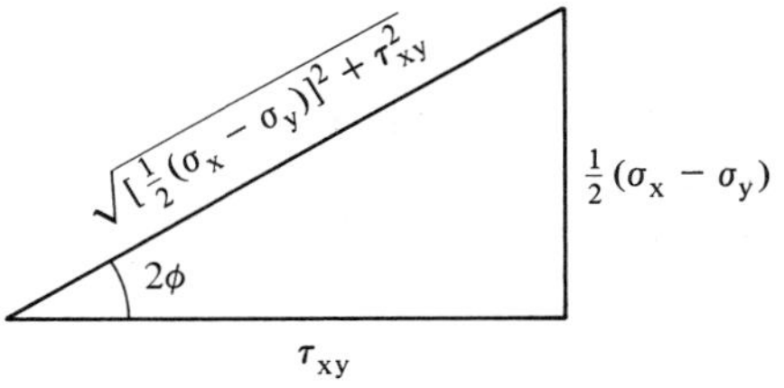

The triangle is defined by Equation (10-7):

$$\tan 2\phi = \frac{\tfrac{1}{2}(\sigma_x - \sigma_y)}{\tau_{xy}}$$

$$\sin 2\phi = \frac{\tfrac{1}{2}(\sigma_x - \sigma_y)}{\sqrt{[\tfrac{1}{2}(\sigma_x - \sigma_y)]^2 + \tau_{xy}^2}}$$

$$\cos 2\phi = \frac{\tau_{xy}}{\sqrt{[\tfrac{1}{2}(\sigma_x - \sigma_y)]^2 + \tau_{xy}^2}}$$

Figure 10-15 Development of $\sin 2\phi$ and $\cos 2\phi$ for maximum shear stress formula.

Here we should check to see if there is a normal stress existing on the element having the maximum shear stress. Substituting the value of ϕ from Equation (10-8) into the general normal stress equation (10-1) gives

$$\sigma = \tfrac{1}{2}(\sigma_x + \sigma_y) \qquad (10\text{-}10)$$

This is the formula for the *average* of the initial normal stresses, σ_x and σ_y. Thus we can conclude:

> *On the element on which the maximum shear stress occurs, there will also be a normal stress equal to the average of the initial normal stresses.*

10-8 MOHR'S CIRCLE FOR STRESS

The use of Equations (10-1) through (10-10) often presents difficulties because of the many possible combinations of signs for the terms σ_x, σ_y, τ_{xy}, and ϕ. Also, two roots of the square root and the fact that the inverse tangent function can result in angles in any of the four quadrants offers difficulties. Fortunately, there is a graphical aid, called *Mohr's circle*, available that can help to overcome these problems. The use of Mohr's circle should give you a better understanding of the general case of stress at a point.

It can be shown that the two equations, (10-1) and (10-2), for the normal and shear stress at a point in any direction can be combined and arranged in the form of the equation for a circle. First presented by Otto Mohr in 1895, the circle allows a rapid and exact computation of:

1. The maximum and minimum principal stresses [Equations (10-5) and (10-6)]
2. The maximum shear stress [Equation (10-9)]
3. The angles of orientation of the principal stress element and the maximum shear stress element [Equations (10-4) and (10-8)]
4. The normal stress that exists along with the maximum shear stress on the maximum shear stress element [Equation (10-10)]
5. The stress condition at any angular orientation of the stress element [Equations (10-1) and (10-2)]

Mohr's circle is drawn on a set of perpendicular axes with shear stress, τ, plotted vertically and normal stresses, σ, horizontally as shown in Figure 10-16. The follow-

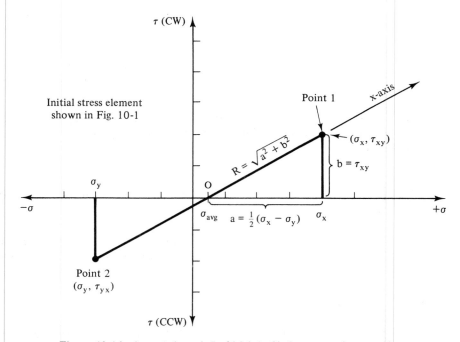

Figure 10-16 Steps 1 through 7 of Mohr's Circle construction procedure.

ing convention is used in this book:

1. Positive normal stresses (tensile) are to the right.
2. Negative normal stresses (compressive) are to the left.
3. Shear stresses that tend to rotate the stress element clockwise (CW) are plotted upward on the τ-axis.
4. Shear stresses that tend to rotate the stress element counterclockwise (CCW) are plotted downward.

The procedure listed next can be used to draw the Mohr's circle. Steps 1–7 are shown in Figure 10-16. The complete stress element as shown in Figure 10-1 is the basis for this example.

1. Identify the stress condition at the point of interest and represent it as the *initial stress element* in the manner shown in Figure 10-1.

2. The combination σ_x and τ_{xy} is plotted as *point 1* on the $\sigma-\tau$ plane.

3. The combination σ_y and τ_{yx} is then plotted as *point 2*. Note that τ_{xy} and τ_{yx} always act in opposite directions. Therefore, one point will be plotted above the σ-axis and one will be below.

4. Draw a straight line between the two points.

5. This line crosses the σ-axis at the center of Mohr's circle, which is also the value of the *average normal stress* applied to the initial stress element. The location of the center can be observed from the data used to plot the points or computed from Equation (10-10), repeated here:

$$\sigma_{avg} = \tfrac{1}{2}(\sigma_x + \sigma_y)$$

For convenience, label the center O.

6. Identify the line from O through point 1 (σ_x, τ_{xy}) as the *x*-axis. This line corresponds to the original *x*-axis and is essential to correlating the data from Mohr's circle to the original *x* and *y* directions.

7. The points O, σ_x, and point 1 form an important right triangle because the distance from O to point 1, the hypotenuse of the triangle, is equal to the radius of the circle, R. Calling the other two sides a and b, the following calculations can be made.

$$a = \tfrac{1}{2}(\sigma_x - \sigma_y)$$
$$b = \tau_{xy}$$
$$R = \sqrt{a^2 + b^2} = \sqrt{[\tfrac{1}{2}(\sigma_x - \sigma_y)]^2 + \tau_{xy}^2}$$

Note that the equation for R is identical to Equation (10-9) for the maximum shear stress on the element. Thus *the length of the radius of Mohr's circle is equal to the magnitude of the maximum shear stress*.

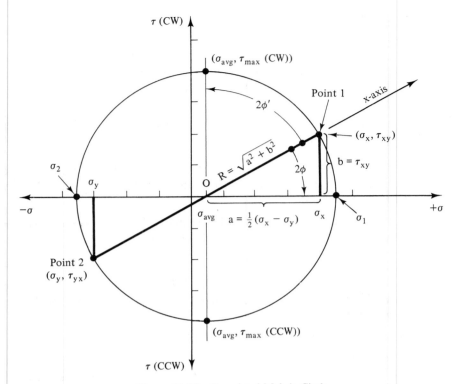

Figure 10-17 Completed Mohr's Circle.

8. Draw the complete circle with the center at O and the radius R. Steps 8–11 are shown in Figure 10-17.

9. Draw the vertical diameter of the circle. The point at the top of the circle has the coordinates $(\sigma_{avg}, \tau_{max})$, where the shear stress has the clockwise (CW) direction. The point at the bottom of the circle represents $(\sigma_{avg}, \tau_{max})$ where the shear stress is counterclockwise (CCW).

10. Identify the points on the σ-axis at the ends of the horizontal diameter as σ_1 at the right (the maximum principal stress) and σ_2 at the left (the minimum principal stress). Note that the shear stress is zero at these points.

11. Determine the values for σ_1 and σ_2 from

$$\sigma_1 = \text{``}O\text{''} + R \tag{10-11}$$

$$\sigma_2 = \text{``}O\text{''} - R \tag{10-12}$$

where "O" represents the coordinate of the center of the circle, σ_{avg}, and R is the radius of the circle. Thus Equations (10-11) and (10-12) are identical to Equations (10-5) and (10-6) for the principal stresses.

The following steps determine the angles of orientation of the principal stress element and the maximum shear stress element. An important concept to remember is that *angles obtained from Mohr's circle are double the true angles*. The reason for this is that the equations on which it is based, Equations (10-1) and (10-2), are functions of 2ϕ.

12. The orientation of the principal stress element is determined by finding the angle *from* the x-axis *to* the "σ_1"-axis, labeled 2ϕ in Figure 10-17. From the data on the circle you can see that

$$2\phi = \tan^{-1}\frac{b}{a}$$

The argument of this inverse tangent function is the same as the absolute value of the argument shown in Equation (10-4). Problems with signs for the resulting angle are avoided by noting the *direction from the x-axis to the σ_1-axis* on the circle, clockwise for the present example. Then the principal stress element is rotated *in the same direction* from the x-axis by an amount ϕ to locate the face on which the maximum principal stress σ_1 acts.

13. Draw the principal stress element in its proper orientation as determined from step 12 with the two principal stresses σ_1 and σ_2 shown (see Figure 10-18, (a) and (b)).

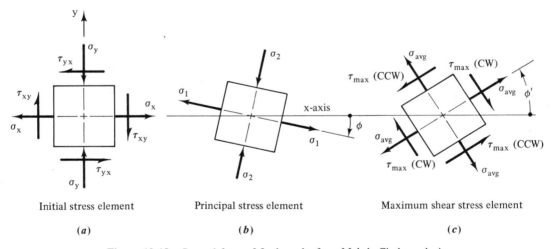

Initial stress element	Principal stress element	Maximum shear stress element
(*a*)	(*b*)	(*c*)

Figure 10-18 General form of final results from Mohr's Circle analysis.

14. The orientation of the maximum shear stress element is determined by finding the angle *from* the x-axis *to* the τ_{max}-axis, labeled $2\phi'$ in Figure 10-17. In the present example,

$$2\phi' = 90° - 2\phi$$

From trigonometry it can be shown that this is equivalent to finding the inverse tangent of a/b, the reciprocal of the argument used to find 2ϕ. Thus it is a true evaluation of

Equation (10-8), derived to find the angle of orientation of the element on which the maximum shear stress occurs.

Again problems with signs for the resulting angle are avoided by noting the *direction from the x-axis to the τ_{max}-axis* on the circle, counterclockwise for the present example. Then the maximum shear stress element is rotated *in the same direction* from the x-axis by an amount ϕ' to locate the face on which the clockwise maximum shear stress acts.

15. Draw the maximum shear stress element in its proper orientation as determined from step 14 with the shear stresses on all four faces and the average normal stress acting on each face (see Figure 10-18(c)). As a whole, Figure 10-18 is the desired result from a Mohr's circle analysis. Shown are the initial stress element that establishes the x and y axes, the principal stress element drawn in proper rotation relative to the x-axis, and the maximum shear stress element also drawn in the proper rotation relative to the x-axis.

Example Problem 10-2

It has been determined that a point in a load-carrying member is subjected to the following stress condition:

$$\sigma_x = 400 \text{ MPa} \qquad \sigma_y = -300 \text{ MPa} \qquad \tau_{xy} = 200 \text{ MPa (CW)}$$

Perform the following:
    (a) Draw the initial stress element.
    (b) Draw the complete Mohr's circle, labeling critical points.
    (c) Draw the complete principal stress element.
    (d) Draw the maximum shear stress element.

Solution The 15-step procedure described before will be used to complete the problem. The initial stress element is drawn in Figure 10-20(a) and the Mohr's circle is in Figure 10-19. The numerical results from steps 1–7 are summarized below.

Center of the circle O is at σ_{avg}:

$$\sigma_{avg} = \tfrac{1}{2}(\sigma_x + \sigma_y) = \tfrac{1}{2}[400 + (-300)] = 50 \text{ MPa}$$

The lower side of the triangle,

$$a = \tfrac{1}{2}(\sigma_x - \sigma_y) = \tfrac{1}{2}[400 - (-300)] = 350 \text{ MPa}$$

The vertical side of the triangle,

$$b = \tau_{xy} = 200 \text{ MPa}$$

The radius of the circle,

$$R = \sqrt{a^2 + b^2} = \sqrt{(350)^2 + (200)^2} = 403 \text{ MPa}$$

Step 8 is the drawing of the circle. The significant data points from steps 9–11 are summarized below.

$$\sigma_{avg} = 50 \text{ MPa (same as the location of } O)$$

$$\tau_{max} = 403 \text{ MPa (same as the value of } R)$$

$$\sigma_1 = O + R = 50 + 403 = 453 \text{ MPa}$$

$$\sigma_2 = O - R = 50 - 403 = -353 \text{ MPa}$$

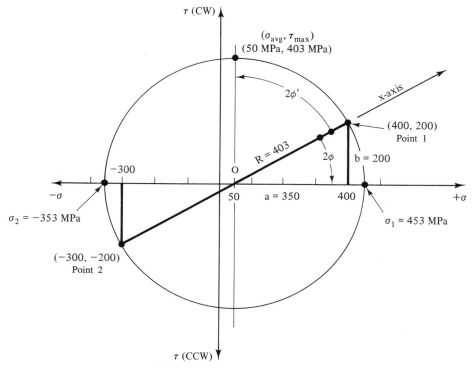

Figure 10-19 Complete Mohr's Circle for Example Problem 10-2.

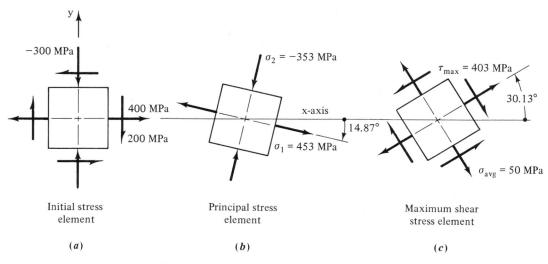

Figure 10-20 Results for Example Problem 10-2.

Steps 12–15 are completed in Figures 10-19 and 10-20. The computations of the angles are summarized below.

$$2\phi = \tan^{-1}\frac{b}{a} = \tan^{-1}\frac{200}{350} = 29.74°$$

Note that 2ϕ is CW from the x-axis to σ_1 on the circle.

$$\phi = \frac{29.74°}{2} = 14.87°$$

Thus, in Figure 10-20(b), the principal stress element is drawn rotated 14.87° CW from the original x axis to the face on which σ_1 acts.

$$2\phi' = 90° - 2\phi = 90° - 29.74° = 60.26°$$

Note that $2\phi'$ is CCW from the x-axis to τ_{max} on the circle.

$$\phi' = \frac{60.26°}{2} = 30.13°$$

Thus, in Figure 10-20(c) the maximum shear stress element is drawn rotated 30.13° CCW from the original x-axis to the face on which τ_{max} acts.

Example Problem 10-2 is completed.

10-9 EXAMPLES OF THE USE OF MOHR'S CIRCLE

The data for Example Problem 10-2 in the preceding section and Example Problems 10-3 through 10-8 to follow, have been selected to demonstrate a wide variety of results. A major variable is the quadrant in which the x-axis lies and the corresponding definition of the angles of rotation for the principal stress element and the maximum shear stress element.

Example Problems 10-6, 10-7, and 10-8 present the special cases of biaxial stress with no shear, uniaxial tension with no shear, and pure shear. These should help you understand the behavior of load-carrying members subjected to such stresses.

The solution for each Example Problem to follow is the Mohr's circle itself along with the stress elements, properly labeled. For each problem, the objectives are to:

(a) Draw the initial stress element
(b) Draw the complete Mohr's circle, labeling critical points
(c) Draw the complete principal stress element
(d) Draw the complete shear stress element.

Example Problem Statements and Summary of Results

Example   Stress condition on the given element; figure showing the resulting
Prob. No.   Mohr's circle; and remarks concerning the results.

10-2 $\sigma_x = 400$ MPa; $\sigma_y = -300$ MPa; $\tau_{xy} = 200$ MPa CW
Figures 10-19 and 10-20. The x-axis is in the first quadrant.

10-3 $\sigma_x = 60$ ksi; $\sigma_y = -40$ ksi; $\tau_{xy} = 30$ ksi CCW
Figure 10-21. The x-axis is in the second quadrant.

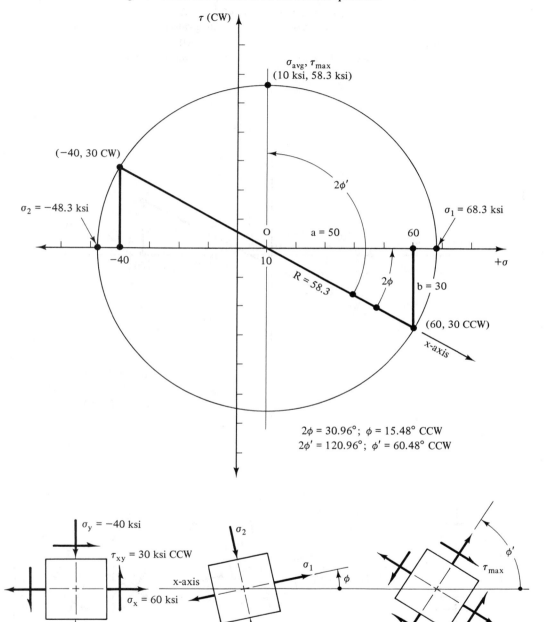

Figure 10-21   Results for Example Problem 10-3. X-axis in the second quadrant.

10-4 $\sigma_x = -120$ MPa; $\sigma_y = 180$ MPa; $\tau_{xy} = 80$ MPa CCW
Figure 10-22. The x-axis is in the third quadrant.

10-5 $\sigma_x = -30$ ksi; $\sigma_y = 20$ ksi; $\tau_{xy} = 40$ ksi CW
Figure 10-23. The x-axis is in the fourth quadrant.

10-6 $\sigma_x = 220$ MPa; $\sigma_y = -120$ MPa; $\tau_{xy} = 0$ MPa
Figure 10-24. The x-axis is coincident with the σ-axis.

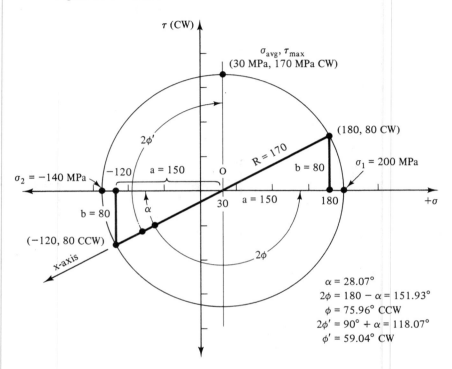

$\alpha = 28.07°$
$2\phi = 180 - \alpha = 151.93°$
$\phi = 75.96°$ CCW
$2\phi' = 90° + \alpha = 118.07°$
$\phi' = 59.04°$ CW

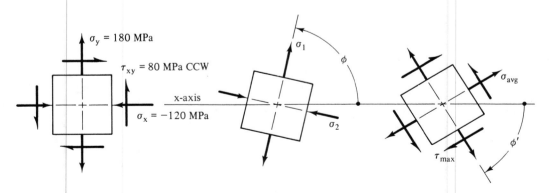

Figure 10-22 Results for Example Problem 10-4. X-axis in the third quadrant.

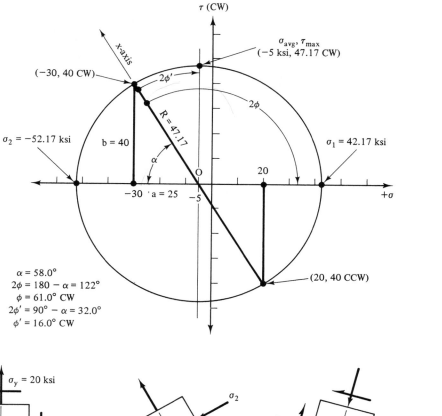

Figure 10-23 Results for Example Problem 10-5. X-axis in the fourth quadrant.

This is the special case of biaxial stress with no shear. The initial stress element and the principal stress element are identical. The maximum shear stress element is rotated 45° from the *x*-axis.

10-7 $\sigma_x = 40$ ksi; $\sigma_y = 0$ ksi; $\tau_{xy} = 0$ ksi
Figure 10-25. The *x*-axis is coincident with the σ-axis.
This is the special case of uniaxial tension with no shear. The initial stress element and the principal stress element are identical. The maximum shear stress element is rotated 45° from the *x*-axis. The maximum shear stress is one-half the given tensile stress.

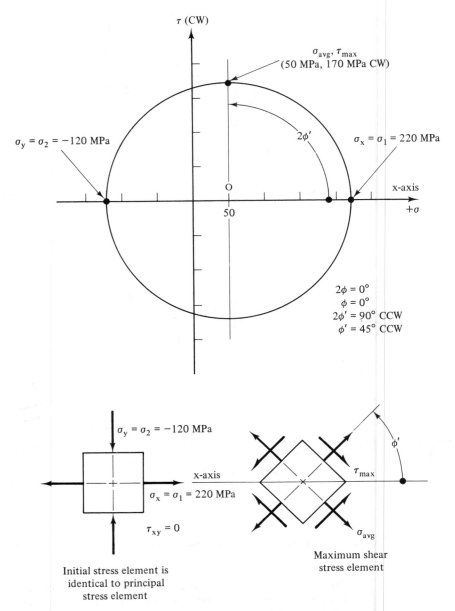

τ (CW)

σ_{avg}, τ_{max}
(50 MPa, 170 MPa CW)

$\sigma_y = \sigma_2 = -120$ MPa

$2\phi'$

$\sigma_x = \sigma_1 = 220$ MPa

x-axis

O

50

$+\sigma$

$2\phi = 0°$
$\phi = 0°$
$2\phi' = 90°$ CCW
$\phi' = 45°$ CCW

$\sigma_y = \sigma_2 = -120$ MPa

x-axis

$\sigma_x = \sigma_1 = 220$ MPa

$\tau_{xy} = 0$

Initial stress element is
identical to principal
stress element

ϕ'

τ_{max}

σ_{avg}

Maximum shear
stress element

Figure 10-24 Results of Example Problem 10-6. Special case of biaxial stress with no shear.

10-8 $\sigma_x = 0$ ksi; $\sigma_y = 0$ ksi; $\tau_{xy} = 40$ ksi CW

Figure 10-26. The x-axis is coincident with the τ-axis.

This is the special case of pure shear. The initial stress element and the maximum shear stress element are identical. The principal stress element is rotated 45° from the x-axis. The principal stresses are equal in magnitude to the given shear stress.

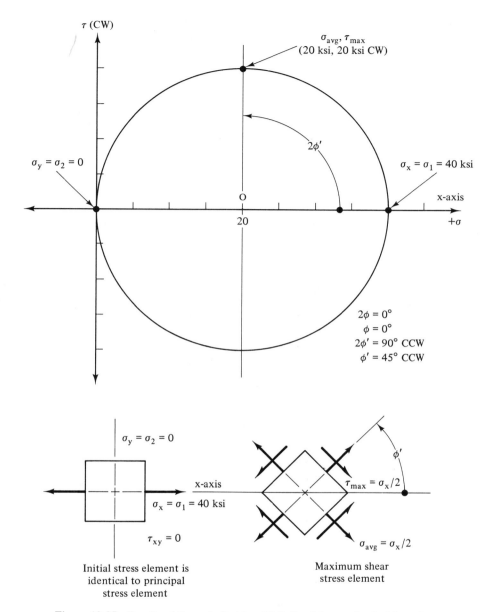

Figure 10-25 Results of Example Problem 10-7. Special case of uniaxial tension.

10-10 STRESS CONDITION ON SELECTED PLANES

There are some cases in which it is desirable to know the stress condition on an element at some selected angle of orientation relative to the reference directions. Figures 10-27 and 10-28 show examples.

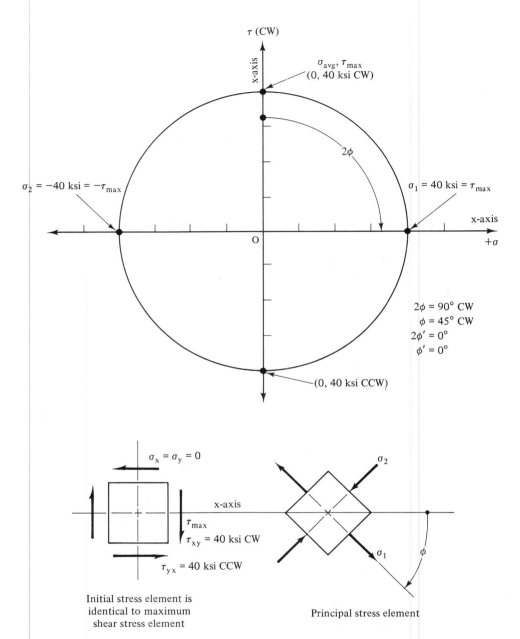

Figure 10-26 Results of Example Problem 10-8. Special case of pure shear.

The wood block in Figure 10-27 shows that the grain of the wood is inclined at an angle of 30° CCW from the given x-axis. Because the wood is very weak in shear parallel to the grain, it is desirable to know the stresses in this direction.

Figure 10-28 shows a member fabricated by welding two components along a seam inclined at an angle relative to the given x-axis. The welding operation could

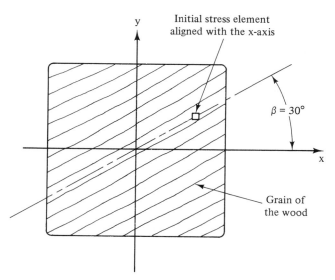

Initial stress element
aligned with the x-axis

$\beta = 30°$

Grain of
the wood

Figure 10-27   Cross section of a wood post with the grain running 30° with respect to the x-axis.

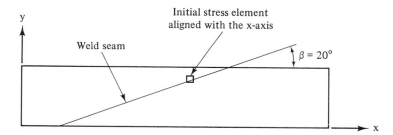

Initial stress element
aligned with the x-axis

Weld seam

$\beta = 20°$

Figure 10-28   Flat bar welded along a 20° inclined seam.

produce a weaker material in the near vicinity of the weld, particularly if the component parts were made from steel that was heat-treated prior to welding. The same is true of many aluminum alloys. For such cases the allowable stresses are somewhat lower along the weld line.

The environmental conditions to which a part is exposed during operation may also affect the material properties. For example, a furnace part may be subjected to local heating by radiant energy along a particular line. The strength of the heated material will be lower than that which remains cool and thus it is desirable to know the stress condition along the angle of the heat-affected zone.

Mohr's circle can be used to determine the stress condition at specified angles of orientation of the stress element. The procedure is demonstrated in Example Problem 10-9.

Example Problem 10-9

For the load-carrying member shown in Figure 10-28, the stress element located parallel to the x and y axes shown is subjected to the following stress condition,

$$\sigma_x = 400 \text{ MPa} \qquad \sigma_y = -300 \text{ MPa} \qquad \tau_{xy} = 200 \text{ MPa CW}$$

Sec. 10-10   Stress Condition on Selected Planes

337

Determine the stress condition that would occur on the element inclined at an angle $\beta = 20°$ CCW from the x-axis, aligned with the direction of the welded seam.

Solution The given data for the x-y directions is the same as that used in Example Problem 10-2. Figure 10-19 shows the completed Mohr's circle for these data. Notice that the x-axis is the line from the center of the circle through the first plotted point at (σ_x, τ_{xy}), in this case at $(400, 200)$. The circle is reproduced in Figure 10-29.

The desired axis is one inclined at $20°$ CCW from the x-axis. Recalling that angles in Mohr's circle are *double* the true angles, we can draw a line through the center of the circle at an angle of $2\beta = 40°$ CCW from the x-axis. The intersection of this line with the circle, labeled A in the figure, locates the point on the circle defining the stress condition of the desired element. The coordinates of this point, (σ_A, τ_A), give the normal and shear stresses acting on one set of faces of the desired stress element.

Simple trigonometry and the basic geometry of the circle can be used to determine σ_A and τ_A by projecting lines vertically and horizontally from point A to the σ and τ axes, respectively. The total angle from the σ-axis to the axis through point A, called η (eta) in the figure, is the sum of 2ϕ and 2β. In Example Problem 10-2 we found $2\phi = 29.74°$. Then

$$\eta = 2\phi + 2\beta = 29.74° + 40° = 69.74°$$

A triangle has been identified in Figure 10-29 with sides labeled d, g, and R. From this triangle, we can compute

$$d = R \cos \eta = (403) \cos 69.74° = 140$$

$$g = R \sin \eta = (403) \sin 69.74° = 378$$

These values enable the computation of

$$\sigma_A = O + d = 50 + 140 = 190 \text{ MPa}$$

$$\tau_A = g = 378 \text{ MPa CW}$$

where O indicates the value of the normal stress at the center of the Mohr's circle.

The stresses on the remaining set of faces of the desired stress element are the coordinates of the point A' located $180°$ from A on the circle and, therefore, $90°$ from the faces on which (σ_A, τ_A) act. Projecting vertically and horizontally from A' to the σ and τ axes locates $\sigma_{A'}$ and $\tau_{A'}$. By similar triangles we can say that $d' = d$ and $g' = g$. Then

$$\sigma_{A'} = O - d' = 50 - 140 = -90 \text{ MPa}$$

$$\tau_{A'} = g' = 378 \text{ MPa CCW}$$

Figure 10-29(c) shows the final stress element inclined at $20°$ from the x-axis. This is the stress condition experienced by the material along the weld line.

10-11 SPECIAL CASE WHEN BOTH PRINCIPAL STRESSES HAVE THE SAME SIGN

In preceding sections involving Mohr's circle, we used the convention that σ_1 is the maximum principal stress and σ_2 is the minimum principal stress. This is true for cases of plane stress (stresses applied in a single plane) if σ_1 and σ_2 have opposite signs, that

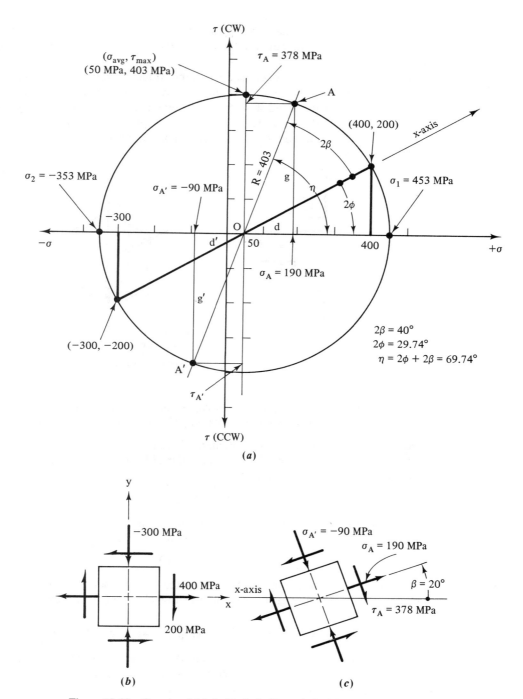

Figure 10-29 Complete Mohr's Circle for Example Problem 10-9 showing stresses on an element inclined at 20° ccw from the x-axis.

is, when one is tensile and one is compressive. Also, in such cases, the shear stress found at the top of the circle (equal to the radius, R) is the true maximum shear stress on the element.

But special care must be exercised when the Mohr's circle indicates that σ_1 and σ_2 have the same sign. Even though we are dealing with plane stress, the true stress element is three-dimensional and should be represented as a cube rather than a square, as shown in Figure 10-30. Faces 1, 2, 3, and 4 correspond to the sides of the square element and faces 5 and 6 are the "front" and "back." For plane stress the stresses on faces 5 and 6 are zero.

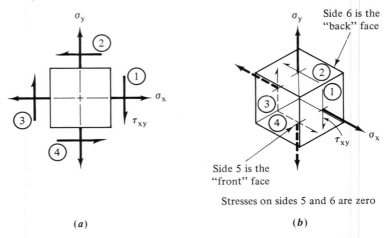

(a)   *(b)*

Figure 10-30   Plane stress shown as two dimensional and three dimensional stress elements. (a) Two dimensional stress element (b) Three dimensional stress element.

For the three-dimensional stress element, three principal stresses exist, called σ_1, σ_2, and σ_3, acting on the mutually perpendicular sides of the stress element. Convention calls for the following order:

$$\sigma_1 > \sigma_2 > \sigma_3$$

Thus σ_3 is the true minimum principal stress and σ_1 is the true maximum principal stress. It can also be shown that the true maximum shear stress can be computed from

$$\tau_{max} = \tfrac{1}{2}(\sigma_1 - \sigma_3) \tag{10-13}$$

Figure 10-31 illustrates a case in which the three-dimensional stress element must be considered. The initial stress element, shown in part (a), carries the following stresses:

$$\sigma_x = 200 \text{ MPa} \qquad \sigma_y = 120 \text{ MPa} \qquad \tau_{xy} = 40 \text{ MPa CW}$$

Part (b) of the figure shows the conventional Mohr's circle, drawn according to the procedure outlined in Section 10-8. Note that both σ_1 and σ_2 are positive or tensile. Then, considering that the stress on the "front" and "back" faces is zero, this is the true minimum principal stress. We can then say that

$$\sigma_1 = 216.6 \text{ MPa}$$

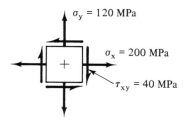

(*a*) Initial stress element

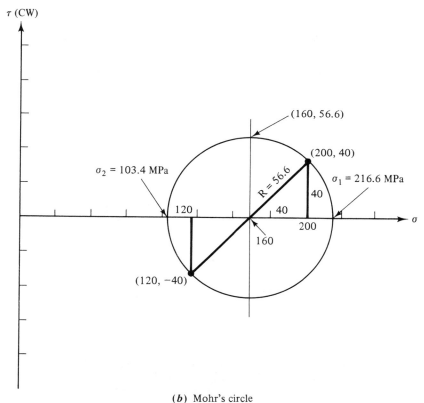

(*b*) Mohr's circle

Figure 10-31 Mohr's Circle for which σ_1 and σ_2 are both positive.

$$\sigma_2 = 103.4 \text{ MPa}$$

$$\sigma_3 = 0 \text{ MPa}$$

From Equation (10-13), the true maximum shear stress is

$$\tau_{max} = \tfrac{1}{2}(\sigma_1 - \sigma_3) = \tfrac{1}{2}(216.6 - 0) = 108.3 \text{ MPa}$$

These concepts can be visualized graphically by creating a set of three Mohr's circles rather than only one. Figure 10-32 shows the circle obtained from the initial

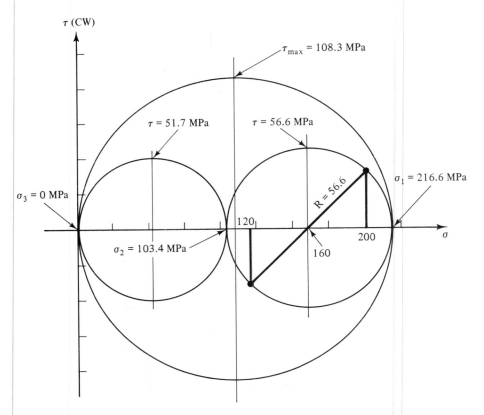

Figure 10-32 Three related Mohr's Circles showing σ_1, σ_2, σ_3, and τ_{max}.

stress element, a second circle encompassing σ_1 and σ_3, and a third encompassing σ_2 and σ_3. Thus each circle represents the plane on which two of the three principal stresses act. The point at the top of each circle indicates the largest shear stress that would occur in that plane. Then the largest circle, drawn for σ_1 and σ_3, produces the true maximum shear stress and its value is consistent with Equation (10-13).

Figure 10-33 illustrates another case in which the principal stresses from the initial stress element have the same sign, both negative in this case. The initial stresses are

$$\sigma_x = -50 \text{ MPa} \qquad \sigma_y = -180 \text{ MPa} \qquad \tau_{xy} = 30 \text{ MPa CCW}$$

Here, too, the supplementary circles must be drawn. But in this case, the zero stress on the "front" and "back" faces of the element becomes the *maximum* principal stress (σ_1). That is,

$$\sigma_1 = 0 \text{ MPa}$$

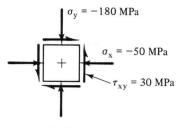

(*a*) Initial stress element

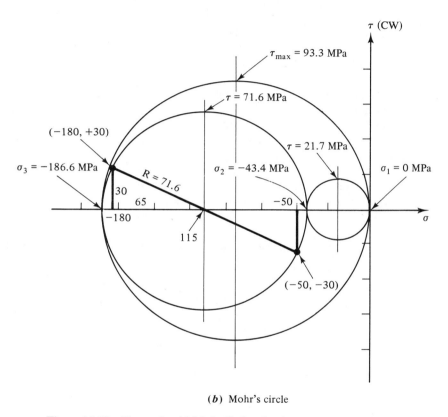

(*b*) Mohr's circle

Figure 10-33 Three related Mohr's Circles showing σ_1, σ_2, σ_3, and τ_{max}.

$$\sigma_2 = -43.4 \text{ MPa}$$

$$\sigma_3 \doteq -186.6 \text{ MPa}$$

and the maximum shear stress is

$$\tau_{max} = \tfrac{1}{2}(\sigma_1 - \sigma_3) = \tfrac{1}{2}[0 - (-186.6)] = 93.3 \text{ MPa}$$

10-12 THE MAXIMUM SHEAR STRESS THEORY OF FAILURE

One of the most widely used principles of design is the *maximum shear stress theory of failure*, which states:

> *A ductile material can be expected to fail when the maximum shear stress to which the material is subjected exceeds the yield strength of the material in shear.*

Of course, to apply this theory, it is necessary to be able to compute the magnitude of the maximum shear stress. If the member is subjected to pure shear, such as torsional shear stress, direct shear stress, or shearing stress in beams in bending, the maximum shear stress can be computed directly from formulas such as those developed in this book. But if a combined stress condition exists, the use of Equation (10-9) or the Mohr's circle should be used to determine the maximum shear stress.

One special case of combined stress that occurs often is one in which a normal stress in only one direction is combined with a shear stress. For example, a round bar could be subjected to a direct axial tension while also being twisted. In many types of mechanical power transmissions, shafts are subjected to bending and torsion simultaneously. Certain fasteners may be subjected to tension combined with direct shear.

A simple formula can be developed for such cases using Mohr's circle or Equation (10-9). If only a normal stress in the x-direction, σ_x, combined with a shearing stress, τ_{xy} exists, the maximum shear stress is

$$\tau_{max} = \sqrt{(\sigma_x/2)^2 + \tau_{xy}^2} \qquad (10\text{-}14)$$

This formula can be developed from Equation (10-9) by letting $\sigma_y = 0$.

Example Problem 10-9

A solid circular bar is 45 mm in diameter and is subjected to an axial tensile force of 120 kN along with a torque of 1150 N · m. Compute the maximum shear stress in the bar.
Solution   The applied normal stress is found from the direct stress formula. Note that 120 kN = 120 000 N.

$$\sigma = F/A$$

$$A = \pi D^2/4 = \pi (45 \text{ mm})^2/4 = 1590 \text{ mm}^2$$

$$\sigma = (120\ 000 \text{ N})/(1590 \text{ mm}^2) = 75.5 \text{ N/mm}^2 = 75.5 \text{ MPa}$$

The shear stress is found from the torsional shear stress formula. Note that 1150 N · m = 1 150 000 N · mm.

$$\tau = T/Z_p$$

$$Z_p = \pi D^3/16 = \pi (45 \text{ mm})^3/16 = 17\ 892 \text{ mm}^3$$

$$\tau = (1\ 150\ 000 \text{ N} \cdot \text{mm})/(17\ 892 \text{ mm}^3) = 64.3 \text{ N/mm}^2 = 64.3 \text{ MPa}$$

Then using Equation (10-14) yields

$$\tau_{max} = \sqrt{\left(\frac{75.5 \text{ MPa}}{2}\right)^2 + (64.3 \text{ MPa})^2} = 74.6 \text{ MPa}$$

REFERENCES

1. Muvdi, B. B., and J. W. McNabb, *Engineering Mechanics of Materials*, Macmillan, New York, 1980.
2. Popov, E. P., *Mechanics of Materials*, 2nd ed. Prentice-Hall, Englewood Cliffs, N.J., 1976.
3. Shigley, J. E., *Mechanical Engineering Design*, 4th ed. McGraw-Hill, New York, 1982.

PROBLEMS

A. For problems 10-1 to 10-28, determine the principal stresses and the maximum shear stress using Mohr's circle. The data sets below give the stresses on the initial stress element. Perform the following operations.

    (a) Draw the complete Mohr's circle, labeling critical points including σ_1, σ_2, τ_{max}, and σ_{avg}.

    (b) On the Mohr's circle, indicate which line represents the x-axis on the initial stress element.

    (c) On the Mohr's circle, indicate the angles from the line representing the x-axis to the σ_1-axis and the τ_{max}-axis.

    (d) Draw the principal stress element and the maximum shear stress element in their proper orientation relative to the initial stress element.

Problem	σ_x	σ_y	τ_{xy}
10-1	300 MPa	−100 MPa	80 MPa CW ↑
10-2	250 MPa	−50 MPa	40 MPa CW
10-3	80 MPa	−10 MPa	60 MPa CW
10-4	150 MPa	10 MPa	100 MPa CW
10-5	20 ksi	−5 ksi	10 ksi CCW
10-6	38 ksi	−25 ksi	18 ksi CCW
10-7	55 ksi	15 ksi	40 ksi CCW
10-8	32 ksi	−50 ksi	20 ksi CCW
10-9	−900 kPa	600 kPa	350 kPa CCW
10-10	−580 kPa	130 kPa	75 kPa CCW
10-11	−840 kPa	−35 kPa	650 kPa CCW
10-12	−325 kPa	50 kPa	110 kPa CCW
10-13	−1800 psi	300 psi	800 psi CW
10-14	−6500 psi	1500 psi	1200 psi CW
10-15	−4250 psi	3250 psi	2800 psi CW
10-16	−150 psi	8600 psi	80 psi CW
10-17	260 MPa	0 MPa	190 MPa CCW
10-18	1450 kPa	0 kPa	830 kPa CW
10-19	22 ksi	0 ksi	6.8 ksi CW
10-20	6750 psi	0 psi	3120 psi CCW
10-21	0 ksi	−28 ksi	12 ksi CW
10-22	0 MPa	440 MPa	215 MPa CW
10-23	0 MPa	260 MPa	140 MPa CCW
10-24	0 kPa	−1560 kPa	810 kPa CCW
10-25	225 MPa	−85 MPa	0 MPa
10-26	6250 psi	−875 psi	0 psi
10-27	775 kPa	−145 kPa	0 kPa
10-28	38.6 ksi	−13.4 ksi	0 ksi

B. For problems that result in the principal stresses from the initial Mohr's circle having the same sign, use the procedures from Section 10-11 to draw supplementary circles and find the following:
   (a) The three principal stresses: σ_1, σ_2, and σ_3.
   (b) The true maximum shear stress.

Problem	σ_x	σ_y	τ_{xy}
10-29	300 MPa	100 MPa	80 MPa CW
10-30	250 MPa	150 MPa	40 MPa CW
10-31	180 MPa	110 MPa	60 MPa CW
10-32	150 MPa	80 MPa	30 MPa CW
10-33	30 ksi	15 ksi	10 ksi CCW
10-34	38 ksi	25 ksi	8 ksi CCW
10-35	55 ksi	15 ksi	5 ksi CCW
10-36	32 ksi	50 ksi	20 ksi CCW
10-37	−840 kPa	−335 kPa	120 kPa CCW
10-38	−325 kPa	−50 kPa	60 kPa CCW
10-39	−1800 psi	−300 psi	80 psi CW
10-40	−6500 psi	−2500 psi	1200 psi CW

C. For the following problems, use the data from the indicated problem for the initial stress element to draw the Mohr's circle. Then determine the stress condition on the element at the specified angle of rotation from the given x-axis. Draw the rotated element in its proper relation to the initial stress element and indicate the normal and shear stresses acting on it.

Problem	Problem for the initial stress data	Angle of rotation from the x-axis
10-41	10-1	30 deg CCW
10-42	10-1	30 deg CW
10-43	10-4	70 deg CCW
10-44	10-6	20 deg CW
10-45	10-8	50 deg CCW
10-46	10-10	45 deg CW
10-47	10-13	10 deg CCW
10-48	10-15	25 deg CW
10-49	10-16	80 deg CW
10-50	10-18	65 deg CW

D. For the following problems, use Equation (10-14) to compute the magnitude of the maximum shear stress for the data from the indicated problem.

10-51. Use data from Problem 10-17.
10-52. Use data from Problem 10-18.
10-53. Use data from Problem 10-19.
10-54. Use data from Problem 10-20.

COMPUTER PROGRAMMING ASSIGNMENTS

1. Write a program for a computer or a programmable calculator to aid in the construction of Mohr's circle. Input the initial stresses, σ_x, σ_y, and τ_{xy}. Have the program compute the radius of the circle, the maximum and minimum principal stresses, the maximum shear stress, and the average stress. Use the program in conjunction with freehand sketching of the circle for the data for Problems 10-1 through 10-24.

2. Enhance the program in Assignment 1 by computing the angle of orientation of the principal stress element and the angle of orientation of the maximum shear stress element.

3. Enhance the program in Assignment 1 by computing the normal and shear stresses on the element rotated at any specified angle relative to the original x-axis.

4. Enhance the program in Assignment 1 by causing it to detect if the principal stresses from the initial Mohr's circle are of the same sign; and in such cases, print out the three principal stresses in the proper order, σ_1, σ_2, σ_3. Also have the program compute the true maximum shear stress from Equation (10-13).

11

Special Cases of Combined Stresses

11-1 OBJECTIVES OF THIS CHAPTER

This chapter can be studied either after completing Chapter 10, or independently from it. There are several practical cases involving combined stresses that can be solved without resorting to the more rigorous and time-consuming procedures presented in Chapter 10, although the techniques discussed here are based on the principles of that chapter.

When a beam is subjected to both bending and a direct axial stress, either tensile or compressive, simple superposition of the applied stresses can be used to determine the combined stress. Many forms of power transmission equipment involve shafts that are subjected to torsional shear stress along with bending stress. Such shafts can be analyzed by using the maximum shear stress theory of failure and employing the equivalent torque technique of analysis.

After completing this chapter, you should be able to:

1. Compute the combined normal stress resulting from the application of bending stress with either direct tensile or compressive stresses using the principle of superposition.

2. Recognize the importance of visualizing the stress distribution over the cross

section of a load-carrying member and considering the stress condition at a point.

3. Recognize the importance of free-body diagrams of components of structures and mechanisms in the analysis of combined stresses.

4. Evaluate the design factor for combined normal stress, including the properties of either isotropic or nonisotropic materials.

5. Optimize the shape and dimensions of a load-carrying member relative to the variation of stress in the member and its strength properties.

6. Analyze members subjected to combined bending and torsion only by computing the resulting maximum shear stress.

7. Use the maximum shear stress theory of failure properly.

8. Apply the equivalent torque technique to analyze members subjected to combined bending and torsion.

9. Consider stress concentration factors when using the equivalent torque technique.

11-2 COMBINED NORMAL STRESSES

The first combination to be considered is bending with direct tension or compression. In any combined stress problem, it is helpful to visualize the stress distribution caused by the various components of the total stress pattern. Notice that bending results in tensile and compressive stresses, as do both direct tension and direct compression. Since the same kind of stresses are produced, a simple algebraic sum of the stresses produced at any point is all that is required to compute the resultant stress at that point. This process is called *superposition*.

An example of a member in which both bending and direct tensile stresses are developed is shown in Figure 11-1. The two horizontal beams support a 10 000-lb load by means of the cable assembly. The beams are rigidly attached to columns, so that they act as cantilever beams. The load at the end of each beam is equal to the tension in the cable. Figure 11-2 shows that the vertical component of the tension in each cable must be 5000 lb. That is,

$$F \cos 60° = 5000 \text{ lb}$$

and the total tension in the cable is

$$F = \frac{5000 \text{ lb}}{\cos 60°} = 10 \ 000 \text{ lb}$$

This is the force applied to each beam, as shown in Figure 11-3. For purposes of analyzing the stresses in the beam, it is convenient to resolve the applied force into horizontal and vertical components. The principle of superposition can now be applied by considering the effect of each component separately. The horizontal component F_h tends to produce a direct tensile stress over the entire cross section of the beam. The

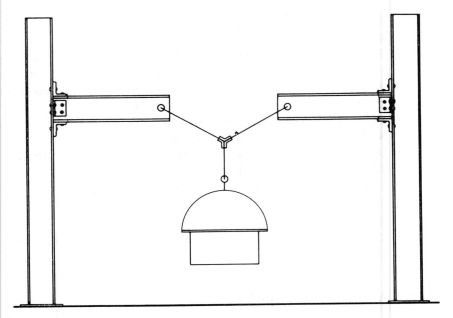

Figure 11-1 Two cantilever beams subjected to combined bending and axial tensile stresses.

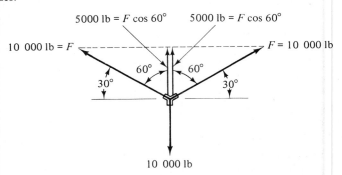

5000 lb = F cos 60°   5000 lb = F cos 60°

10 000 lb = F F = 10 000 lb

60°   60°

30°   30°

10 000 lb

Figure 11-2 Force analysis on cable system.

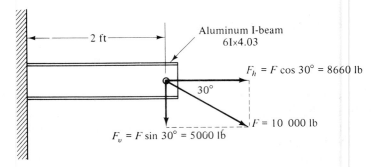

2 ft

Aluminum I-beam
6I×4.03

$F_h = F$ cos 30° = 8660 lb

30°

$F = 10\ 000$ lb

$F_v = F$ sin 30° = 5000 lb

Figure 11-3 Forces applied to each beam.

vertical component F_v produces bending in such a way that the upper part of the beam would be in tension and the lower part would be in compression.

Let's compute the magnitudes of the direct tensile stress and the bending stress separately. Considering *only* the horizontal force F_h, we find resulting tensile stress to be

$$\sigma_1 = \frac{F_h}{A}$$

Here the subscript 1 is merely indicating the first stress to be computed. The area of the aluminum 6I × 4.03 beam shape is 3.427 in², given in Appendix A-11. Then

$$\sigma_1 = \frac{8660 \text{ lb}}{3.427 \text{ in}^2} = 2527 \text{ psi}$$

Considering the beam to be a cantilever with a concentrated load of 5000 lb downward at its end, we find that the maximum bending moment would occur at the left end where it is attached to the column.

$$M = F_v(2 \text{ ft}) = 5000 \text{ lb } (2 \text{ ft})(12 \text{ in./ft}) = 120\ 000 \text{ lb} \cdot \text{in.}$$

From the flexure formula, using a section modulus of 7.33 in³ for the beam, the stress due to bending at the left end of the beam is

$$\sigma_2 = \frac{M}{S} = \frac{120\ 000 \text{ lb} \cdot \text{in.}}{7.33 \text{ in}^3} = 16\ 371 \text{ psi}$$

When combining the stresses, it is advisable to make a judgment as to the points in the member where the maximum tensile and compressive stresses would probably occur. In this case, the direct tensile stress σ_1 is the same throughout the beam. However, the bending stress σ_2 is zero at the right end, where the bending moment is zero, and reaches a maximum at the left end. Thus the left end appears to be the critical location, that is, where the maximum stresses occur.

Figure 11-4 shows a series of three sketches that illustrate the principle of superposition. Part (b) shows the uniform tensile stress $\sigma_1 = 2527$ psi acting on the beam cross section. Part (a) shows the bending stress distribution. Establishing a sign

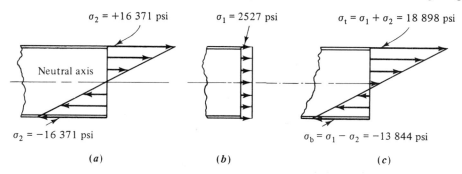

Figure 11-4 Diagram of superposition principle applied to the beams of Figure 11-1. (a) Bending stress distribution (b) Direct tensile stress distribution (c) Combined stress distribution

convention which denotes tensile stresses as positive and compressive stresses as negative, we can say that at the top of the beam the bending stress is $+16\ 371$ psi. At the bottom, the stress is $-16\ 371$ psi. Now since both σ_1 and σ_2 are produced at the same time, the combined stress on the section has the distribution shown in Figure 11-4(c). At the top,

$$\sigma_t = \sigma_1 + \sigma_2 = 2527 + 16\ 371 = 18\ 898 \text{ psi} \qquad \text{(tension)}$$

At the bottom,

$$\sigma_b = \sigma_1 - \sigma_2 = 2527 - 16\ 371 = -13\ 844 \text{ psi} \qquad \text{(compression)}$$

These are the maximum tensile and compressive stresses in the beam.

The analysis of members carrying loads which produce a combination of bending stress with direct tensile and compressive stresses can be summarized by the equation

$$\sigma = \pm\frac{M}{S} \pm \frac{F}{A} \tag{11-1}$$

The dual signs ($\pm$) indicate that it is necessary to determine the sense of the stresses (tensile or compressive) *at the point* where the combined stress is to be calculated.

Example Probllem 11-1

A picnic table in a park is made by supporting a circular top on a pipe which is rigidly held in concrete in the ground. Figure 11-5 shows the arrangement. Compute the maximum stress in the pipe if a person having a mass of 135 kg sits on the edge of the table. The pipe is made of an aluminum alloy and has an outside diameter of 170 mm and an inside diameter of 163 mm. If the aluminum has a yield strength of 241 MPa, what is the design factor for this loading?

Solution The pipe is subjected to combined bending and direct compression as illustrated in Figure 11-6, the free-body diagram of the pipe. The effect of the load is to

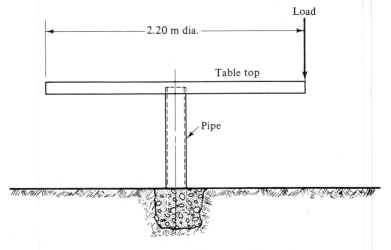

Figure 11-5 Picnic table supported by a pipe.

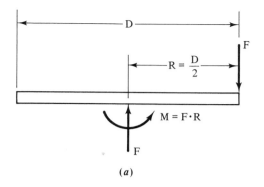

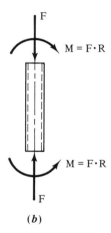

Figure 11-6 Free body diagrams for table top and pipe for Example Problem 11-1. (a) Free body diagram of table top (b) Free body diagram of pipe

produce a downward force on the top of the pipe while exerting a moment in the clockwise direction. The moment is the product of the load times the radius of the tabletop. The reaction at the bottom of the pipe, supplied by the concrete, is an upward force combined with a counterclockwise moment.

The force is the gravitational attraction of the 135-kg mass.

$$F = m \cdot g = 135 \text{ kg} \cdot 9.81 \text{ m/s}^2 = 1324 \text{ N}$$

Since this force acts at a distance of 1.1 m from the axis of the pipe, the moment is

$$M = (1324 \text{ N})(1.1 \text{ m}) = 1456 \text{ N} \cdot \text{m}$$

Now the direct compressive stress in the pipe is

$$\sigma_1 = -\frac{F}{A}$$

But

$$A = \frac{\pi(D_o^2 - D_i^2)}{4} = \frac{\pi(170^2 - 163^2) \text{ mm}^2}{4} = 1831 \text{ mm}^2$$

Then

$$\sigma_1 = -\frac{1324 \text{ N}}{1831 \text{ mm}^2} = -0.723 \text{ MPa}$$

This stress is a uniform, compressive stress across any cross section of the pipe.
The bending stress computation requires the application of the flexure formula,

$$\sigma_2 = \frac{Mc}{I}$$

where $c = \dfrac{D_o}{2} = \dfrac{170 \text{ mm}}{2} = 85 \text{ mm}$

$$I = \frac{\pi}{64}(D_0^4 - D_i^4) = \frac{\pi}{64}(170^4 - 163^4) \text{ mm}^4$$
$$I = 6.347 \times 10^6 \text{ mm}^4$$

Then

$$\sigma_2 = \frac{(1456 \text{ N} \cdot \text{m})(85 \text{ mm})}{6.347 \times 10^6 \text{ mm}^4} \times \frac{10^3 \text{ mm}}{\text{m}} = 19.5 \text{ MPa}$$

The bending stress σ_2 produces compressive stress on the right side of the pipe and tension on the left side. Since the direct compression stress adds to the bending stress on the right side, that is where the maximum stress would occur. The combined stress would then be

$$\sigma_t = -\sigma_1 - \sigma_2 = (-0.723 - 19.5) \text{ Mpa} = -20.22 \text{ MPa}$$

Based on a yield strength of 241 MPa, the design factor would be

$$N = \frac{S_y}{\sigma_t} = \frac{241 \text{ MPa}}{20.22 \text{ MPa}} = 11.9$$

11-3 COMBINED NORMAL AND SHEAR STRESSES

Rotating shafts in machines transmitting power represent good examples of members loaded in such a way as to produce combined bending and torsion. Figure 11-7 shows a shaft carrying two chain sprockets. Power is delivered to the shaft through the sprocket at C and transmitted down the shaft to the sprocket at B, which in turn delivers it to another shaft. Because it is transmitting power, the shaft between B and C is subject to a torque and torsional shear stress, as you learned in Chapter 5. In order for the sprockets to transmit torque, they must be pulled by one side of the chain. At C, the back side of the chain must pull down with the force F_1 in order to drive the sprocket clockwise. Since the sprocket at B drives a mating sprocket, the front side of the chain would be in tension under the force F_2. The two forces, F_1 and F_2, acting downward cause bending of the shaft. Thus the shaft must be analyzed for both torsional shear stress and bending stress. Then, since both stresses act at the same place on the shaft, their combined effect must be determined. The method of analysis to be used is called the *maximum shear stress theory of failure*, which is described below. Then example problems will be shown.

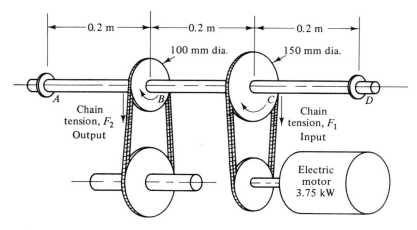

Figure 11-7 Power transmission shafts.

Maximum shear stress theory of failure. When the tensile or compressive stress caused by bending occurs at the same place that a shearing stress occurs, the two kinds of stress combine to produce a larger shearing stress. The maximum stress can be computed from

$$\tau_{max} = \sqrt{\left(\frac{\sigma}{2}\right)^2 + \tau^2} \qquad (11\text{-}2)$$

In Equation (11-2), σ refers to the magnitude of the tensile or compressive stress at the point, and τ is the shear stress at the same point. The result τ_{max} is the maximum shear stress at the point. The basis of Equation (11-2) can be shown by using Mohr's circle.

The maximum shear stress theory of failure states that a member fails when the maximum shear stress exceeds the yield strength of the material in shear. This failure theory shows good correlation with test results for ductile metals such as most steels.

Equivalent torque. Equation (11-2) can be expressed in a simplified form for the particular case of a circular shaft subjected to bending and torsion. Evaluating the bending stress separately, the maximum tensile or compressive stress would be

$$\sigma = \frac{M}{S}$$

where $S = \dfrac{\pi D^3}{32}$ = section modulus

D = diameter of the shaft
M = bending moment on the section

The maximum stress due to bending occurs at the outside surface of the shaft, as shown in Figure 11-8.

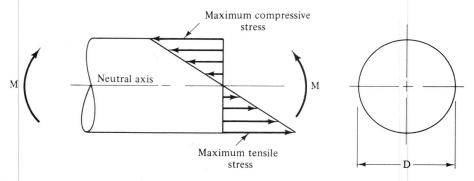

Figure 11-8 Bending stress distribution in a round shaft.

Now consider the torsional shear stress separately. In Chapter 5 the torsional shear stress equation was derived:

$$\tau = \frac{T}{Z_p}$$

where $Z_p = \dfrac{\pi D^3}{16}$ = polar section modulus

T = torque on the section

The maximum shear stress occurs at the outer surface of the shaft all the way around the diameter, as shown in Figure 11-9.

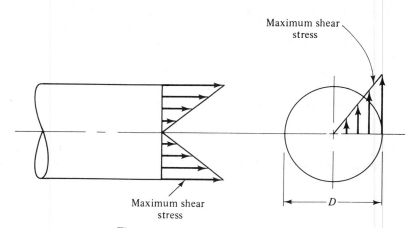

Figure 11-9   Shear stress distribution in a round shaft.

Thus the maximum tensile stress and the maximum torsional shear stress both occur at the same point in the shaft. Now let's use Equation (11-2) to get an expression for the combined stress in terms of the bending moment M, the torque T, and the shaft diameter D.

$$\tau_{max} = \sqrt{\left(\frac{\sigma}{2}\right)^2 + \tau^2}$$

$$\tau_{max} = \sqrt{\left(\frac{M}{2S}\right)^2 + \left(\frac{T}{Z_p}\right)^2} \qquad (11\text{-}3)$$

Notice that, from the definitions of S and Z_p given above,

$$2S = Z_p$$

Substituting this into Equation (11-3) gives

$$\tau_{max} = \sqrt{\left(\frac{M}{Z_p}\right)^2 + \left(\frac{T}{Z_p}\right)^2}$$

Factoring out Z_p yields

$$\tau_{max} = \frac{1}{Z_p} \sqrt{M^2 + T^2}$$

Sometimes the term $\sqrt{M^2 + T^2}$ is called the *equivalent torque* because it represents the amount of torque which would have to be applied to the shaft by itself to cause the equivalent magnitude of shear stress as the combination of bending and torsion. Calling the equivalent torque T_e,

$$T_e = \sqrt{M^2 + T^2} \qquad (11\text{-}4)$$

and

$$\tau_{max} = \frac{T_e}{Z_p} \qquad (11\text{-}5)$$

Equations (11-4) and (11-5) greatly simplify the calculation of the maximum shear stress in a circular shaft subjected to bending and torsion.

In the design of circular shafts subjected to bending and torsion, a design stress can be specified giving the maximum allowable shear stress. This was done in Chapter 5.

$$\tau_d = \frac{s_{ys}}{N}$$

where s_{ys} is the yield strength of the material in shear. Since s_{ys} is seldom known, the approximate value found from $s_{ys} = s_y/2$ can be used. Then

$$\tau_d = \frac{s_y}{2N} \qquad (11\text{-}6)$$

where s_y is the yield strength in tension, as reported in most material property tables, such as those in the Appendix. It is recommended that the value of the design factor be *no less than 4*. A rotating shaft subjected to bending is a good example of a repeated and reversed load. With each revolution of the shaft, a particular point on the surface is subjected to the maximum tensile and then the maximum compressive stress. Thus fatigue is the expected mode of failure, and $N = 4$ or greater is recommended, based on yield strength.

Stress concentrations.   In shafts, stress concentrations are created by abrupt changes in geometry such a key seats, shoulders, and grooves. The proper application of stress concentration factors to the equivalent torque Equations (11-4) and (11-5) should be considered carefully. If the value of K_t at a section of interest is equal for both bending and torsion, then it can be applied directly to Equation (11-5). That is,

$$\tau_{max} = \frac{T_e K_t}{Z_p} \tag{11-7}$$

The form of Equation (11-7) can also be applied as a conservative calculation of τ_{max} by selecting K_t as the largest value for either torsion or bending.

To account for the proper K_t for both torsion and bending, Equation (11-4) can be modified as

$$T_e = \sqrt{(K_{tB}M)^2 + (K_{tT}T)^2} \tag{11-8}$$

Then Equation (11-5) can be used directly to compute the maximum shear stress.

Example Problem 11-2

Specify a suitable material for the shaft shown in Figure 11-7. The shaft has a uniform diameter of 55 mm and rotates at 120 rpm while transmitting 3.75 kW of power. The chain sprockets at B and C are keyed to the shaft with sled-runner key seats. Sprocket C receives the power, and sprocket B delivers it to another shaft. The bearings at A and D provide simple supports for the shaft.

Solution   The several steps to be used to solve this problem are outlined below.

1. The torque in the shaft will be computed for the known power and rotational speed from $T = P/n$, as developed in Chapter 5.
2. The tensions in the chains for sprockets B and C will be computed. These are the forces which produce bending in the shaft.
3. Considering the shaft to be a beam, the shear and bending moment diagrams will be drawn for it.
4. At the section where the maximum bending moment occurs, the equivalent torque T_e will be computed from Equation (11-4).
5. The polar section modulus Z_p and the stress concentration factor K_t will be determined.
6. The maximum shear stress will be computed from Equation (11-7).
7. The required yield strength of the shaft material will be computed by letting $\tau_{max} = \tau_d$ in Equation (11-6) and solving for s_y. Remember, let $N = 4$ or more.
8. A steel that has a sufficient yield strength will be selected from Appendix A-13.

Now let's follow the steps.

Step 1.   The desirable unit for torque is N · m. Then it is most convenient to observe that the power unit of kilowatts is equivalent to the units of kN · m/s. Also, rotational speed must be expressed in rad/s.

$$n = \frac{120 \text{ rev}}{\text{min}} \times \frac{2\pi \text{ rad}}{\text{rev}} \times \frac{1 \text{ min}}{60 \text{ s}} = 12.57 \text{ rad/s}$$

We can now compute torque.

$$T = \frac{P}{n} = \frac{3.75 \text{ kN} \cdot \text{m}}{\text{s}} \times \frac{1}{12.57 \text{ rad/s}} = 0.298 \text{ kN} \cdot \text{m}$$

Step 2. The tensions in the chains are indicated in Figure 11-7 by the forces F_1 and F_2. For the shaft to be in equilibrium, the torque on both sprockets must be the same in magnitude but opposite in direction. On either sprocket the torque is the product of the chain force times the *radius* of the pulley. That is,

$$T = F_1 R_1 = F_2 R_2$$

The forces can now be computed.

$$F_1 = \frac{T}{R_1} = \frac{0.298 \text{ kN} \cdot \text{m}}{75 \text{ mm}} \times \frac{10^3 \text{ mm}}{\text{m}} = 3.97 \text{ kN}$$

$$F_2 = \frac{T}{R_2} = \frac{0.298 \text{ kN} \cdot \text{m}}{50 \text{ mm}} \times \frac{10^3 \text{ mm}}{\text{m}} = 5.96 \text{ kN}$$

Step 3. Figure 11-10 shows the complete shear and bending moment diagrams found by the methods of Chapter 6. The maximum bending moment is $1.06 \text{ kN} \cdot \text{m}$ at section B, where one of the sprockets is located.

Step 4. At section B, the torque in the shaft is $0.298 \text{ kN} \cdot \text{m}$ and the bending

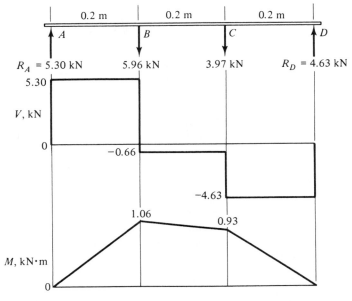

Figure 11-10 Shearing force and bending moment diagrams for Example Problem 11-2

moment is 1.06 kN · m. Then

$$T_e = \sqrt{M^2 + T^2} = \sqrt{(1.06)^2 + (0.298)^2} = 1.10 \text{ kN} \cdot \text{m}$$

Step 5.

$$Z_p = \frac{\pi D^3}{16} = \frac{\pi (55 \text{ mm})^3}{16} = 32.67 \times 10^3 \text{ mm}^3$$

For the key seat at section B securing the sprocket to the shaft, we will use $K_t = 1.6$, as reported in Figure 5-7.

Step 6.

$$\tau_{max} = \frac{T_e K_t}{Z_p} = \frac{1.10 \times 10^3 \text{ N} \cdot \text{m} \, (1.6)}{32.67 \times 10^3 \text{ mm}^3} \times \frac{10^3 \text{ mm}}{\text{m}} = 53.9 \text{ MPa}$$

Step 7. Let

$$\tau_{max} = \tau_d = \frac{s_y}{2N}$$

Then

$$s_y = 2N\tau_{max} = (2)(4)(53.9 \text{ MPa}) = 431 \text{ MPa}$$

Step 8. Referring to Appendix A-13, we find that several alloys could be used. For example, AISI 1040, cold drawn, has a yield strength of 490 MPa. Alloy AISI 1141 OQT 1300 has a yield strength of 469 MPa and also a very good ductility, as indicated by the 28% elongation. Either of these would be reasonable choices.

PROBLEMS

11-1. A $2\frac{1}{2}$-inch schedule 40 steel pipe is used as a support for a basketball backboard, as shown in Figure 11-11. It is securely fixed into the ground. Compute the stress which would be developed in the pipe if a 230-lb player hung on the base of the rim of the basket.

11-2. The bracket shown in Figure 11-12 has a rectangular cross section 18 mm wide by 75 mm high. It is securely attached to the wall. Compute the maximum stress in the bracket.

11-3. The beam shown in Figure 11-13 carries a 6000-lb load attached to a bracket below the beam. Compute the stress at points M and N, where it is attached to the column.

11-4. For the beam shown in Figure 11-13, compute the stress at points M and N if the 6000-lb load acts vertically downward instead of at an angle.

11-5. For the beam shown in Figure 11-13, compute the stress at points M and N if the 6000-lb load acts back toward the column at an angle of 40 deg below the horizontal instead of as shown.

11-6. Compute the maximum stress in the top portion of the coping saw frame shown in Figure 11-14 if the tension in the blade is 125 N.

11-7. Compute the maximum stress in the crane beam shown in Figure 11-15 when a load of 12 kN is applied at the middle of the beam.

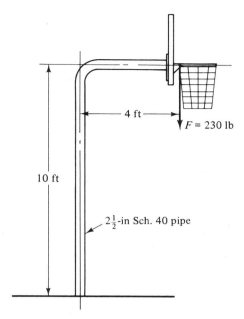

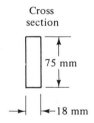

Figure 11-11 Basketball backboard for Problem 11-1.

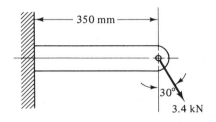

Figure 11-12 Bracket for Problem 11-2.

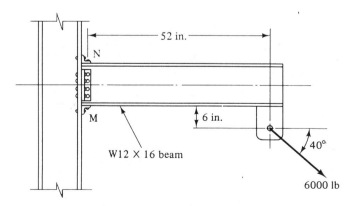

Figure 11-13 Beam for Problems 11-3, 11-4, and 11-5.

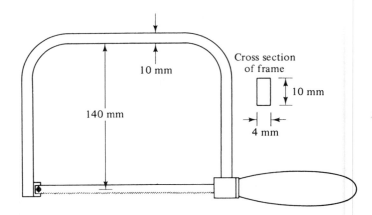

Figure 11-14 Coping saw frame for Problem 11-6.

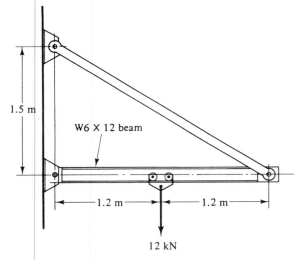

Figure 11-15 Crane beam for Problem 11-7.

12 kN

11-8. Figure 11-16 shows a metal-cutting hacksaw. Its frame is made of hollow tubing having an outside diameter of 12 mm and a wall thickness of 1.0 mm. The blade is pulled taut by the wing nut so that a tensile force of 160 N is applied to the blade. Compute the maximum stress in the top section of the tubular frame.

Figure 11-16 Hack saw frame for Problem 11-8.

11-9. The C-clamp in Figure 11-17 is made of cast zinc, ASTM B669-80. Determine the allowable clamping force which the clamp can exert if it is desired to have a design factor of 4 based on ultimate strength in either tension or compression.

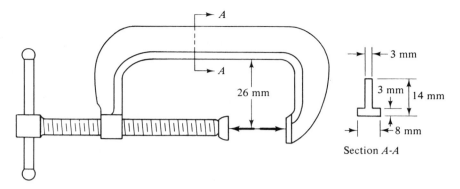

Figure 11-17 C-clamp for Problem 11-9.

11-10. The C-clamp shown in Figure 11-18 is made of cast malleable iron, ASTM A220 grade 45008. Determine the allowable clamping force which the clamp can exert if it is desired to have a design factor of 4 based on ultimate strength in either tension or compression.

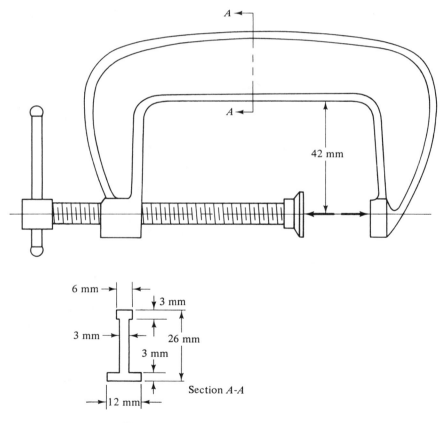

Figure 11-18 C-clamp for Problem 11-10.

11-11. A tool used for compressing a coil spring to allow its installation in a car is shown in Figure 11-19. A force of 1200 N is applied near the ends of the extended lugs, as shown. Compute the maximim tensile stress in the threaded rod. Assume a stress concentration factor of 3.0 at the thread root for both tensile and bending stresses. Then, using a design factor of 2 based on yield strength, specify a suitable material for the rod.

11-12. Figure 11-20 shows a portion of the steering mechanism for a car. The steering arm has the detailed design shown. Notice that the arm has a constant thickness of 10 mm so

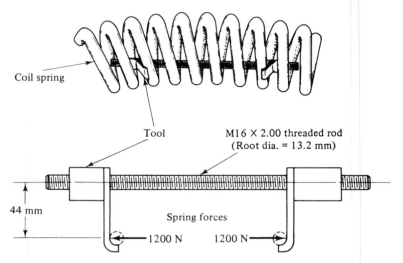

Figure 11-19 Tool for compressing springs for Problem 11-11.

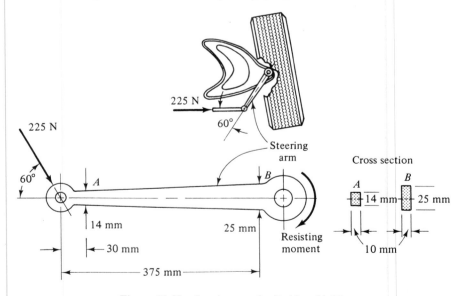

Figure 11-20 Steering arm for Problem 11-12.

Special Cases of Combined Stresses   Chap. 11

that all cross sections between the end lugs are rectangular. The 225-N force is applied to the arm at a 60-deg angle. Compute the stress in the arm at sections A and B. Then, if the arm is made of ductile iron, ASTM A536, grade 80-55-6, compute the minimum design factor at these two points based on ultimate strength.

11-13. An S3 × 5.7 American Standard I-beam is subjected to the forces shown in Figure 11-21. The 4600-lb force acts directly in line with the axis of the beam. The 500-lb downward force at A produces the reactions shown at the supports B and C. Compute the maximum tensile and compressive stresses in the beam.

11-14. The horizontal crane boom shown in Figure 11-22 is made of a hollow rectangular steel tube. Compute the stress in the boom just to the left of point B when a mass of 1000 kg is supported at the end.

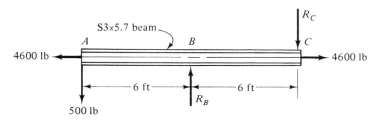

Figure 11-21 Beam for Problem 11-13.

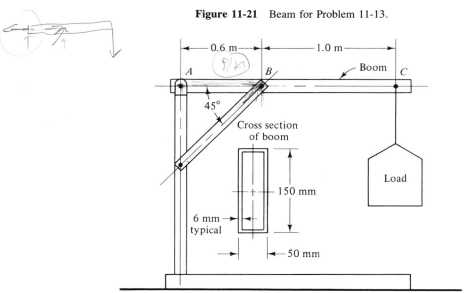

Figure 11-22 Crane boom for Problems 11-14 and 11-15.

11-15. For the crane boom shown in Figure 11-22, compute the load that could be supported if a design factor of 3 based on yield strength is desired. The boom is made of AISI 1040 hot-rolled steel. Analyze only sections where the full box section carries the load, assuming that sections at connections are adequately reinforced.

11-16. A television antenna is mounted on a hollow aluminum tube as sketched in Figure 11-23. During installation, a force of 20 lb is applied to the end of the antenna as shown.

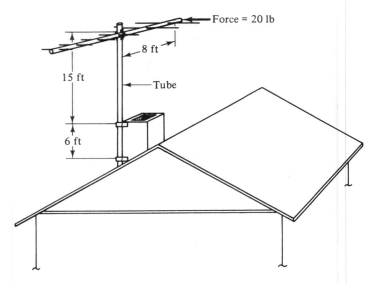

Figure 11-23 Antenna mounting tube for Problem 11-16.

Calculate the torsional shear stress in the tube and the stress due to bending. Consider the tube to be simply supported against bending at the clamps, but assume that rotation is not permitted. If the tube is made of 6061-T6 aluminum, would it be safe under this load? The tube has an outside diameter of 1.50 in. and a wall thickness of $\frac{1}{16}$ in.

11-17. Figure 11-24 shows a crank to which a force F of 1200 N is applied. Compute the maximum stress in the circular portion of the crank.

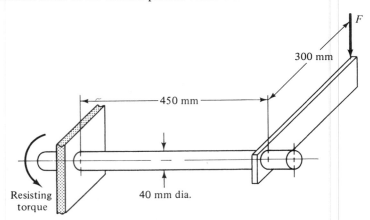

Figure 11-24 Crank for Problem 11-17.

11-18. A standard steel pipe is to be used to support a bar carrying four loads, as shown in Figure 11-25. Specify a suitable pipe which would keep the maximum combined shear stress to 8000 psi.

11-19. A 1.0-in.-diameter solid round shaft will carry 25 hp while rotating at 1150 rpm in the arrangement shown in Figure 11-26. The total bending loads at the gears A and C are

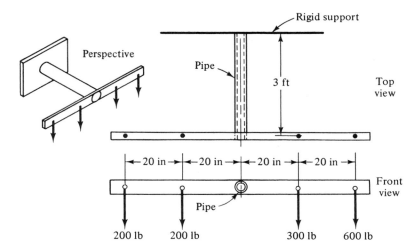

Figure 11-25 Bracket for Problem 11-18.

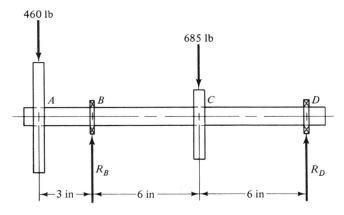

Figure 11-26 Shaft for Problem 11-19.

shown, along with the reactions at the bearings B and D. Use a design factor of 6 for the maximum shear stress theory of failure, and determine a suitable steel for the shaft.

11-20. Three pulleys are mounted on a rotating shaft, as shown in Figure 11-27. Belt tensions are as shown in the end view. All power is received by the shaft through pulley C from below. Pulleys A and B deliver power to mating pulleys above.

(a) Compute the torque at all points in the shaft.

(b) Compute the bending stress and the torsional shear stress in the shaft at the point where the maximum bending moment occurs if the shaft diameter is 1.75 in.

(c) Compute the maximum shear stress at the point used in step (b). Then specify a suitable material for the shaft.

11-21. The vertical shaft shown in Figure 11-28 carries two belt pulleys. The tensile forces in the belts under operating conditions are shown. Also, the shaft carries an axial compressive load of 6.2 kN. Considering torsion, bending, and axial compressive stresses, compute the maximum shear stress using Equation (11-2).

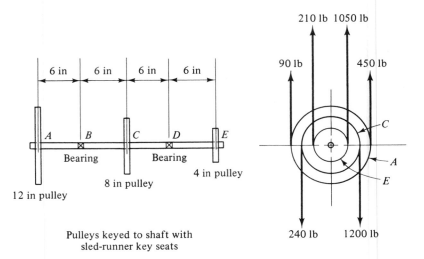

Pulleys keyed to shaft with
sled-runner key seats

Figure 11-27 Shaft for Problem 11-20.

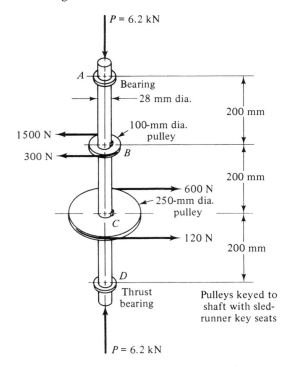

Figure 11-28 Shaft for Problem 11-21.

11-22. For the shaft in Problem 11-21, specify a suitable steel that would provide a design factor of 4 based on yield strength in shear.

11-23. The member *EF* in the truss shown in Figure 11-29 carries an axial tensile load of 54 000 lb in addition to the two 1200-lb loads shown. It is planned to use two steel

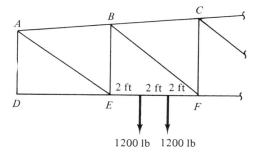

Figure 11-29 Truss for Problem 11-23.

angles for the member, arranged back to back. Specify a suitable size for the angles if they are to be made of ASTM A36 structural steel with an allowable stress of 0.6 times the yield strength.

11-24. A solid circular bar is 40 mm in diameter and is subjected to an axial tensile force of 150 kN along with a torque of 500 N · m. Compute the maximum shear stress in the bar.

11-25. A solid circular bar is 2.25 in. in diameter and is subjected to an axial tensile force of 47 000 lb along with a torque of 8500 lb · in. Compute the maximum shear stress in the bar.

11-26. A short solid circular bar is 4.00 in. in diameter and is subjected to an axial compressive force of 40 000 lb along with a torque of 25 000 lb · in. Compute the maximum shear stress in the bar.

11-27. A short hollow circular post is made from a 12-in. schedule 40 steel pipe and it carries an axial compressive load of 250 000 lb along with a torque of 180 000 lb · in. Compute the maximum shear stress in the post.

11-28. A short hollow circular support is made from a 3-in. schedule 40 steel pipe and it carries an axial compressive load of 25 000 lb along with a torque of 15 500 lb · in. Compute the maximum shear stress in the support.

11-29. A machine screw has Number 8-32 UNC American Standard threads (see Appendix A-3). The screw is subjected to an axial tensile force that produces a direct tensile stress in the threads of 15 000 psi based on the tensile stress area. There is a section under the head without threads that has a diameter equal to the major diameter of the threads. This section is also subjected to a direct shearing force of 120 lb. Compute the maximum shear stress in this section.

11-30. Repeat Problem 11-29 except the screw threads are $\frac{1}{4}$-20 UNC American Standard and the shearing force is 775 lb.

11-31. Repeat Problem 11-29 except the screw threads are No. 4-48 UNF American Standard and the shearing force is 50 lb.

11-32. Repeat Problem 11-29 except the screw threads are $1\frac{1}{4}$-12 UNF and the shearing force is 2500 lb.

11-33. A machine screw has metric threads with a major diameter of 16 mm and a pitch of 2.0 mm (see Appendix A-3). The screw is subjected to an axial tensile force that produces a direct tensile stress in the threads of 120 MPa based on the tensile stress area. There is a section under the head without threads that has a diameter equal to the major diameter of the threads. This section also is subjected to a direct shearing force of 8.0 kN. Compute the maximum shear stress in this section.

11-34. A machine screw has metric threads with a major diameter of 48 mm and a pitch of 5.0 mm (see Appendix A-3). The screw is subjected to an axial tensile force that produces a direct tensile stress in the threads of 120 MPa based on the tensile stress area. There is a section under the head without threads that has a diameter equal to the major diameter of the threads. This section also is subjected to a direct shearing force of 80 kN. Compute the maximum shear stress in this section.

11-35. A rectangular bar is used as a beam carrying a concentrated load of 5500 lb at the middle of its 60-in. length. The cross section is 2.00 in. wide and 6.00 in. high with the 6.00-in. dimension oriented vertically. Compute the maximum shear stress that occurs in the bar near the load at the following points within the section:
   (a) At the bottom of the bar.
   (b) At the top of the bar.
   (c) At the neutral axis.
   (d) At a point 1.0 in. above the bottom of the bar.
   (e) At a point 2.0 in. above the bottom of the bar.

11-36. Repeat Problem 11-35 except the beam is an aluminum I-beam, I6 × 4.692.

11-37. Repeat Problem 11-35 except the load is a uniformly distributed load of 100 lb/in. over the entire length. Consider cross sections near the middle of the beam, near the supports, and 15 in. from the left support.

11-38. A circular shaft carries the load shown in Figure 11-30. The shaft carries a torque of 1500 N · m between sections B and C. Compute the maximum shear stress in the shaft near section B.

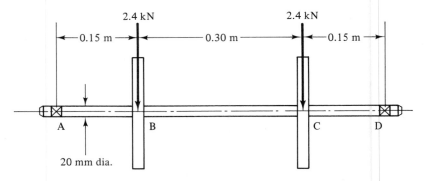

Figure 11-30   Shaft for Problem 11-38.

11-39. A circular shaft carries the load shown in Figure 11-31. The shaft carries a torque of 4500 N · m between sections B and C. Compute the maximum shear stress in the shaft near section B.

11-40. A square bar is 25 mm on a side and carries an axial tensile load of 75 kN along with a torque of 245 N · m. Compute the maximum shear stress in the bar. (*Note*: Refer to Section 5-9 and Figure 5-15.)

11-41. A rectangular bar is 30 mm by 50 mm in cross section and is subjected to an axial tensile force of 175 kN along with a torque of 525 N · m. Compute the maximum shear stress in the bar. (*Note*: Refer to Section 5-9 and Figure 5-15.)

11-42. A bar has a cross section in the form of an equilateral triangle, 50 mm on a side. It carries an axial tensile force of 115 kN along with a torque of 775 N · m. Compute the maximum shear stress in the bar. (*Note*: Refer to Section 5-9 and Figure 5-15.)

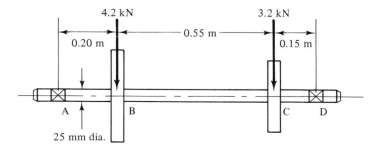

Figure 11-31 Shaft for Problem 11-39.

11-43. A link in a large mechanism is made from a square structural tube, $3 \times 3 \times \frac{1}{4}$ (see Appendix A-9). It is originally designed to carry an axial tensile load that produces a design factor of 3 based on the yield strength of ASTM A500 cold-formed structural steel, shaped, grade C.

    (a) Determine this load and the maximum shear stress that results in the tube under this load.

    (b) In operation, a torque of 950 lb · ft is experienced by the tube in addition to the axial load. Compute the maximum shear stress under this combined loading and compute the resulting design factor based on the yield strength of the steel in shear.

For Problems 11-44 through 11-47, specify a suitable size for the horizontal portion of the member that will maintain the combined tensile or compressive stress to 6000 psi if the data are in the U.S. Customary unit system and 42 MPa if the SI metric system is used. The cross section is to be square. The loading for these members is the same as that used in problems in Chapter 6 based on Figure 6-54 for which the free-body diagrams and the shearing force and bending moment diagrams were to be determined.

11-44. Use Figure 6-54(a).
11-45. Use Figure 6-54(b).
11-46. Use Figure 6-54(c).
11-47. Use Figure 6-54(d).

12

Deflection of Beams

12-1 OBJECTIVES OF THIS CHAPTER

The proper performance of machine parts, the structural rigidity of buildings, vehicles, and machine frames, and the tendency for a part to vibrate are all dependent on deformation in beams. Therefore, the ability to analyze beams for deflections under load is very important.

The spindle of a lathe or drill press and the arbor of a milling machine carry cutting tools for machining metals. Deflection of the spindle or arbor would have an adverse effect on the accuracy that the machine could produce. The manner of loading and support of these machine elements indicate that they are beams, and the approach to computing their deflection will be discussed in this chapter.

Precision measuring equipment must also be designed to be very rigid. Deflection caused by the application of measuring forces reduces the precision of the desired measurement.

Power-transmission shafts carrying gears must have sufficient rigidity to ensure that the gear teeth mesh properly. Excessive deflection of the shafts would tend to separate the mating gears, resulting in a movement away from the most desirable point of contact between the gear teeth. Noise generation, decreased power-transmitting capability, and increased wear would result. For straight spur gears, it is recommended that the movement between two gears not exceed 0.005 in. (0.13 mm). This limit is the *sum* of the movement of the two shafts carrying the mating gears at the location of the gears.

The floors of buildings must have sufficient rigidity to carry expected loads. Occupants of the building should not notice floor deflections. Machines and other equipment require a stable floor support for proper operation. Beams carrying plastered ceilings must not deflect excessively so as not to crack the plaster. A limit of $\frac{1}{360}$ times the span of the beam carrying a ceiling is often used for deflection.

Frames of vehicles, metal-forming machines, automation devices, and process equipment must also possess sufficient rigidity to ensure satisfactory operation of the equipment carried by the frame. The bed of a lathe, the crown of a punch press, the structure of an automatic assembly device, and the frame of a truck are examples.

Vibration is caused by the forced oscillations of parts of a structure or machine. The tendency to vibrate at a certain frequency and the severity of the vibrations are functions of the flexibility of the parts. Of course, flexibility is a term used to describe how much a part deflects under load. Vibration problems can be solved by either *increasing or decreasing* the stiffness of the part, depending on the circumstances. In either case, an understanding of how to compute deflections of beams is important.

In this chapter we present the principles on which the computation of the deflection of beams is based, along with four popular methods of deflection analysis: the *successive integration method*, the *moment-area method*, the *formula method*, and the *superposition method*.

Each method has its advantages and disadvantages, and your choice of which method to use depends on the nature of the problem. The formula method is the simplest but it depends on the availability of a suitable formula to match the application. The superposition method, a modest extension of the formula method, dramatically expands the number of practical problems that can be handled without a significant increase in the complexity of your work. The moment-area method is fairly quick and simple, but it is typically used for computing the deflections of only one or a few points on the beam. Its use requires a high level of understanding of the principle of moments and the techniques of preparing bending moment diagrams. The successive integration method is perhaps the most general and it can be used to solve virtually any combination of loading and support conditions for statically determinate beams. But its use requires the ability to write the equations for the shearing force and bending moment diagrams and to derive equations for the slope and deflection of the beam using integral calculus. The successive integration method results in equations for the slope and deflection for the entire beam and enables the direct determination of the point of maximum deflection. Published formulas were developed using the successive integration or the moment-area method.

Several computer-assisted beam analysis programs are available to reduce the time and computation required to determine the deflection of beams. Although they can relieve the designer of much work, it is recommended that the principles on which they are based be understood before using them.

After completing this chapter, you should be able to:

1. Understand the need for considering beam deflections.

2. Understand the development of the relationships between the manner of loading and support for a beam and the deflection of the beam.
3. Graphically show the relationships among the load, shearing force, bending moment, slope, and deflection curves for beams.
4. Develop formulas for the deflection of beams for certain cases using the successive integration method.
5. Apply the method of successive integration to beams having a variety of loading and support conditons.
6. Use standard formulas to compute the deflection of beams at selected points.
7. Use the principle of superposition along with standard formulas to solve problems of greater complexity.
8. Use the moment-area method to solve for the slope and deflection for beams.
9. Write computer programs to assist in using the several methods of beam analysis described in this chapter.

The organization of the chapter allows selective coverage. In general, all the information necessary to use each method is included within that part of the chapter. An exception is that the understanding of the formula method is necessary before using the superposition method.

12-2 DEFINITION OF TERMS

To describe graphically the condition of a beam carrying a pattern of loading, five diagrams are used as shown in Figure 12-1. We have used the first three diagrams in earlier work. The *load diagram* is the free-body diagram on which all external loads and support reactions are shown. From that, the *shearing force diagram* was developed which enables the calculation of shearing stresses in the beam at any section. The *bending moment diagram* is a plot of the variation of bending moment with position on the beam with the results used for computing the stress due to bending. The horizontal axis of these plots is the position on the beam, called x. It is typical to measure x relative to the left end of the beam, but any reference point can be used.

Deflection diagram. The last two diagrams are related to the deformation of the beam under the influence of the loads. It is convenient to begin discussion with the last diagram, the *deflection diagram*, because this shows the shape of the deflected beam. Actually, it is the plot of the position of the neutral axis of the beam relative to its initial position. The initial position is taken to be the straight line between the two support points. The amount of deflection will be called y, with positive values measured upward. Typical beams carrying downward loads, such as that shown in Figure 12-1, will result in downward (negative) deflections of the beam.

Slope diagram. A line drawn tangent to the deflection curve at a point of interest would define the slope of the deflection curve at that point. The slope is indicated as the angle, θ, measured in radians, relative to the horizontal as shown in

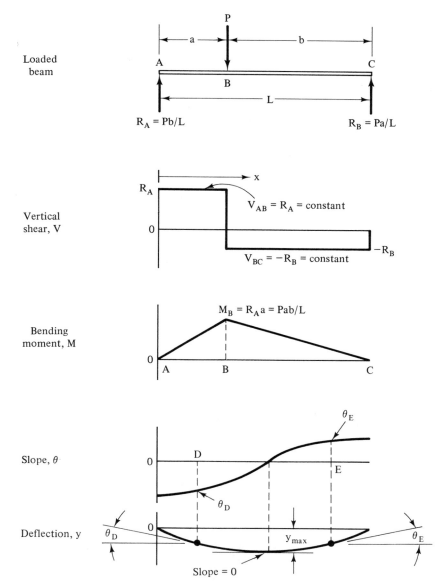

Figure 12-1 Five beam diagrams.

Figure 12-1. The plot of the slope as a function of position on the beam is the *slope curve*, drawn below the bending moment curve and above the deflection curve. Note on the given beam that the slope of the left portion of the deflection curve is negative and the right portion has a positive slope. The point where the tangent line is itself horizontal is the point of zero slope and defines the location of the maximum deflection. This observation will be used in the discussion of the moment-area method and the successive integration method later in this chapter.

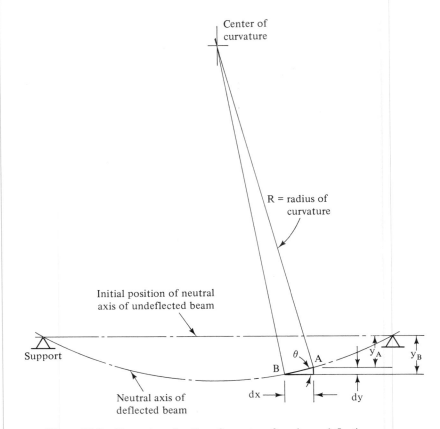

Figure 12-2 Illustration of radius of curvature for a beam deflection curve.

Radius of curvature. Figure 12-2 shows the radius of curvature, R, at a particular point. For practical beams, the curvature is very slight, resulting in a very large value for R. For convenience, the shape of the deflection curve is exaggerated to aid in visualizing the principles and the variables involved in the analysis. Remember from analytic geometry that the radius of curvature at a point is perpendicular to the line drawn tangent to the curve at that point.

The relationship between slope and deflection is also illustrated in Figure 12-2. Over a small distance dx, the deflection changes by a small amount dy. A small portion of the deflection curve itself completes the right triangle from which we can define.

$$\tan \theta = \frac{dy}{dx} \tag{12-1}$$

The absolute value of θ will be very small because the curvature of the beam is very slight. We can then take advantage of the observation that for small angles, $\tan \theta = \theta$. Then

$$\theta = \frac{dy}{dx} \tag{12-2}$$

Thus *the slope of the deflection curve at a point is equal to the derivative of the deflection with respect to position on the beam.*

Beam stiffness. It will be shown later that the amount of deflection for a beam is inversely proportional to the *beam stiffness,* indicated by the product *EI,* where

E = modulus of elasticity of the material of the beam

I = moment of inertia of the cross section of the beam
with respect to the neutral axis

12-3 BASIC PRINCIPLES FOR BEAM DEFLECTION

In this section we show the mathematical relationships between the moment, slope, and deflection curves from which you can solve for the actual equations for a given beam with a given loading and support condition.

Figure 12-3 shows a small segment of a beam in its initial straight shape and its deflected shape. The sides of the segment remain straight as the beam deflects, but they rotate with respect to a point at the neutral axis. This results in compression in

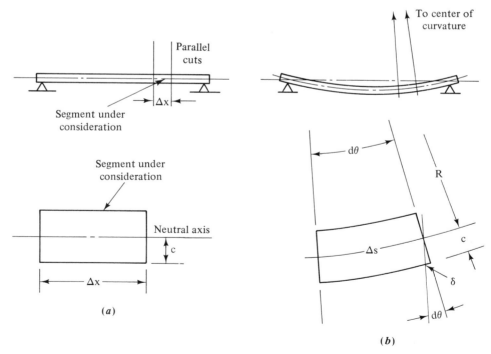

Figure 12-3 Relation between radius R and deformation δ. (a) Segment of a straight beam (b) Segment of deflected beam

the top portion of the segment and tension in the bottom portion, a fact used in the development of the flexure formula in Chapter 8.

The rotated sides of the segment intersect at the center of curvature and form the small angle $d\theta$. Note also the radius of curvature, R, measured from the center of curvature to the neutral axis. From the geometry shown in the figure,

$$\Delta s = R(d\theta) \tag{12-3}$$

and

$$\delta = c(d\theta) \tag{12-4}$$

where Δs is the length of the segment at the neutral axis and δ is the elongation of the bottom line of the segment that occurs as the beam deflects. The term c has the same meaning as in the flexure formula, the distance from the neutral axis to the outer fiber of the section.

Recall that the definition of the neutral axis states that no strain occurs there. Then the length Δs in the deflected beam segment equals the length Δx in the undeflected segment, and Equation (12-3) can be written

$$\Delta x = R(d\theta) \tag{12-5}$$

Now both Equations (12-4) and (12-5) can be solved for $d\theta$.

$$d\theta = \frac{\delta}{c}$$

$$d\theta = \frac{\Delta x}{R}$$

Equating these values of $d\theta$ to each other gives

$$\frac{\delta}{c} = \frac{\Delta x}{R}$$

Another form of this equation is

$$\frac{c}{R} = \frac{\delta}{\Delta x}$$

The right side of this equation conforms to the definition of strain, ϵ. Then

$$\epsilon = \frac{c}{R} \tag{12-6}$$

Earlier it was shown that

$$\epsilon = \frac{\sigma}{E}$$

where σ, the stress due to bending, can be computed from the flexure formula,

$$\sigma = \frac{Mc}{I}$$

Then

$$\epsilon = \frac{\sigma}{E} = \frac{Mc}{EI}$$

Combining this with Equation (12-6) gives

$$\frac{c}{R} = \frac{Mc}{EI}$$

Dividing both sides by c gives

$$\frac{1}{R} = \frac{M}{EI} \tag{12-7}$$

Equation (12-7) is useful in developing the moment-area method for finding beam deflections.

In analytic geometry, the reciprocal of the radius of curvature, $1/R$, is defined as the *curvature* and denoted by κ, the lowercase Greek letter kappa. Then

$$\kappa = \frac{M}{EI} \tag{12-8}$$

Equation (12-8) indicates that the curvature gets greater as the bending moment increases, which stands to reason. Similarly, the curvature decreases as the beam stiffness, EI, increases.

Another principle of analytic geometry states that if the equation of a curve is expressed as $y = f(x)$, that is, y is a function of x, then the curvature is

$$\kappa = \frac{d^2y}{dx^2} \tag{12-9}$$

Combining Equations (12-8) and (12-9) gives

$$\frac{M}{EI} = \frac{d^2y}{dx^2} \tag{12-10}$$

or

$$M = EI\frac{d^2y}{dx^2} \tag{12-11}$$

Equations (12-10) and (12-11) are useful in developing the successive integration method for finding beam deflections.

12-4 BEAM DEFLECTIONS—SUCCESSIVE INTEGRATION METHOD— GENERAL APPROACH

A general approach will now be presented which allows the determination of deflection at any point on the beam. The advantages of this approach are listed below.

1. The result is a set of equations for deflection at all parts of the beam. Deflection at any point can then be found by substitution of the beam stiffness properties of E and I, and the position on the beam.

2. Data are easily obtained from which a plot of the shape of the deflection curve may be made.

3. The equations for the *slope* of the beam at any point are generated, as are deflections. This is important in some machinery applications such as shafts at bearings and shafts carrying gears. An excessive slope of the shaft would result in poor performance and reduced life of the bearings or gears.

4. The fundamental relationships between loads, manner of support, beam stiffness properties, slope, and deflections are emphasized in the solution procedure. The designer who understands these relationships can make more efficient designs.

5. The method requires the application of only simple mathematical concepts.

6. The point of maximum deflection can be found directly from the resulting equations.

The basis for the successive integration method has been developed in Sections 12-2 and 12-3. The five beam diagrams will be prepared, in a manner similar to that shown in Figure 12-1, to relate the loads, shearing forces, bending moments, slopes, and deflections over the entire length of the beam.

The load, shearing force, and bending moment diagrams can be drawn using the principles from Chapter 6. Then equations for the bending moment are derived for all segments of the bending moment diagram.

Equation (12-11) is then used to develop the slope and deflection equations from the moment equations by integrating twice with respect to the position on the beam, x, as follows.

$$M = EI \frac{d^2y}{dx^2} \tag{12-11}$$

Now, integrating once with respect to x gives,

$$\int M \, dx = EI \int \frac{d^2x}{dx^2} \, dx = EI \frac{dy}{dx} \tag{12-12}$$

Earlier we showed that $dy/dx = \theta$, the slope of the deflection curve. Then,

$$\int M \, dx = EI\theta \tag{12-13}$$

Equation (12-12) can be integrated again, giving

$$\int EI\theta \, dx = EI \int \frac{dy}{dx} \, dx = EI \, y \tag{12-14}$$

After the final values for $EI\theta$ and $EI \, y$ have been determined, they will be divided by the beam stiffness, EI, to obtain the values for slope, θ, and deflection, y.

The steps indicated by Equations (12-11) through (12-14) are to be completed for each segment of the beam for which the moment diagram is continuous. Also, because our objective is to obtain discrete equations for slope and deflection for particular beam loading patterns, we will need to evaluate a constant of integration for each integration performed.

The development of the equations for the bending moment versus position is often accomplished by integrating the equations for the shearing force versus x as shown in Chapter 6. This follows from the rule that the change in bending moment between two points on a beam is equal to the area under the shearing force curve between the same two points.

The step-by-step method used to find the deflection of beams using the general approach is as follows.

1. Determine the reactions at the supports for the beam.
2. Draw the shear and bending moment diagrams using the same procedures presented in Chapter 6, and identify the magnitudes at critical points.
3. Divide the beam into segments in which the shear diagram is continuous by naming points at the places where abrupt changes occur with the letters A, B, C, D, etc.
4. Write equations for the shear curve in each segment. In most cases, these will be equations of straight lines, that is, equations involving x to the first power. Sometimes, as for beams carrying concentrated loads, the equation will be simply of the form

$$V = \text{constant}$$

5. For each segment, perform the process,

$$M = \int V \, dx + C$$

To evaluate the constant of integration which ties the moment equation to the particular values already known for the moment diagram, insert known boundary conditions and solve for C.

6. For each segment, perform the process,

$$\theta EI = \int M \, dx + C$$

The constant of integration generated here cannot be evaluated directly right away. So each constant should be identified separately by a subscript such as C_1, C_2, C_3, etc. Then when they are evaluated, they can be put in their proper places.

7. For each segment, perform the process,

$$yEI = \int \theta EI \, dx + C$$

Here again, the constants should be labeled with subscripts.

8. Establish *boundary conditions* for the slope and deflection diagrams. As many boundary conditions must be identified as there are unknown constants from

steps 6 and 7. Boundary conditions express mathematically the special values of slope and deflection at certain points and the fact that both the slope curve and the deflection curve are continuous. Typical boundary conditions are:

a. The deflection of the beam at each support is zero.
b. The deflection of the beam at the end of one segment is equal to the deflection of the beam at the beginning of the next segment. This follows from the fact that the deflection curve is continuous; that is, it has no abrupt changes.
c. The slope of the beam at the end of one segment is equal to the slope at the beginning of the next segment. The slope has no abrupt changes.
d. For the special case of a cantilever, the slope of the beam at the support is also zero.

9. Combine all the boundary conditions to evaluate all the constants of integration.
10. Substitute the constants of integration back into the slope and deflection equations, thus completing them. The value of the slope or deflection at any point can then be evaluated by simply placing the proper value of the position on the beam in the equation.

The method will now be illustrated with an example problem.

Example Problem 12-1

Figure 12-4 shows a beam used as a part of a special structure for a machine. The 20 000-lb load at A and the 30 000-lb load at C represent places where heavy equipment is supported. Between the two supports at B and D, the uniformly distributed load of 2000 lb/ft is due to stored bulk materials which are in a bin supported by the beam. To maintain accuracy of the product produced by the machine, the maximum allowable deflection of the beam is 0.05 in. Specify an acceptable wide-flange steel beam, and also check the stress in the beam.

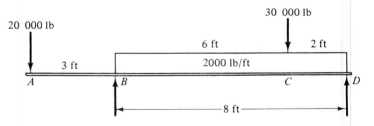

Figure 12-4 Beam for Example Problem 12-1.

Solution The beam will be analyzed to determine where the maximum deflection will occur. Then the required moment of inertia will be determined to limit the deflection to 0.05 in. A wide flange beam which has the required moment of inertia will then be selected. The ten-step procedure discussed earlier will be used. The solution is shown in a programmed format. You should work through the problem yourself before looking at the given solution.

In order to reduce the size of numbers to a manageable number of digits, the loads should be restated in the unit of *kilopounds,* sometimes called kips. One kip equals 1000 lb. Then the 20 000-lb load becomes 20 kips, the 30 000-lb load is 30 kips, and the distributed load is 2 kips/ft.

Steps 1 and 2 call for drawing the shear and bending moment diagrams. Do this now before checking the result below.

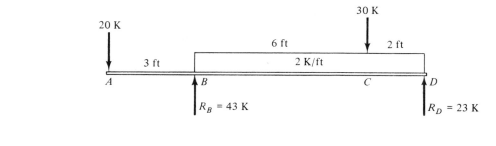

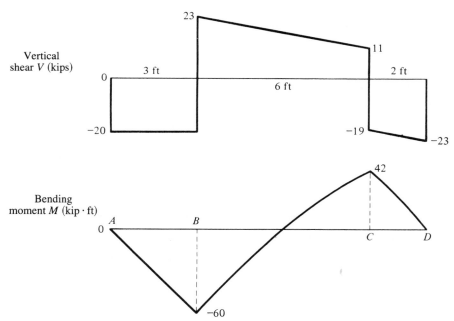

Figure 12-5 Load, shear and bending moment diagrams for Example Problem 12-1.

Figure 12-5 shows the results. Now do *step 3*.

Three segments are required, *AB*, *BC*, and *CD*. These are the segments over which the shear diagram is continuous. Now do *step 4* to get the shear curve equations.

The results are:

$$V_{AB} = -20 \tag{a}$$

$$V_{BC} = -2x + 29 \tag{b}$$

$$V_{CD} = -2x - 1 \tag{c}$$

In the segments BC and CD, the shear curve is a straight line with a slope of -2 kips/ft, the same as the load. Any method of writing the equation of a straight line can be used to derive these equations.

Now do *step 5* of the procedure.

You should have the following for the moment equations. First,

$$M_{AB} = \int V_{AB}\, dx + C = \int -20\, dx + C = -20x + C$$

At $x = 0$, $M_{AB} = 0$. Therefore, $C = 0$ and

$$M_{AB} = -20x \qquad\qquad\qquad \text{(d)}$$

Next,

$$M_{BC} = \int V_{BC}\, dx + C = \int (-2x + 29)\, dx + C = -x^2 + 29x + C$$

At $x = 3$, $M_{BC} = -60$. Therefore, $C = -138$ and

$$M_{BC} = -x^2 + 29x - 138 \qquad\qquad\qquad \text{(e)}$$

Finally,

$$M_{CD} = \int V_{CD}\, dx + C = \int (-2x - 1)\, dx + C = -x^2 - x + C$$

At $x = 9$, $M_{CD} = 42$. Therefore, $C = 132$ and

$$M_{CD} = -x^2 - x + 132 \qquad\qquad\qquad \text{(f)}$$

Now do *step 6* to get equations for θEI.

By integrating the moment equations,

$$\theta_{AB}EI = \int M_{AB}\, dx + C = \int -20x\, dx + C$$

$$\theta_{AB}EI = -10x^2 + C_1 \qquad\qquad\qquad \text{(g)}$$

$$\theta_{BC}EI = \int M_{BC}\, dx + C = \int (-x^2 + 29x - 138)\, dx + C$$

$$\theta_{BC}EI = -0.333x^3 + 14.5x^2 - 138x + C_2 \qquad\qquad\qquad \text{(h)}$$

$$\theta_{CD}EI = \int M_{CD}\, dx + C = \int (-x^2 - x + 132)\, dx + C$$

$$\theta_{CD}EI = -0.333x^3 - 0.5x^2 + 132x + C_3 \qquad\qquad\qquad \text{(i)}$$

Now in *step 7*, integrate Equations (g), (h), and (i) to get the yEI equations.

You should have

$$y_{AB}EI = \int \theta_{AB}EI\, dx + C$$

$$y_{AB}EI = -3.33x^3 + C_1x + C_4 \tag{j}$$

$$y_{BC}EI = \int \theta_{BC}EI \, dx + C$$

$$y_{BC}EI = -0.0833x^4 + 4.83x^3 - 69x^2 + C_2x + C_5 \tag{k}$$

$$y_{CD}EI = \int \theta_{CD}EI \, dx + C$$

$$y_{CD}EI = -0.0833x^4 - 0.167x^3 + 66x^2 + C_3x + C_6 \tag{l}$$

Step 8 calls for identifying boundary conditions. Six are required since there are six unknown constants of integration in Equations (g) through (l). Write them now.

Considering zero deflection points and the continuity of the slope and deflection curves, we can say

1. At $x = 3$, $y_{AB}EI = 0$
2. At $x = 3$, $y_{BC}EI = 0$ } (zero deflection at supports).
3. At $x = 11$, $y_{CD}EI = 0$
4. At $x = 9$, $y_{BC}EI = y_{CD}EI$ (continuous deflection curve at C).
5. At $x = 3$, $\theta_{AB}EI = \theta_{BC}EI$ } (continuous slope curve at B and C).
6. At $x = 9$, $\theta_{BC}EI = \theta_{CD}EI$

We can now substitute the values of x above into the proper equations and solve for C_1 through C_6. First make the substitutions and reduce the resulting equations to the form involving the constants.

For the six conditions listed above, the following equations result.

1. $3C_1 + C_4$ $= 90$
2. $3C_2 + C_5$ $= 497.25$
3. $11C_3 + C_6$ $= -6544$
4. $9C_2 - 9C_3 + C_5 - C_6 = 7290$
5. $C_1 - C_2$ $= -202.5$
6. $C_2 - C_3$ $= 1215$

Now solve the six equations simultaneously for the values of C_1 through C_6.

The results are:

$$C_1 = 132.5, \qquad C_4 = -307.5$$
$$C_2 = 335, \qquad C_5 = -507$$
$$C_3 = -880, \qquad C_6 = 3138$$

The final equations for θ and y can now be written by substituting the constants into Equations (g) through (l).

$$\theta_{AB}EI = -10x^2 + 132.5$$

$$\theta_{BC}EI = -0.333x^3 + 14.5x^2 - 138x + 335$$

$$\theta_{CD}EI = -0.333x^3 - 0.5x^2 + 132x - 880$$

$$y_{AB}EI = -3.33x^3 + 132.5x - 307.5$$

$$y_{BC}EI = -0.0833x^4 + 4.83x^3 - 69x^2 + 335x - 507$$

$$y_{CD}EI = -0.0833x^4 - 0.167x^3 + 66x^2 - 880x + 3138$$

Having the completed equations, we can now determine the point of maximum deflection, which is the primary objective of the analysis. Based on the loading, the probable shape of the deflected beam would be like that shown in Figure 12-6. Therefore, the maximum deflection could occur at point A at the end of the overhang, at a point to the right of B (upward), or at a point near the load at C (downward). It is probable that there are two points of zero slope at the points E and F, as shown in Figure 12-6. We would need to know where the slope equation $\theta_{BC}EI$ equals zero to determine where the maximum deflections occur. Notice that this equation is a third-degree equation and that

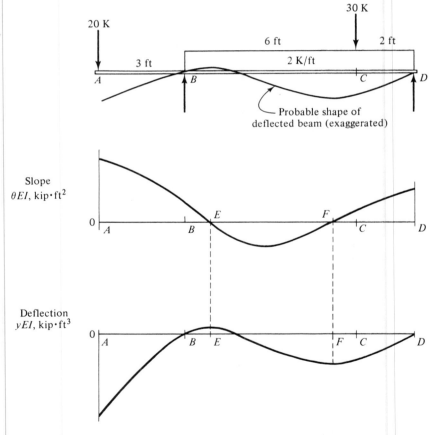

Figure 12-6 Slope and deflection curves for Example Problem 12-1.

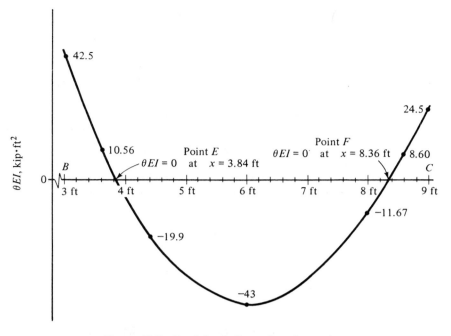

Figure 12-7 Graph for finding points of zero slope.

it would be difficult to solve for the value of x where θEI equals zero. A graphical approach is easier. We can substitute a few values of x into the equation at points near where θEI is likely to be zero. Then plotting the resulting values on a graph would reveal where θEI equals zero. Figure 12-7 shows such a graph, which locates point E at $x = 3.84$ ft and point F at $x = 8.36$ ft.

We can now determine the values for yEI at points A, E, and F to find out which is larger.

Point A.   At $x = 0$,

$$y_A EI = -307.5 \text{ kip} \cdot \text{ft}^3$$

Point E.   At $x = 3.84$ ft in segment BC,

$$y_E EI = -0.0833(3.84)^4 + 4.83(3.84)^3 - 69(3.84)^2 + 335(3.84) - 507$$

$$= +17.5 \text{ kip} \cdot \text{ft}^3$$

Point F.   At $x = 8.36$ ft in segment BC,

$$y_F EI = -0.0833(8.36)^4 + 4.83(8.36)^3 - 69(8.36)^2 + 335(8.36) - 507$$

$$= -111.8 \text{ kip} \cdot \text{ft}^3$$

The largest value occurs at point A; so that is the critical point. We must choose a beam that limits the deflection at A to 0.05 in. or less.

$$y_A EI = -307 \text{ kip} \cdot \text{ft}^3$$

Let $y_A = -0.05$ in. Then the required I is

$$I = \frac{-307 \text{ kip} \cdot \text{ft}^3}{Ey_A} \times \frac{1000 \text{ lb}}{\text{kip}} \times \frac{(12 \text{ in.})^3}{\text{ft}^3}$$

$$= \frac{(-307)(1000)(1728) \text{ lb} \cdot \text{in}^3}{(30 \times 10^6 \text{ lb/in}^2)(-0.05 \text{ in.})} = 354 \text{ in}^4$$

Consult the table of wide-flange beams and select a suitable beam.

A W18 $\times$ 40 is the best choice from Appendix A-7 since it is the lightest beam that has a large enough value for I. For this beam, $I = 612$ in^4, and the section modulus is $S = 68.4$ in^3. Now compute the maximum bending stress in the beam.

In Figure 12-5 we find the maximum bending moment to be 60 kip·ft. Then

$$\sigma = \frac{M}{S} = \frac{60 \text{ kip} \cdot \text{ft}}{68.4 \text{ in}^3} \times \frac{1000 \text{ lb}}{\text{kip}} \times \frac{12 \text{ in.}}{\text{ft}} = 10\ 526 \text{ psi}$$

Since the allowable stress for structural steel is approximately 22000 psi, the selected beam is safe.

12-5 BEAM DEFLECTIONS USING THE FORMULA METHOD

For many practical configurations of beam loading and support, formulas have been derived that allow the computation of the deflection at any point on the beam. The method of successive integration or the moment-area method, described later, may be used to develop the equations. Appendixes A-22, A-23, and A-24 include many examples of formulas for beam deflections.

The deflection formulas are valid only for the cases where the cross section of the beam is uniform for its entire length. Example problems will demonstrate the application of the formulas.

Example Problem 12-2

Determine the maximum deflection of a simply supported beam carrying a hydraulic cylinder in a machine used to press bushings into a casting, as shown in Figure 12-8. The force exerted during the pressing operation is 15 kN. The beam is rectangular, 25 mm thick and 100 mm high, and made of steel.

Solution Using the formula from Appendix A-22-a we find the maximum deflection to be

$$y = \frac{PL^3}{48EI}$$

where $P = 15$ kN $= 15 \times 10^3$ N
$\quad\quad L = 1.6$ m

From Appendix A-13, $E = 207$ GPa $= 207 \times 10^9$ N/m^2. For the rectangular beam,

$$I = \frac{(25)(100)^3}{12} = 2.083 \times 10^6 \text{ mm}^4$$

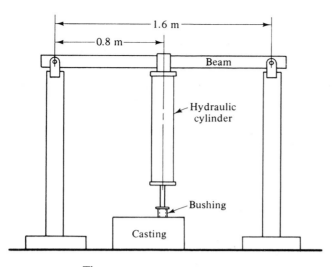

Casting

Figure 12-8 Beam for Example Problem 12-2.

Then

$$y = \frac{PL^3}{48EI} = \frac{(15 \times 10^3 \text{ N})(1.6 \text{ m})^3}{48(207 \times 10^9 \text{ N/m}^2)(2.083 \times 10^6 \text{ mm}^4)} \times \frac{(10^3 \text{ mm})^5}{\text{m}^5}$$

$$= 2.97 \text{ mm}$$

Example Problem 12-3

A round shaft, 45 mm in diameter, carries a 3500-N load as shown in Figure 12-9. If the shaft is steel, compute the deflection at the load and at the point C, 100 mm from the right end of the shaft.

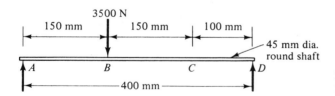

Figure 12-9 Shaft for Example Problem 12-3.

Solution The shaft conforms to the case depicted in Appendix A-22-b. Then, at the load, point B,

$$y_B = \frac{-Pa^2b^2}{3EIL}$$

But

$$I = \frac{\pi D^4}{64} = \frac{\pi(45 \text{ mm})^4}{64} = 0.201 \times 10^6 \text{ mm}^4$$

Also, in this case it would be convenient to express the modulus of elasticity E in the units of N/mm² instead of N/m² for consistency. The deflection y will then be in mm.

$$E = \frac{207 \times 10^9 \text{ N}}{\text{m}^2} \times \frac{1 \text{ m}^2}{(10^3 \text{ mm})^2} = 207 \times 10^3 \text{ N/mm}^2$$

Now

$$y_B = \frac{-Pa^2b^2}{3EIL} = \frac{-(3500)(150)^2(250)^2}{3(207 \times 10^3)(0.201 \times 10^6)(400)} = -0.0985 \text{ mm}$$

We can now compute the deflection at point C. In the set of formulas given, the one for deflections in the segment of the beam of length a applies because it is the longer segment.

$$y_C = \frac{-Pbx}{6EIL}(L^2 - b^2 - x^2)$$

where $P = 3500N$
$ b = 150$ mm
$ L = 400$ mm
$ x = 100$ mm $\quad$ (from right support to C)

Then

$$y_C = \frac{-(3500)(150)(100)}{6(207 \times 10^3)(0.201 \times 10^6)(400)}(400^2 - 150^2 - 100^2)$$

$$= -0.0670 \text{ mm}$$

Example Problem 12-4

Determine the deflection at the middle of an aluminum I10 $\times$ 8.646 beam when it carries a total load of 55 800 lb distributed over a length of 6.0 ft. The beam is simply supported at its ends.

Solution The deflection can be computed using the formula in Appendix A-22-d,

$$y = \frac{-5WL^3}{384EI}$$

Consistency of units must be carefully observed. Express W in *pounds*, length in *inches*, E in *psi*, and I in *inches* to the fourth power. From the Appendix,

$$I = 132.09 \text{ in}^4$$

$$E = 10 \times 10^6 \text{ psi}$$

and

$$L = 6.0 \text{ ft} \times 12 \text{ in./ft} = 72 \text{ in.}$$

Then

$$y = \frac{-5(55\ 800)(72)^3}{384(10 \times 10^6)(132.09)} \text{ in.} = -0.205 \text{ in.}$$

12-6 SUPERPOSITION USING DEFLECTION FORMULAS

Formulas such as those used in the preceding section are available for a large number of cases of loading and support conditions. Obviously, these cases would allow the solution of many practical beam deflection problems. An even larger number of situations can be handled by the use of the *principle of superposition*.

If a particular loading and support pattern can be broken into components such that each component is like one of the cases for which a formula is available, then the total deflection at a point on the beam is equal to the sum of the deflections caused by each component. The deflection due to one component load is *superposed* on deflections due to the other loads, thus giving the name *superposition*.

An example of where superposition can be applied is shown in Figure 12-10. Represented in the figure is a roof beam carrying a uniformly distributed roof load of 800 lb/ft and also supporting a portion of a piece of process equipment which provides a concentrated load at the middle. Figure 12-11 shows how the loads are considered separately. Each component load produces a maximum deflection at the middle. Therefore, the maximum total deflection will occur there also. Let the subscript 1 refer to the concentrated load case and the subscript 2 refer to the distributed load case. Then

$$y_1 = \frac{-PL^3}{48EI}$$

$$y_2 = \frac{-5}{384} \frac{WL^3}{EI}$$

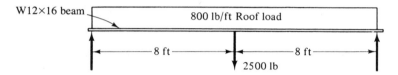

Figure 12-10   Roof beam.

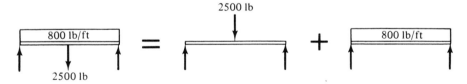

Figure 12-11   Illustration of a superposition principle.

The total deflection will be

$$y_T = y_1 + y_2$$

The terms L, E, and I will be the same for both cases.

$$L = 16 \text{ ft} \times 12 \text{ in./ft} = 192 \text{ in.}$$

$$E = 30 \times 10^6 \text{ psi} \qquad \text{for steel}$$

$$I = 103 \text{ in}^4 \qquad \text{for W12} \times 16 \text{ beam}$$

To compute y_1, let $P = 2500$ lb.

$$y_1 = \frac{-2500(192)^3}{48(30 \times 10^6)(103)} \text{ in.} = -0.119 \text{ in.}$$

To compute y_2, W is the total resultant of the distributed load.

$$W = (800 \text{ lb/ft})(16 \text{ ft}) = 12\,800 \text{ lb}$$

Then

$$y_2 = \frac{-5(12\,800)(192)^3}{384(30 \times 10^6)(103)} \text{ in.} = -0.382 \text{ in.}$$

and

$$y_T = y_1 + y_2 = -0.119 \text{ in.} - 0.382 \text{ in.} = -0.501 \text{ in.}$$

Since this is the total deflection, we could check to see if it meets the recommendation that the maximum deflection should be less than $\frac{1}{360}$ times the span of the beam. The span is L.

$$\frac{L}{360} = \frac{192 \text{ in.}}{360} = 0.533 \text{ in.}$$

The computed deflection was 0.501 in., which is satisfactory.

The superposition principle is valid for any place on the beam, not just at the loads. The following example problem illustrates this.

Example Problem 12-5

Figure 12-12 shows a shaft carrying two gears and simply supported at its ends by bearings. The mating gears above exert downward forces which tend to separate the gears. Other forces acting horizontally are not considered in this analysis. Determine the total deflection at the gears B and C if the shaft is steel and has a diameter of 45 mm.

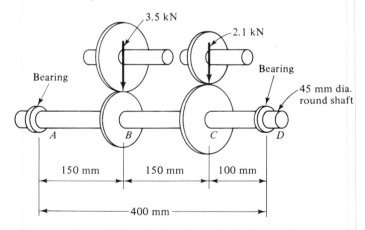

Figure 12-12   Shaft for Example Problem 12-5.

Solution   The two unsymmetrically placed, concentrated loads constitute a situation for which none of the given cases for beam deflection formulas is valid. However, it can be solved by using the formulas of Appendix A-22-b twice. Considering each load separately, the deflections at B and C can be found for each. The total, then, would be the

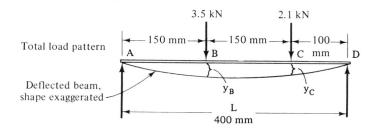

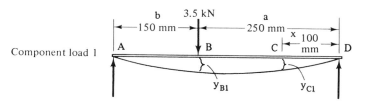

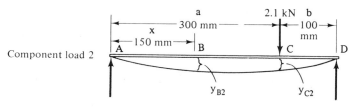

Figure 12-13 Superposition logic for deflection of shaft in Figure 12-12.

sum of the component deflections. Figure 12-13 shows the logic from which we can say

$$y_B = y_{B1} + y_{B2}$$

$$y_C = y_{C1} + y_{C2}$$

where y_B = total deflection at B
y_C = total deflection at C
y_{B1} = deflection at B due to 3.5-kN load alone
y_{C1} = deflection at C due to 3.5-kN load alone
y_{B2} = deflection at B due to 2.1-kN load alone
y_{C2} = deflection at C due to 2.1-kN load alone

For all the calculations, the values of E, I, and L will be needed. These are the same as those used in Example Problem 12-3.

$$E = 207 \times 10^3 \text{ N/mm}^2$$

$$I = 0.201 \times 10^6 \text{ mm}^4$$

$$L = 400 \text{ mm}$$

The product of EIL is present in all the formulas.

$$EIL = (207 \times 10^3)(0.201 \times 10^6)(400) = 16.64 \times 10^{12}$$

Now the individual component deflections will be computed. For component 1:

$$y_{B1} = \frac{-Pa^2b^2}{3EIL} = \frac{-(3.5 \times 10^3)(250)^2(150)^2}{3(16.64 \times 10^{12})} = -0.0985 \text{ mm}$$

$$y_{C1} = \frac{-Pbx}{6EIL}(L^2 - b^2 - x^2)$$

$$y_{C1} = \frac{-(3.5 \times 10^3)(150)(100)}{6(16.64 \times 10^{12})}(400^2 - 150^2 - 100^2) = -0.0670 \text{ mm}$$

For component 2, the load will be 2.1 kN at point C. Then

$$y_{B2} = \frac{-Pbx}{6EIL}(L^2 - b^2 - x^2)$$

$$y_{B2} = \frac{-(2.1 \times 10^3)(100)(150)}{6(16.64 \times 10^{12})}(400^2 - 100^2 - 150^2) = -0.0402 \text{ mm}$$

$$y_{C2} = \frac{-Pa^2b^2}{3EIL} = \frac{-(2.1 \times 10^3)(300)^2(100)^2}{3(16.64 \times 10^{12})} = -0.0378 \text{ mm}$$

Now, by superposition,

$$y_B = y_{B1} + y_{B2} = -0.0985 \text{ mm} - 0.0402 \text{ mm} = -0.1387 \text{ mm}$$

$$y_C = y_{C1} + y_{C2} = -0.0670 \text{ mm} - 0.0378 \text{ mm} = -0.1048 \text{ mm}$$

In Section 12-1 it was observed that a recommended limit for the movement of one gear relative to its mating gear is 0.13 mm. Thus this shaft is too flexible since the deflection at B exceeds 0.13 mm without even considering the deflection of the mating shaft.

12-7 BEAM DEFLECTIONS—MOMENT-AREA METHOD

The semigraphical procedure for finding beam deflections, called the *moment-area method*, is useful for problems in which a fairly complex loading pattern occurs or when the beam has a varying cross section along its length. Such cases are difficult to handle with the other methods presented in this chapter.

Shafts for mechanical drives are examples where the cross section varies throughout the length of the beam. Figure 12-14 shows a shaft designed to carry two gears where the changes in diameter provide shoulders against which to seat the gears and bearings to provide axial location. Note, also, that the bending moment decreases toward the ends of the shaft, allowing smaller sections to be safe with regard to bending stress.

In structural applications of beams, varying cross sections are often used to make more economical members. Larger sections having higher moments of inertia are used

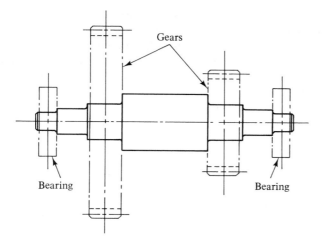

Figure 12-14 Shaft with varying cross sections.

at sections where the bending moment is high while decreased section sizes are used in places where the bending moment is lower. Figure 12-15 shows an example.

The moment-area method uses the quantity M/EI, the bending moment divided by the stiffness of the beam, to determine the deflection of the beam at selected points.

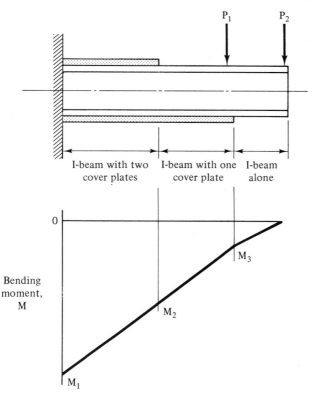

Figure 12-15 Cantilever with varying cross sections.

Then it is convenient to prepare such a diagram as a part of the beam analysis procedure. If the beam has the same cross section over its entire length, the M/EI diagram looks similar to the familiar bending moment diagram except that its values have been divided by the quantity EI. However, if the moment of inertia of the cross section varies along the length of the beam, the shape of the M/EI diagram will be different. This is shown in Figure 12-16.

Recall Equation (12-10) from Section 12-3,

$$\frac{M}{EI} = \frac{d^2y}{dx^2} \tag{12-10}$$

This formula relates the deflection of the beam, y, as a function of position, x, the bending moment, M, and the beam stiffness, EI. The right side of Equation (12-10) can be rewritten as

$$\frac{d^2y}{dx^2} = \frac{d}{dx}\left(\frac{dy}{dx}\right)$$

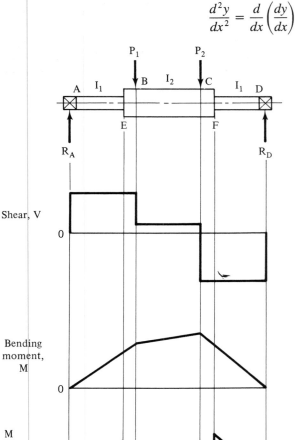

Figure 12-16   Illustration of M/EI diagrams for a beam with varying cross sections.

But note that dy/dx is defined as the slope of the deflection curve, θ; that is, $dy/dx = \theta$. Then

$$\frac{d^2y}{dx^2} = \frac{d\theta}{dx}$$

Equation (12-10) can then be written

$$\frac{M}{EI} = \frac{d\theta}{dx}$$

Solving for $d\theta$ gives

$$d\theta = \frac{M}{EI}\,dx \qquad (12\text{-}15)$$

The interpretation of Equation (12-15) can be seen in Figure 12-17 in which the right side, $(M/EI)\,dx$, is the area under the M/EI diagram over a small length dx. Then $d\theta$ is the change in the angle of the slope over the same distance dx. If tangent lines

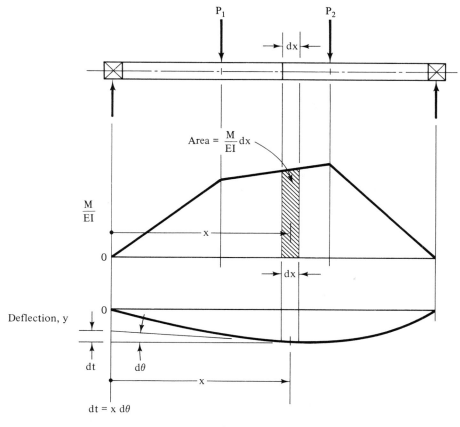

Figure 12-17 Principles of the moment-area method for beam deflections.

are drawn to the deflection curve of the beam at the two points marking the beginning and the end of the segment dx, the angle between them is $d\theta$.

The change in angle $d\theta$ causes a change in the vertical position of a point at some distance x from the small element dx as shown in Figure 12-17. Calling the change in vertical position dt, it can be found from

$$dt = x\, d\theta = \frac{M}{EI}\, x\, dx \tag{12-16}$$

To determine the effect of the change in angle over a larger segment of the beam, Equations (12-15) and (12-16) must be integrated over the length of the segment. For example, over the segment A-B shown in Figure 12-18 from Equation (12-15),

$$\int_A^B d\theta = \theta_B - \theta_A = \int_A^B \frac{M}{EI}\, dx \tag{12-17}$$

The last part of this equation is the area under the M/EI curve between A and B. This is equal to the change in the angle of the tangents at A and B, $\theta_B - \theta_A$.

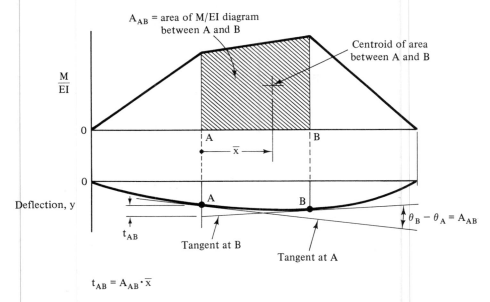

$$t_{AB} = A_{AB} \cdot \overline{x}$$

Figure 12-18 Illustrations of the two theorems of the moment-area method for beam deflections.

From Equation (12-16),

$$\int_A^B dt = t_{AB} = \int_A^B \frac{M}{EI}\, x\, dx \tag{12-18}$$

Here the term t_{AB} represents the tangential deviation of the point A from the tangent to point B, as shown in Figure 12-18. Also, the right side of Equation (12-18) is the *moment of the area of the M/EI diagram between points A and B*. In practice, the moment of the area is computed by multiplying the area under the M/EI curve by the distance to the centroid of the area, also shown in Figure 12-18.

Equations (12-17) and (12-18) form the basis for the two *theorems of the moment-area method for finding beam deflections*. They are:

Theorem 1. The change in angle, in radians, between tangents drawn at two points A and B on the deflection curve for a beam, is equal to the area under the M/EI diagram between A and B.

Theorem 2. The vertical deviation of point A on the deflection curve for a beam from the tangent through another point B on the curve is equal to the moment of the area under the M/EI curve with respect to point A.

12-8 APPLICATIONS OF THE MOMENT-AREA METHOD

In this section we show several examples of the use of the moment-area method for finding the deflection of beams. Procedures are developed for each class of beam according to the manner of loading and support. Considered are;

1. Cantilevers with a variety of loads
2. Symmetrically loaded simply supported beams
3. Beams with varying cross section
4. Unsymmetrically loaded simply supported beams

Cantilevers. The definition of a cantilever includes the requirement that it is rigidly fixed to a support structure so that no rotation of the beam can occur at the support. Therefore, the tangent to the deflection curve at the support is always in line with the original position of the neutral axis of the beam in the unloaded state. If the beam is horizontal, as we usually picture it, this tangent is also horizontal.

The procedure for finding the deflection of any point on a cantilever, listed below, uses the two theorems developed in the Section 12-7 along with the observation that the tangent to the deflection curve at the support is horizontal.

1. Draw the load, shear, and bending moment diagrams.
2. Divide the bending moment values by the beam stiffness, EI, and draw the M/EI diagram.
3. Compute the area of the M/EI diagram and locate its centroid. If the diagram is not a simple shape, break it into parts and find the area and centroid for each part separately. If the deflection at the end of the cantilever is desired, the area of the entire M/EI diagram is used. If the deflection of some other point is desired, only the area between the support and the point of interest is used.
4. Use Theorem 2 to compute the vertical deviation of the point of interest from the tangent to the beam's neutral axis at the support. Because the tangent is coincident with the original position of the neutral axis, the deviation thus found is

the actual deflection of the beam at the point of interest. If all loads are in the same direction, the maximum deflection occurs at the end of the cantilever.

Example Problem 12-6

Use the moment-area method to determine the deflection at the end of the steel cantilever shown in Figure 12-19.

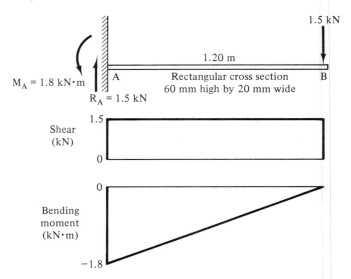

Figure 12-19 Load, shear and bending moment diagrams for Example Problem 12-6 and 12-7.

Solution

Step 1. Figure 12-19 shows the load, shear, and bending moment diagrams.

Step 2. The stiffness is computed here:

$$E = 207 \text{ GPa} = 207 \times 10^9 \text{ N/m}^2$$

$$I = \frac{th^3}{12} = \frac{(0.02 \text{ m})(0.06 \text{ m})^3}{12} = 3.60 \times 10^{-7} \text{ m}^4$$

$$EI = (207 \times 10^9 \text{ N/m}^2)(3.60 \times 10^{-7} \text{ m}^4) = 7.45 \times 10^4 \text{ N} \cdot \text{m}^2$$

The M/EI diagram is drawn in Figure 12-20. Note that the only changes from the bending moment diagram are the units and the values because the beam has a constant stiffness along its length.

Step 3. The desired area is that of the entire triangular shape of the M/EI diagram, called A_{BA} here to indicate that it is used to compute the deflection of point B relative to A.

$$A_{BA} = (0.5)(24.15 \times 10^{-3} \text{ m}^{-1})(1.20 \text{ m}) = 14.5 \times 10^{-3} \text{ rad}$$

The centroid of this area is two-thirds of the distance from B to A, 0.80 m.

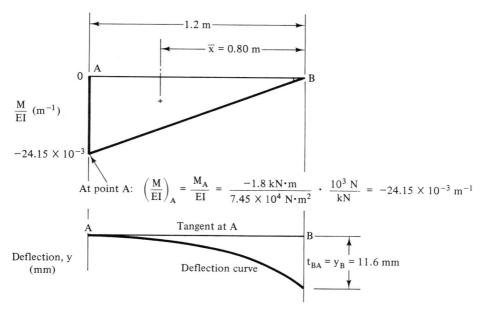

$$\text{At point A:} \quad \left(\frac{M}{EI}\right)_A = \frac{M_A}{EI} = \frac{-1.8 \text{ kN}\cdot\text{m}}{7.45 \times 10^4 \text{ N}\cdot\text{m}^2} \cdot \frac{10^3 \text{ N}}{\text{kN}} = -24.15 \times 10^{-3} \text{ m}^{-1}$$

Figure 12-20 M/EI curve and deflection curve for Example Problem 12-6.

Step 4.   To implement Theorem 2, we need to compute the moment of the area found in step 3. This is equal to t_{BA}, the vertical deviation of point B from the tangent drawn to the deflection curve at point A.

$$t_{BA} = A_{BA} \cdot \bar{x} = (14.5 \times 10^{-3} \text{ rad})(0.80 \text{ m})$$
$$= 11.6 \times 10^{-3} \text{ m} = 11.6 \text{ mm}$$

Because the tangent to point A is horizontal, t_{BA} is equal to the deflection of the beam at its end, point B.

Example Problem 12-7

For the same beam used in Example Problem 12-6 and shown in Figure 12-19 compute the deflection at a point 1.0 m from the support.

Solution   Part of the solution for Example Problem 12-6 can be used here. Steps 1 and 2 are identical, resulting in the load, shear, bending moment, and *M/EI* diagrams shown in Figures 12-19 and 12-20.

Step 3.   This step changes because only the area of the *M/EI* diagram between the support and the point 1.0 m out on the beam is used, as shown in Figure 12-21. Calling the point of interest, point C, we need to compute t_{CA}; that is, the vertical deviation of point C relative to the tangent drawn to point A at the support. The required area is a trapezoid and it is convenient to break it into a triangle and a rectangle and treat them as simple shapes. The calculation then takes the form

$$t_{CA} = A_{CA} \cdot \bar{x} = A_{CA1} \cdot x_1 + A_{CA2} \cdot x_2$$

where the areas and x-distances are shown in Figure 12-21. Note that the distances are

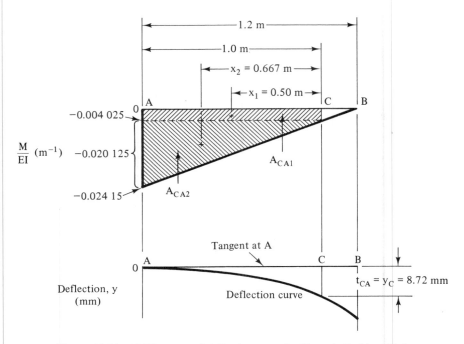

Figure 12-21 M/EI curve and deflection curve for Example Problem 12-7.

measured *from the point C to the centroid of the component area.* Then

$$A_{CA1} \cdot x_1 = (0.004\ 025\ \text{m}^{-1})(1.0\ \text{m})(0.50\ \text{m}) = 2.01 \times 10^{-3}\ \text{m}$$

$$A_{CA2} \cdot x_2 = (0.5)(0.020\ 125\ \text{m}^{-1})(1.0\ \text{m})(0.667\ \text{m}) = 6.71 \times 10^{-3}\ \text{m}$$

$$t_{CA} = (2.01 + 6.71)(10^{-3})\ \text{m} = 8.72\ \text{mm}$$

As before, because the tangent to point A is horizontal, the vertical deviation, t_{CA}, is the true deflection of point C.

Symmetrically loaded simply supported beams. This class of problems has the advantage that it is known that the maximum deflection occurs at the middle of the span of the beam. An example is shown in Figure 12-22, in which the beam carries two identical loads equally spaced from the supports. Of course, any loading for which the point of maximum deflection can be predicted can be solved by the procedure illustrated here.

1. Draw the load, shear, and bending moment diagrams.
2. Divide the bending moment values by the beam stiffness, EI, and draw the M/EI diagram.
3. If the maximum deflection, at the middle of the span, is desired, use that part of the M/EI diagram between the middle and one of the supports, that is, half of the diagram.

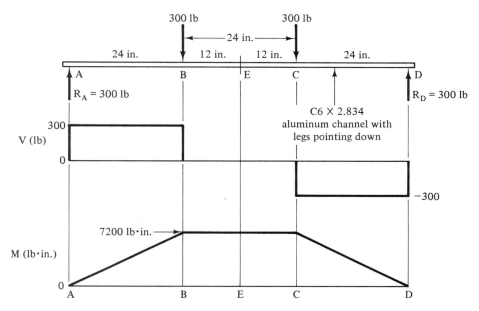

Figure 12-22 Load, shear and bending moment diagrams for Example Problems 12-8 and 12-9.

4. Use Theorem 2 to compute the vertical deviation of the point at one of the supports from the tangent to the beam's neutral axis at its middle. Because the tangent is horizontal and because the deflection at the support is actually zero, the deviation thus found is the actual deflection of the beam at its middle.

Example Problem 12-8

Determine the maximum deflection of the beam shown in Figure 12-22. The beam is an aluminum channel, C6 × 2.834, made from 6061-T6 aluminum, positioned with the legs downward.

Solution Because the loading pattern is symmetrical, the maximum deflection will occur at the middle and the tangent drawn at that point will be horizontal. The procedure listed above is used to determine the deflection.

Step 1. The load, shear, and bending moment diagrams are shown in Figure 12-22, prepared in the traditional manner. The maximum bending moment is 7200 lb · in. between *B* and *C*.

Step 2. The stiffness of the beam, *EI*, is found using data from the Appendix. From Appendix A-17, *E* for 6061-T6 aluminum is 10×10^6 psi. From Appendix A-10, the moment of inertia of the channel, with respect to the *Y-Y* axis is 1.53 in⁴. Then

$$EI = (10 \times 10^6 \text{ lb/in}^2)(1.53 \text{ in}^4) = 1.53 \times 10^7 \text{ lb} \cdot \text{in}^2$$

Because the stiffness of the beam is uniform over its entire length, the *M/EI* diagram has the same shape as the bending moment diagram, but the values are different as shown in Figure 12-23. The maximum value of *M/EI* is 4.71×10^{-4} in.⁻¹.

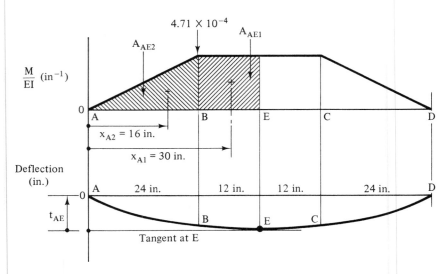

Figure 12-23 M/EI diagram and deflection curve for Example Problem 12-8.

Step 3. To determine the deflection at the middle of the beam, one-half of the *M/EI* diagram is used. For convenience, this is separated into a rectangle and a triangle and the centroid of each part is shown.

Step 4. We need to find t_{AE}, the vertical deviation of point *A* from the tangent drawn to the deflection curve at point *E*, the middle of the beam. By Theorem 2,

$$t_{AE} = A_{AE1} \cdot x_{A1} + A_{AE2} \cdot x_{A2}$$

The symbols x_{A1} and x_{A2} indicate that the distances to the centroids of the areas must be measured *from point A*.

$$A_{AE1} \cdot x_{A1} = (4.71 \times 10^{-4} \text{ in.}^{-1})(12 \text{ in.})(30 \text{ in.}) = 0.170 \text{ in.}$$

$$A_{AE2} \cdot x_{A2} = (0.5)(4.71 \times 10^{-4} \text{ in.}^{-1})(24 \text{ in.})(16 \text{ in.}) = 0.090 \text{ in.}$$

$$t_{AE} = 0.170 + 0.090 = 0.260 \text{ in.}$$

This is the vertical deviation of point *A* from the tangent to point *E*. Because the tangent is horizontal and because the actual deflection of point *A* is zero, this represents the true deflection of point *E* relative to the original position of the neutral axis of the beam.

Example Problem 12-9

For the same beam used for Example Problem 12-8, shown in Figure 12-22, determine the deflection at point *B* under one of the loads.

Solution In this case we can use the moment-area method to determine the vertical deviation, t_{BE}, of point *B* from the tangent to point *E* at the middle of the beam. Then, subtracting that from the value of t_{AE} found in Example Problem 12-8 gives the true deflection of point *B*. *Steps 1 and 2* are the same as shown in Figures 12-22 and 12-23. Shown in Figure 12-24 are the data needed to compute t_{BE}.

$$t_{BE} = A_{BE1} \cdot x_{B1} = (4.71 \times 10^{-4} \text{ in.}^{-1})(12 \text{ in.})(6 \text{ in.}) = 0.034 \text{ in.}$$

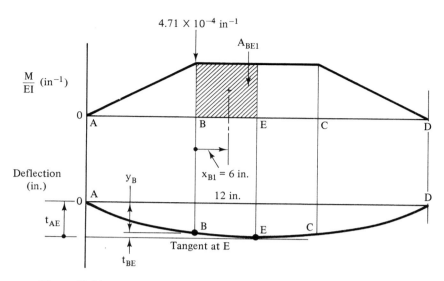

Figure 12-24 M/EI diagram and deflection curve for Example Problem 12-9.

Note that the distance x_{B1} must be measured from point B. Then the deflection of point B is

$$y_B = t_{AE} - t_{BE} = 0.260 - 0.034 = 0.226 \text{ in.}$$

Beams with varying cross section. One of the major uses of the moment-area method is to find the deflection of a beam having a varying cross section along its length. Only one additional step is required as compared with beams having a uniform cross section, like those considered thus far.

An example of such a beam is illustrated in Figure 12-25. Note that it is a modification of the beam used in Example Problems 12-8 and 12-9 and shown in Figure 12-22. Here we have added a rectangular plate, 0.25 in. by 6.0 in., to the underside of the original channel over the middle 48 in. of the length of the beam. The box shape would provide a significant increase in stiffness, thereby reducing the deflection of the beam. The stress in the beam would also be reduced.

The change in the procedure for analyzing the deflection of the beam is in the preparation of the *M/EI* diagram. Figure 12-26 shows the load, shear, and bending moment diagrams as before. In the first and last 12 in. of the *M/EI* diagram, the stiffness of the plain channel is used as before. For each segment over the middle 48 in., the stiffness of the box shape must be used. The *M/EI* diagram then includes the effect of the change in stiffness along the length of the beam.

Example Problem 12-10

Determine the deflection at the middle of the reinforced beam shown in Figure 12-25.
Solution

Step 1. The load, shear, and bending moment diagrams are prepared in the traditional manner as shown in Figure 12-26.

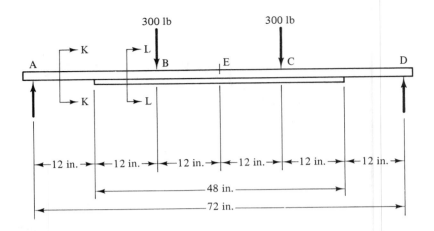

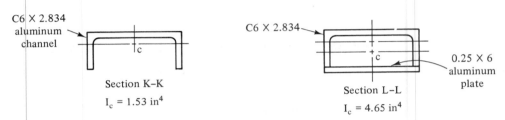

Figure 12-25 Beam for Example Problem 12-10.

Step 2. To prepare the M/EI diagram, two values for the stiffness, EI, are needed. The plain channel has the same value used in previous problems, 1.53×10^7 lb·in². For the box shape,

$$EI = (10 \times 10^6 \text{ lb/in}^2)(4.65 \text{ in}^4) = 4.65 \times 10^7 \text{ lb·in}^2$$

Then, at the point on the beam just before 12 in. from A, where the bending moment is 3600 lb·in.

$$\frac{M}{EI} = \frac{3600 \text{ lb·in.}}{1.53 \times 10^7 \text{ lb·in}^2} = 2.35 \times 10^{-4} \text{ in.}^{-1}$$

Just beyond 12 in. from A,

$$\frac{M}{EI} = \frac{3600 \text{ lb·in.}}{4.65 \times 10^7 \text{ lb·in}^2} = 7.74 \times 10^{-5} \text{ in.}^{-1}$$

At point B, where $M = 7200$ lb·in. and $EI = 4.65 \times 10^7$ lb·in²,

$$\frac{M}{EI} = \frac{7200 \text{ lb·in.}}{4.65 \times 10^7 \text{ lb·in}^2} = 1.55 \times 10^{-4} \text{ in.}^{-1}$$

These values establish the critical points on the M/EI diagram.

Step 3. The moment area for the left half of the M/EI diagram will be used to determine the value of t_{AE}, as done in Example Problem 12-8. For convenience, the total

Deflection of Beams   Chap. 12

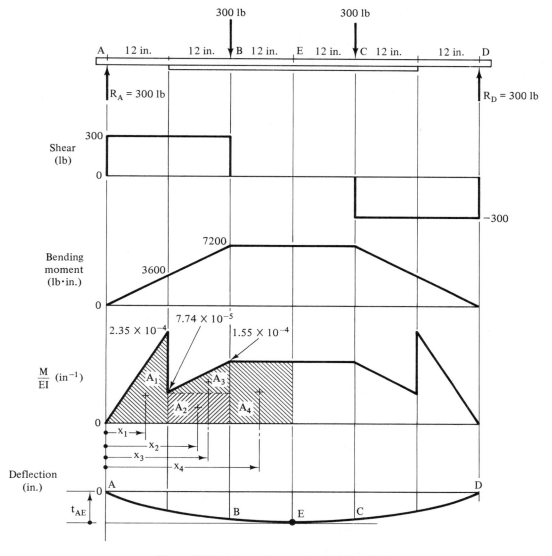

Figure 12-26 Beam diagrams for Problem 12-10.

area is divided into four parts as shown in Figure 12-26 with the locations of the centroids indicated relative to point A. The distances are

$$x_1 = (\tfrac{2}{3})(12 \text{ in.}) = 8 \text{ in.}$$

$$x_2 = (\tfrac{1}{2})(12 \text{ in.}) + 12 \text{ in.} = 18 \text{ in.}$$

$$x_3 = (\tfrac{2}{3})(12 \text{ in.}) + 12 \text{ in.} = 20 \text{ in.}$$

$$x_4 = (\tfrac{1}{2})(12 \text{ in.}) + 24 \text{ in.} = 30 \text{ in.}$$

Step 4. We can now use Theorem 2 to compute the value of t_{AE}, the deviation of point A from the tangent to point E, by computing the moment of each of the four areas shown crosshatched in the M/EI diagram of Figure 12-26.

$$t_{AE} = A_1x_1 + A_2x_2 + A_3x_3 + A_4x_4$$

$$A_1x_1 = (0.5)(2.35 \times 10^{-4} \text{ in.}^{-1})(12 \text{ in.})(8 \text{ in.}) = 1.128 \times 10^{-2} \text{ in.}$$

$$A_2x_2 = (7.74 \times 10^{-5} \text{ in.}^{-1})(12 \text{ in.})(18 \text{ in.}) = 1.672 \times 10^{-2} \text{ in.}$$

$$A_3x_3 = (0.5)(7.74 \times 10^{-5} \text{ in.}^{-1})(12 \text{ in.})(20 \text{ in.}) = 9.293 \times 10^{-3} \text{ in.}$$

$$A_4x_4 = (1.55 \times 10^{-4} \text{ in.}^{-1})(12 \text{ in.})(30 \text{ in.}) = 5.580 \times 10^{-2} \text{ in.}$$

Then

$$t_{AE} = \Sigma (A_ix_i) = 9.309 \times 10^{-2} \text{ in.} = 0.093 \text{ in.}$$

As before, this value is equal to the deflection of point E at the middle of the beam. Comparing it to the deflection of 0.260 in. found in Example Problem 12-8, the addition of the cover plate reduced the maximum deflection by approximately 64%.

Unsymmetrically loaded simply supported beams. The major difference between this type of beam and the ones considered earlier is that the point of maximum deflection is unknown. Special care is needed to describe the geometry of the M/EI diagram and the deflection curve for the beam. The general procedure used is illustrated in an example problem.

Example Problem 12-11

Determine the deflection of the beam shown in Figure 12-27 at its middle, 1.0 m from the supports. The beam is an American Standard steel beam, S3 × 5.7.

Solution The first steps are to prepare the load, shear, and bending moment diagrams, as shown in Figure 12-27. Then the M/EI diagram is drawn by dividing the values for M by the beam stiffness, EI. We will use $E = 207$ GPa for steel. From the Appendix we find $I = 2.52$ in⁴, and this must be converted to metric units.

$$I = \frac{(2.52 \text{ in}^4)(0.0254 \text{ m})^4}{1.0 \text{ in}^4} = 1.049 \times 10^{-6} \text{ m}^4$$

Then the beam stiffness is

$$EI = (207 \times 10^9 \text{ N/m}^2)(1.049 \times 10^{-6} \text{ m}^4) = 2.17 \times 10^5 \text{ N} \cdot \text{m}^2$$

The value of M/EI at point B on the beam can now be computed. Note that $M_B = 2.40$ kN·m = 2400 N·m. Then

$$\left(\frac{M}{EI}\right)_B = \frac{2400 \text{ N} \cdot \text{m}}{2.17 \times 10^5 \text{ N} \cdot \text{m}^2} = 0.0111 \text{ m}^{-1}$$

The M/EI diagram is drawn in Figure 12-27 along with the exaggerated sketch of the deflection curve for the beam. It is desired to compute the deflection of the beam at its middle, labeled point D. The location of point D on the deflection curve is called D'. Then $y_D = DD'$.

One technique for finding the deflection at D is illustrated in Figure 12-28. The following steps are used.

Deflection of Beams   Chap. 12

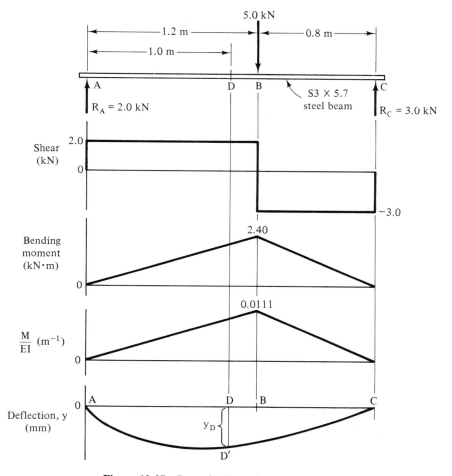

Figure 12-27 Beam for Example Problem 12-11.

Step 1. Draw the tangent to the deflection curve at A and compute the deviation of point C from this tangent, t_{CA}, using Theorem 2. The entire M/EI diagram is used.

$$t_{CA} = A_{CA1}x_{C1} + A_{CA2}x_{C2}$$

$$A_{CA1}x_{C1} = (0.5)(0.0111 \text{ m}^{-1})(0.8 \text{ m})(0.533 \text{ m}) = 0.00237 \text{ m}$$

$$A_{CA2}x_2 = (0.5)(0.0111 \text{ m}^{-1})(1.2 \text{ m})(1.2 \text{ m}) = 0.00799 \text{ m}$$

Then

$$t_{CA} = 0.00237 + 0.00799 = 0.01036 \text{ m} = 10.36 \text{ mm}$$

Step 2. Use the principle of proportion to determine the distance DD'' from D to the tangent line.

$$\frac{t_{CA}}{CA} = \frac{DD''}{AD}$$

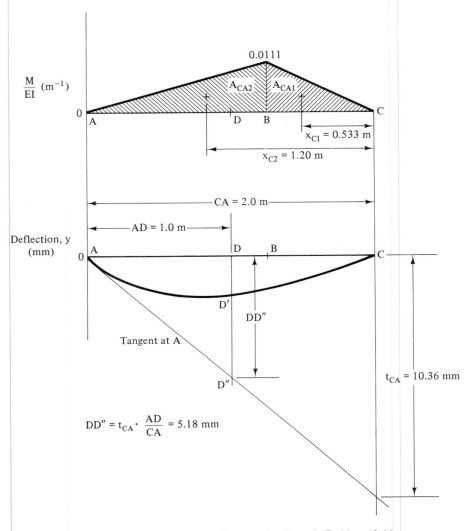

Figure 12-28 Moment-area diagrams for Example Problem 12-11.

or

$$DD'' = t_{CA} \cdot \frac{AD}{CA} = (10.36 \text{ mm}) \cdot \frac{1.0 \text{ m}}{2.0 \text{ m}} = 5.18 \text{ mm}$$

Step 3. Compute the deviation, $t_{D'A}$, of point D' from the tangent line drawn to point A using Theorem 2. The part of the M/EI diagram between D and A is used as shown in Figure 12-29.

$$t_{D'A} = A_{D'A}x_{D1} = (0.5)(0.0092 \text{ m}^{-1})(1.0 \text{ m})(0.333 \text{ m}) = 0.00153 \text{ m}$$

$$= 1.53 \text{ mm}$$

Deflection of Beams   Chap. 12

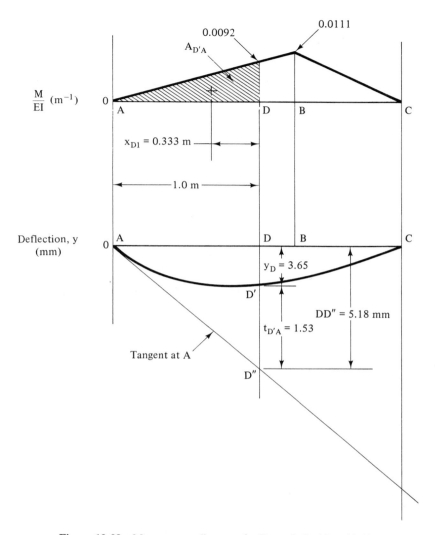

Figure 12-29   Moment area diagrams for Example Problem 12-11.

Step 4.   From the geometry of the deflection diagram shown in Figure 12-29 the deflection at point D, y_D, is

$$y_D = DD' = DD'' - t_{D'A} = 5.18 - 1.53 = 3.65 \text{ mm}$$

12-9 BEAMS WITH DISTRIBUTED LOADS—MOMENT-AREA METHOD

The general procedure for determining the deflection of beams carrying distributed loads is the same as that shown for beams carrying concentrated loads. However, the shape of the bending moment and M/EI curves is different, requiring the use of other formulas for computing the area and centroid location for use in the moment-area method. The following example shows the kind of differences to be expected.

Example Problem 12-12

Determine the deflection at the end of the cantilever carrying a uniformly distributed load shown in Figure 12-30. The beam is a $6 \times 2 \times \frac{1}{4}$ hollow rectangular steel tube with the 6.0-in. dimension horizontal.

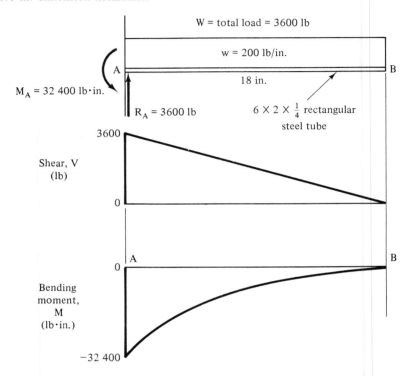

Figure 12-30 Load, shear and bending moment diagrams for Example Problem 12-12.

Solution As before, the solution starts with the preparation of the load, shear, and bending moment diagrams, shown in Figure 12-30. Then the M/EI curve will have the same shape as the bending moment curve because the stiffness of the beam is uniform. From Appendix A-9 we find $I = 2.31$ in^4. Then, using $E = 30 \times 10^6$ psi for steel,

$$EI = (30 \times 10^6 \text{ lb/in}^2)(2.31 \text{ in}^4) = 6.93 \times 10^7 \text{ lb} \cdot \text{in}^2$$

Now Figure 12-31 shows the M/EI curve along with the deflection curve for the beam. We can use the same procedure developed for Example Problem 12-6 to complete the problem. The horizontal line on the deflection diagram is the tangent drawn to the beam shape at point A, where the beam is fixed to the support. Then, at the right end of the beam, the deviation of the beam deflection curve from this tangent, t_{BA}, is equal to the deflection of the beam itself.

Using Theorem 2, the deviation t_{BA} is equal to the product of the area of the M/EI curve between B and A times the distance from point B to the centroid of the area. That

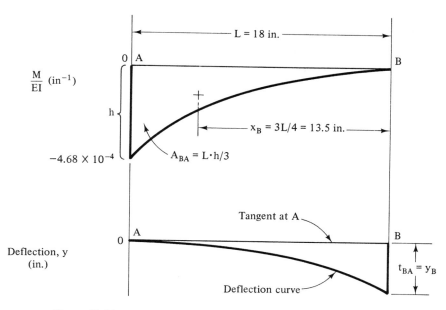

Figure 12-31 M/EI and deflection curves for Example Problem 12-12.

is,

$$t_{BA} = A_{BA} \cdot x_B$$

Recalling that the load, shear, and bending moment diagrams are related to each other in such a way that the higher curve is the derivative of the curve below it, the following conclusions can be drawn:

1. The shear curve is a first-degree curve (straight line with constant slope). Its equation is of the form

$$V = m \cdot x + b$$

where m is the slope of the line and b is its intercept with the vertical axis. The variable x is the position on the beam.

2. The bending moment curve is a second-degree curve, a parabola. The general equation of the curve is of the form,

$$M = a \cdot x^2 + b$$

Appendix A-1 shows the relationships for computing the area and the location of the centroid for areas bounded by second-degree curves. For an area having the shape of the bending moment or *M/EI* curves,

$$\text{area} = \frac{L \cdot h}{3}$$

$$x = \frac{L}{4}$$

where L = length of the base of the area

h = height of the area

x = distance from the side of the area to the centroid

Note that the corresponding distance from the vertex of the curve to the centroid is

$$x' = \frac{3L}{4}$$

Now, for the data shown in Figure 12-31

$$A_{BA} = \frac{L \cdot h}{3} = \frac{(18 \text{ in.})(-4.68 \times 10^{-4} \text{ in.}^{-1})}{3} = 2.808 \times 10^{-3}$$

$$x_B = \frac{3L}{4} = \frac{3(18 \text{ in.})}{4} = 13.5 \text{ in.}$$

Theorem 2 can now be used:

$$t_{BA} = A_{BA}x_B = (2.808 \times 10^{-3})(13.5 \text{ in.}) = 0.0379 \text{ in.}$$

This is equal to the deflection of the beam at its end, y_B.

PROBLEMS

Formula Method and Superposition

12-1. A round shaft having a diameter of 32 mm is 700 mm long and carries a 3.0-kN load at its center. If the shaft is steel and simply supported at its ends, compute the deflection at the center.

12-2. For the shaft in Problem 12-1, compute the deflection if the shaft is 6061-T6 aluminum instead of steel.

12-3. For the shaft in Problem 12-1, compute the deflection if the ends are fixed against rotation instead of simply supported.

12-4. For the shaft in Problem 12-1, compute the deflection if the shaft is 350 mm long rather than 700 mm.

12-5. For the shaft in Problem 12-1, compute the deflection if the diameter is 25 mm instead of 32 mm.

12-6. For the shaft in Problem 12-1, compute the deflection if the load is placed 175 mm from the left support rather than in the center. Compute the deflection both at the load and at the center of the shaft.

12-7. A wide flange steel beam, W12 × 16, carries the load shown in Figure 6-36(d). Compute the deflection at the loads and at the center of the beam.

12-8. A standard steel $1\frac{1}{2}$ in. schedule 40 pipe carries a 650-lb load at the center of a 28-in. span, simply supported. Compute the deflection of the pipe at the load.

12-9. An Aluminum Association Standard I-beam, 8I × 6.181, carries a uniformly distributed load of 1125 lb/ft over a 10-ft span. Compute the deflection at the center of the span.

12-10. For the beam in Problem 12-9, compute the deflection at a point 3.5 ft from the left support of the beam.

12-11. A wide flange steel beam, W12 × 30 carries the load shown in Figure 6-38(d). Compute the deflection at the load.

12-12. For the beam in Problem 12-11, compute the deflection at the load if the left support is moved 2.0 ft toward the load.

12-13. For the beam in Problem 12-11, compute the maximum upward deflection and determine its location.

12-14. The load shown in Figure 6-37(b) is being carried by an extruded aluminum (6061-T6) beam. Compute the deflection of the beam at each load. The beam shape is shown in Figure 7-22(b).

12-15. The loads shown in Figure 6-37(a) represent the feet of a motor on a machine frame. The frame member has the cross section shown in Figure 7-22(c) which has a moment of inertia of 16 956 mm^4. Compute the deflection at each load. The aluminum alloy 2014-T4 is used for the frame.

12-16. Compute the maximum deflection of a steel W18 × 55 beam when it carries the load shown in Figure 6-37(c).

12-17. A 1-in. schedule 40 steel pipe is used as a cantilever beam 8 in. long to support a load of 120 lb at its end. Compute the deflection of the pipe at the end.

12-18. A 1-in. schedule 40 steel pipe carries the two loads shown in Figure 6-40(b). Compute the deflection of the pipe at each load.

12-19. A cantilever beam carries two loads as shown in Figure 6-41(a). If a rectangular steel bar 20 mm wide by 80 mm high is used for the beam, compute the deflection at the end of the beam.

12-20. For the beam in Problem 12-19, compute the deflection if the bar is aluminum 2014-T4 rather than steel.

12-21. For the beam in Problem 12-19, compute the deflection if the bar is magnesium, ASTM AZ 63A-T6, instead of steel.

12-22. The load shown in Figure 6-49(a) is carried by a round steel bar having a diameter of 0.800 in. Compute the deflection of the bar at the right end.

12-23. A round steel bar is to be used to carry a single concentrated load of 3.0 kN at the center of a 700-mm-long span on simple supports. Determine the required diameter of the bar if its deflection must not exceed 0.12 mm.

12-24. For the bar designed in Problem 12-23, compute the stress in the bar and specify a suitable steel to provide a design factor of 8 based on ultimate strength.

12-25. Specify a standard wide flange steel beam which would carry the loading shown in Figure 6-37(c) with a maximum deflection less than $\frac{1}{360}$ times the length of the beam.

12-26. It is planned to use an Aluminum Association channel with its legs down to carry the loads shown in Figure 6-46(a) so that the flat face can contact the load. The maximum allowable deflection is 0.080 in. Specify a suitable channel.

12-27. A flat strip of steel 0.100 in. wide and 1.200 in. long is clamped at one end and loaded at the other like a cantilever beam (as in case a in Appendix A-23). What should be the thickness of the strip if it is to deflect 0.15 in. under a load of 0.52 lb?

12-28. A wood joist in a commercial building is 14 ft 6 in. long and carries a uniformly distributed load of 50 lb/ft. It is 1.50 in. wide and 9.25 in. high. If it is made of southern pine, compute the maximum deflection of the joist. Also compute the stress in the joist due to bending and horizontal shear, and compare them with the allowable stresses for No. 2 grade southern pine.

Successive Integration Method

For the following problems, 12-29 through 12-37 use the general approach outlined in Section 12-4 to determine the equations for the shape of the deflection curves of the beams. Unless otherwise noted, compute the maximum deflection of the beam from the equations and tell where it occurs.

12-29. The loading is shown in Figure 12-32(a). The beam is a rectangular steel bar, 1.0 in. wide and 2.0 in. high.

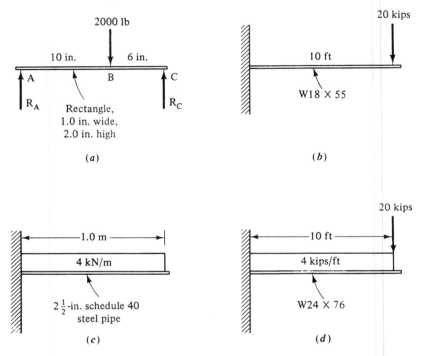

(a)

(b)

(c)

(d)

Figure 12-32   Beams for Problems 12-29 through 12-32.

12-30. The loading is shown in Figure 12-32(b). The beam is a steel wide-flange shape W18 × 55.

12-31. The loading is shown in Figure 12-32(c). The beam is a 2½-in. schedule 40 steel pipe.

12-32. The loading is shown in Figure 12-32(d). The beam is a steel wide-flange shape W24 × 76.

12-33. The load is shown in Figure 12-33(a). Design a round steel bar that will limit the deflection at the end of the beam to 5.0 mm.

12-34. The load is shown in Figure 12-33(b). Design a steel beam that will limit the maximum deflection to 1.0 mm. Use any shape, including those listed in the Appendix.

12-35. The load is shown in Figure 12-33(c). Select an aluminum I-beam that will limit the stress to 120 MPa; then compute the maximum deflection in the beam.

12-36. A steel wide-flange beam W14 × 26 carries the loading shown in Figure 12-33(d). Compute the maximum deflection between the supports and at each end.

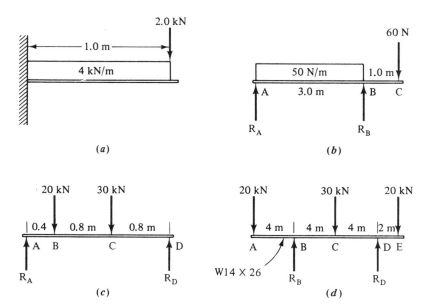

Figure 12-33   Beams for Problems 12-33 through 12-36.

12-37. The loading shown in Figure 12-34 represents a steel shaft for a large machine. The loads are due to gears mounted on the shaft. Assuming that the entire shaft will be the same diameter, determine the required diameter to limit the deflection at any gear to 0.13 mm.

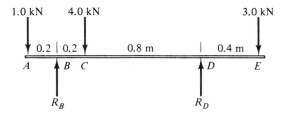

Figure 12-34   Shaft for Problem 12-37.

Moment-Area Method

Use the moment-area method to solve the following problems.

12-38. For the beam shown in Figure 12-32(a), compute the deflection at the load. The beam is a rectangular steel bar 1.0 in. wide and 2.0 in. high.

12-39. For the beam shown in Figure 12-32(a), compute the deflection at the middle, 8.0 in. from either support. The beam is a rectangular steel bar 1.0 in. wide and 2.0 in. high.

12-40. For the beam shown in Figure 12-32(b), compute the deflection at the end. The beam is a steel wide-flange shape W18 × 55.

12-41. For the beam shown in Figure 12-32(c), compute the deflection at the end. The beam is a $2\frac{1}{2}$-in. schedule 40 steel pipe.

12-42. For the beam shown in Figure 12-32(d), compute the deflection at the end. The beam is a steel wide-flange shape W24 × 76.

12-43. For the beam shown in Figure 12-33(a), compute the deflection at the end. The beam is a round aluminum bar 6061-T6 with a diameter of 100 mm.

12-44. For the beam shown in Figure 12-33(b), compute the deflection at the right end, point C. The beam is a square steel structural tube $2 \times 2 \times \frac{1}{4}$.

12-45. For the beam shown in Figure 12-33(c), compute the deflection at point C. The beam is an aluminum I-beam I7 $\times$ 5.800 made from 6061-T6.

12-46. For the beam shown in Figure 12-33(d), compute the deflection at point A. The beam is a steel wide-flange shape W14 $\times$ 26.

12-47. Figure 12-35 shows a stepped round steel shaft carrying a single concentrated load at its middle. Compute the deflection under the load.

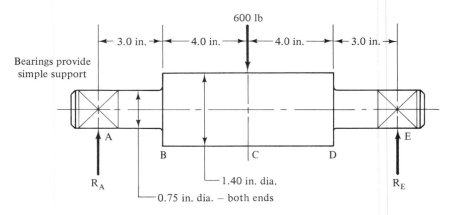

Figure 12-35 Shaft for Problem 12-47.

12-48. Figure 12-36 shows a composite beam made from steel structural tubing. Compute the deflection at the load.

12-49. Figure 12-37 shows a cantilever beam made from a steel wide-flange shape W18 $\times$ 55 with cover plates welded to the top and bottom of the beam over the first 6.0 ft. Compute the deflection at the end of the beam.

12-50. The top member for a gantry crane is made as shown in Figure 12-38. The two end pieces are $3 \times 3 \times \frac{1}{4}$ steel structural tubing. The middle piece is $4 \times 4 \times \frac{1}{2}$. They are combined for a distance of 10.0 in. with the 3-in. tube fitted tightly inside the 4-in. tube and brazed together. Compute the deflection at the middle of the beam.

12-51. A crude diving board is made by nailing two wooden planks together as shown in Figure 12-39. Compute the deflection at the end if the diver exerts a force of 1.80 kN at the end. The wood is No. 2 southern pine.

COMPUTER PROGRAMMING ASSIGNMENTS

1. Write a program to evaluate the deflection of a simply supported beam carrying one concentrated load between the supports using the formulas given in case b of Appendix A-22. The program should accept input for the length, position of the load, length of the overhang, if any, the beam stiffness values (E and I), and the point for which the deflection is to be computed.

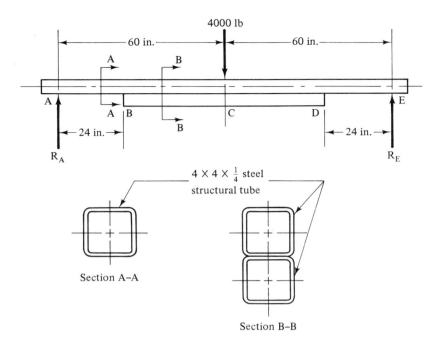

Figure 12-36 Beam for Problem 12-48.

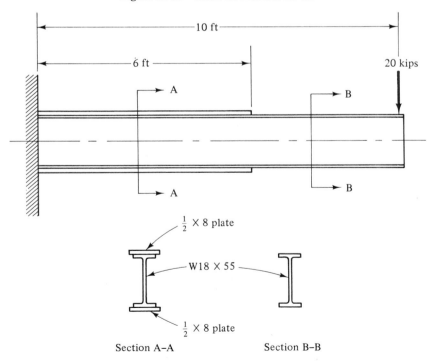

Figure 12-37 Cantilever beam for Problem 12-49.

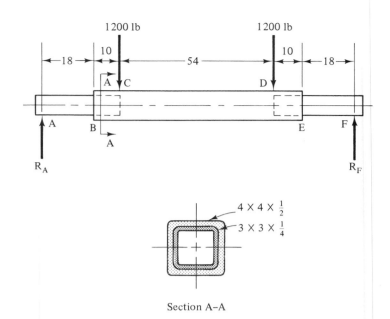

Figure 12-38 Gantry crane beam for Problem 12-50.

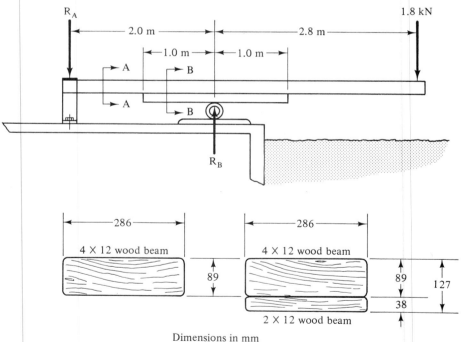

Dimensions in mm

Section A–A   Section B–B

Figure 12-39 Diving board for Problem 12-51.

ENHANCEMENTS

    (a) Design the program so that it computes the deflection at a series of points which could permit the plotting of the complete deflection curve.

    (b) In addition to computing the deflection of the series of points, have the program plot the deflection curve on a plotter or printer.

    (c) Have the program compute the maximum deflection and the point at which it occurs.

2. Repeat Assignment 1 for any of the loading and support patterns shown in Appendix A-22.

3. Write a program similar to that of Assignment 1 for case b of Appendix A-22, but have it accept two or more concentrated loads at any point on the beam and compute the deflection at specified points using the principle of superposition.

4. Combine two or more programs that solve for the deflection of beams for a given loading pattern so that superposition can be used to compute the deflection at any point due to the combined loading. For example, combine cases b and d of Appendix A-22 to handle any beam with a combination of concentrated loads and a complete uniformly distributed load. Or, add case g to include a distributed load over only part of the beam's length.

5. Repeat Assignments 1 through 4 for the cantilevers from Appendix A-23.

6. In the general approach to computing beam deflections, a series of constants of integration are developed that have to be evaluated by solving a set of simultaneous equations. One convenient way of doing this is to use determinants. Write a program that can solve for the constants using determinants.

13

Statically Indeterminate Beams

13-1 OBJECTIVES OF THIS CHAPTER

The beams considered in earlier chapters include beams with two and only two simple supports or cantilevers that have one fixed end and the other end free. It was shown that all the unknown reaction forces and bending moments could be found using the classic equations of equilibrium:

$$\sum F = 0 \qquad \text{in any direction}$$

$$\sum M = 0 \qquad \text{about any point}$$

Such beams are called *statically determinate*.

This chapter deals with beams that do not fall into the categories listed above and thus they are called *statically indeterminate*. Different methods of analyzing such beams are required and will be shown in this chapter. Also, a comparison will be made for the behavior of beams designed to perform a similar function but implemented with different support systems, some of which are statically determinate and some of which are statically indeterminate.

After completing this chapter, you should be able to:

1. Define *statically determinate* and *statically indeterminate.*
2. Recognize statically indeterminate beams from given descriptions of the loading and support conditions.
3. Define *continuous beam.*
4. Define *supported cantilever.*
5. Define *fixed-end beam.*
6. Use established formulas to analyze certain types of statically indeterminate beams.
7. Use superposition to combine simple cases for which formulas are available to solve more complex loading cases.
8. Use the *theorem of three moments* to analyze continuous beams having three or more supports and carrying any combination of concentrated and distributed loads.
9. Compare the relative strength and stiffness of beams having different support systems for given loadings.

13-2 EXAMPLES OF STATICALLY INDETERMINATE BEAMS

Beams with more than two simple supports, cantilevers with a second support, or beams with two fixed ends are important examples of statically indeterminate beams. Figure 13-1 shows the conventional method of representing these kinds of beams. The typical shapes for the deflection curves of the beams are also shown. Significant differences should be noted between these and the curves for beams in the preceding chapter.

Figure 13-1(a) is called a *continuous beam,* the name coming from the fact that the beam is continuous over several supports. It is important to note that the shape of the deflection curve of the beam is also continuous through the supports. This fact will be useful in analyzing such beams. Continuous beams occur frequently in building structures and highway bridges. Many ranch-type homes with basements contain such a beam running the length of the house to carry loads from floor joists and partitions. Expressway bridges carrying local traffic over through lanes are frequently supported at the ends on each side of the roadway and also at the middle in the median strip. Notice that the beams of such bridges are usually one piece or are connected to form a rigid *continuous* beam.

The *fixed-end* beam shown in Figure 13-1(b) is used in building structures and also in machine structures because of the high degree of rigidity provided. The creation of the fixed type of end condition requires that the connections at the ends restrain the beam from rotation as well as provide support for vertical loads. Figure 13-2 shows one way in which the fixed end condition can be achieved. Welding of the cross beam to the supporting columns would achieve the same result. Care must be used in evaluating fixed-ended beams to ensure that the connections can indeed

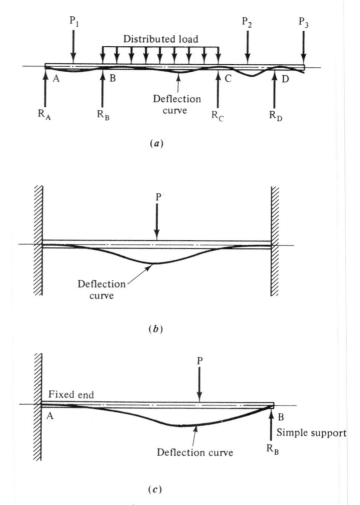

Figure 13-1 Examples of statically indeterminant beams. (a) Continuous beam—a beam on three or more simple supports (b) Beam with two fixed ends (c) Supported cantilever.

restrain the beam from rotating at the support and resist the moments created because of the restraint. Without due care, a condition *between* that of fixed ends and simple supports could result, making analysis difficult and subject to error.

The supported cantilever could be constructed as shown in Figure 13-3. The load on the flat roof is carried by a beam rigidly fixed to a column at one end and resting simply on another column at the other end.

Continuous beams and beams with fixed ends have certain advantages over simply supported beams: A greater rigidity is provided and generally lower stresses would be produced in beams of equivalent size; alternatively, smaller and lighter beams can be used to give the same degree of strength and rigidity. The disadvantages

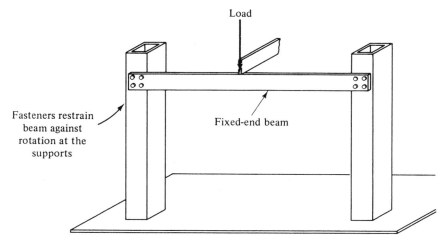

Figure 13-2 Fixed-end beam.

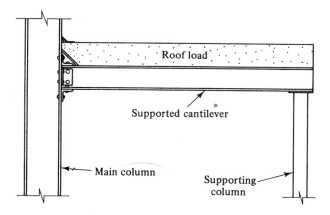

Figure 13-3 Supported cantilever.

are that the analyses are more difficult and the effect of small variations in support conditions can be significant.

13-3 FORMULAS FOR STATICALLY INDETERMINATE BEAMS

Appendix A-24 shows several examples of statically indeterminate beams along with formulas for computing the reactions at supports and the deflection at any point on the beams. These formulas can be applied directly in the manner demonstrated in Chapter 12 for statically determinate beams. Also shown are plots of the shearing force and bending moment diagrams along with the necessary formulas for computing the values at critical points.

Note the following general characteristics of these beams:

1. For the fixed-end beams the bending moments at the ends are *not* zero as they are for simply supported beams.
2. The bending moments at the fixed ends are *negative,* indicating that the deflection curve near the ends is *concave downward.*
3. The bending moments in the middle portion of fixed-end beams are *positive,* indicting that the deflection curve there is *concave upward.*
4. There are typically two points of *zero bending moment* on the fixed-end beams.
5. The means of fastening a fixed-end beam to its supports must be capable of resisting the bending moments and the shear forces at those points, often the *maximum* values.
6. The deflection curve for fixed-end beams has a zero slope at the fixed ends because of the restraint provided there.
7. For the *supported cantilever* beams, the bending moment at the fixed end is typically negative and is frequently the maximum value on the beam.
8. The *support* for the supported cantilever is taken to be a *simple support,* meaning that it cannot resist bending. Thus the bending moment at the simple support is zero, provided that there is no overhang beyond the support.
9. There is a point of zero bending moment on a supported cantilever, generally near the fixed end.
10. If the supported cantilever has an overhang, such as that shown in case d in Appendix A-24, the largest bending moment typically occurs at the simple support. The shape of the bending moment curve is generally opposite that for the supported cantilever with the load placed between the fixed end and the support.
11. For continuous beams, the bending moment is typically maximum at the interior supports.
12. The design of the beam's shape and cross-sectional dimensions can be customized to accommodate the variation in bending moments.

Example Problem 13-1

Compare the behavior of the four beams shown in Figures 13-4 through 13-7 with regard to shearing force, bending moment distribution, and maximum deflection. In each case, the beam is designed to support a single concentrated load at a given distance from a support on the left. A second support is provided at the same distance to the right for some of the proposed designs, with the supports being either simple or fixed.

Solution   Part of the results are shown in Figures 13-4 through 13-7, in which the shearing force and bending moment diagrams are drawn using the approaches developed in Chapter 6 and the information from Appendixes A-22 through A-24. Also shown are the expressions for computing the maximum deflection of the beams in terms of the primary variables; load, P, distance to the support, a, and the beam stiffness, EI.

The approach used here to compare the four methods of supporting the load is to consider the simply supported beam, [Figure 13-4], as the basic case. Then the other proposed designs are compared to it. First, the beam shape required to withstand the

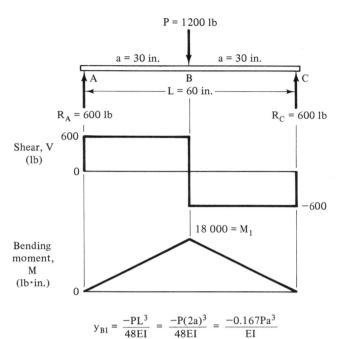

$$y_{B1} = \frac{-PL^3}{48EI} = \frac{-P(2a)^3}{48EI} = \frac{-0.167Pa^3}{EI}$$

Figure 13-4 Bending moment and deflection for simply supported beam.

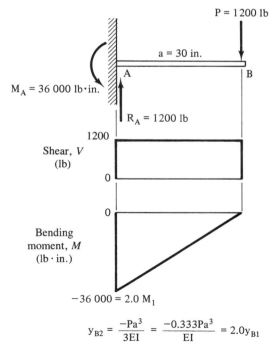

$$y_{B2} = \frac{-Pa^3}{3EI} = \frac{-0.333Pa^3}{EI} = 2.0y_{B1}$$

Figure 13-5 Bending moment and deflection for cantilever.

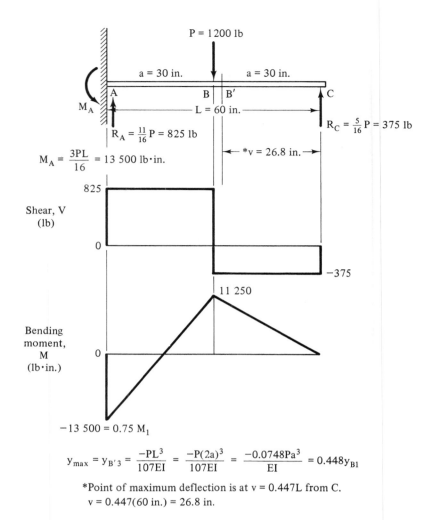

$$M_A = \frac{3PL}{16} = 13\ 500\ lb\cdot in.$$

$$y_{max} = y_{B'3} = \frac{-PL^3}{107EI} = \frac{-P(2a)^3}{107EI} = \frac{-0.0748Pa^3}{EI} = 0.448y_{B1}$$

*Point of maximum deflection is at v = 0.447L from C.
v = 0.447(60 in.) = 26.8 in.

Figure 13-6 Bending moment and deflection for supported cantilever.

bending moment safely according to the AISC code is determined. From the discussion of design stresses in Chapter 3,

$$\sigma_d = 0.66s_y$$

Let's choose to specify a standard steel channel with the legs down made from ASTM A36 structural steel for the beam. Then the design stress is

$$\sigma_d = 0.66(36\ 000\ psi) = 23\ 760\ psi$$

We can now compute the required section modulus for the beam. From Figure 13-4 we find the maximum bending moment to be 18 000 lb · in. Then

$$S = \frac{M}{\sigma_d} = \frac{18\ 000\ lb\cdot in.}{23\ 760\ lb/in^2} = 0.758\ in^3$$

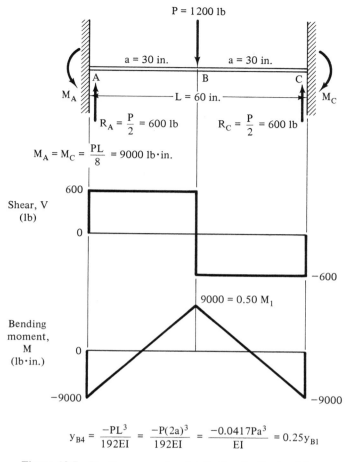

$$y_{B4} = \frac{-PL^3}{192EI} = \frac{-P(2a)^3}{192EI} = \frac{-0.0417Pa^3}{EI} = 0.25y_{B1}$$

Figure 13-7 Bending moment and deflection for fixed-end beam.

Referring to Appendix A-6 and noting that the channel is to be applied with the legs down, the section properties with respect to the *Y-Y* axis are used. A C8 × 11.5 channel has a section modulus of 0.781 in³ and is the lightest beam shape that is suitable.

Now we can consider the deflection of the beam. As shown in Figure 13-4, the maximum deflection occurs at the load, point *B*, and has the value

$$y_{B1} = \frac{-0.167\,Pa^3}{EI}$$

Using $P = 1200$ lb, $a = 30$ in., $E = 30 \times 10^6$ psi, and $I = 1.32$ in⁴,

$$y_{B1} = \frac{(-0.167)(1200\ \text{lb})(30\ \text{in.})^3}{(30 \times 10^6\ \text{lb/in}^2)(1.32\ \text{in}^4)} = -0.136\ \text{in.}$$

One traditional measure of the upper limit for the deflection of a beam is that it should not deflect more than $\frac{1}{360}$ of its length. For this beam,

$$\frac{L}{360} = \frac{60 \text{ in.}}{360} = 0.167 \text{ in.}$$

Because the actual deflection is less, it is acceptable.

Now the other three proposed beam types will be designed in a similar manner and the required channel size compared to the C8 × 11.5 required for the basic case. That is:

1. The maximum bending moment for the beam loading will be found from the bending moment diagram and compared with the maximum bending moment, $M_1 = 18\,000$ lb · in., for the basic case.
2. The required section modulus for the beam cross section will be computed to limit the bending stress to 23 760 psi.
3. The lightest standard channel will be selected from Appendix A-6 that has the required section modulus.
4. The required moment of inertia for the beam cross section will be computed to limit the maximum deflection of the beam to 0.136 in., that computed for the beam in the basic case. This value for I will be compared to that for the basic case, $I_1 = 1.32$ in^4.
5. The lightest standard channel will be selected from Appendix A-6 that has the required moment of inertia.
6. The selected channel for each case will be compared, primarily with regard to its weight, a measure of the efficient use of material to support the load.

Cantilever

1. $M_{\max} = 36\,000$ lb · in. $= 2.0M_1$.
2. $S = M/\sigma_d = 36\,000$ lb · in./23 760 lb/in^2 = 1.52 in^3.
3. The C12 × 25 channel, with $S = 1.88$ in^3, is the lightest acceptable channel. Note that its weight is 13.5 lb/ft *more* than that required for the basic case.
4. The required moment of inertia is found from

$$y_{\max} = \frac{-0.333Pa^3}{EI}$$

Letting $y_{\max} = -0.136$ in. yields

$$I = \frac{-0.333Pa^3}{E(y_{\max})}$$

$$= \frac{(-0.333)(1200 \text{ lb})(30 \text{ in.})^3}{(30 \times 10^6 \text{ lb/in}^2)(-0.136 \text{ in.})} = 2.65 \text{ in}^4 = 2.0I_1$$

5. The required channel is C10 × 20, but Step 3 governs.

Supported Cantilever

1. $M_{\max} = 13\,500$ lb · in. $= 0.750M_1$

2. $S = M/\sigma_d = 13\ 500\ \text{lb}\cdot\text{in.}/23\ 760\ \text{lb/in}^2 = 0.568\ \text{in}^3$.

3. The C8 × 11.5 channel, with $S = 0.781\ \text{in}^3$, is the lightest acceptable channel.

4. The required moment of inertia is found from

$$y_{\text{max}} = \frac{-0.0748Pa^3}{EI}$$

Letting $y_{\text{max}} = -0.136$ in. yields

$$I = \frac{-0.0748Pa^3}{E(y_{\text{max}})}$$

$$= \frac{(-0.0748)(1200\ \text{lb})(30\ \text{in.})^3}{(30 \times 10^6\ \text{lb/in}^2)(-0.136\ \text{in.})} = 0.594\ \text{in}^4 = 0.45I_1$$

5. The required channel is the C6 × 8.2, with $I = 0.693\ \text{in}^4$.

Fixed-End Beam

1. $M_{\text{max}} = 9\ 000\ \text{lb}\cdot\text{in.} = 0.50M_1$.

2. $S = M/\sigma_d = 9\ 000\ \text{lb}\cdot\text{in.}/23\ 760\ \text{lb/in}^2 = 0.379\ \text{in}^3$.

3. The C6 × 8.2 channel, with $S = 0.492\ \text{in}^3$, is the lightest acceptable channel. Note that its weight is 4.7 lb/ft *less* than that required for the basic case.

4. The required moment of inertia is found from

$$y_{\text{max}} = \frac{-0.0417Pa^3}{EI}$$

Letting $y_{\text{max}} = -0.136$ in. yields

$$I = \frac{-0.0417Pa^3}{E(y_{\text{max}})}$$

$$I = \frac{(-0.0417)(1200\ \text{lb})(30\ \text{in.})^3}{(30 \times 10^6\ \text{lb/in}^2)(-0.136\ \text{in.})} = 0.331\ \text{in}^4 = 0.25I_1$$

5. The required channel is C5 × 6.7, with $I = 0.479\ \text{in}^4$.

Summary Comparison

Beam type	M_{max} (lb · in.)	Required S (in)^3	Required I (in^4)	Channel size
Simply supported (basic case)	$18\ 000 = M_1$	0.758	$1.32 = I_1$	C8 × 11.5
Cantilever	$36\ 000 = 2.0M_1$	1.52	$2.65 = 2.0I_1$	C12 × 25
Supported cantilever	$13\ 500 = 0.75M_1$	0.568	$0.594 = 0.45I_1$	C8 × 11.5
Fixed-end	$9\ 000 = 0.50M_1$	0.379	$0.331 = 0.25I_1$	C6 × 8.2

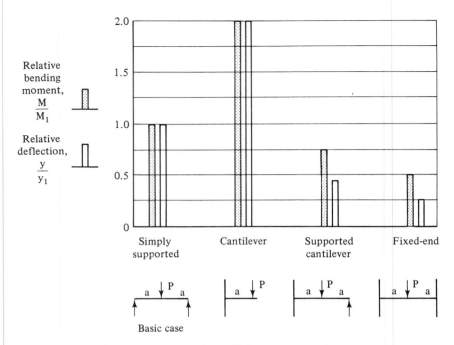

Figure 13-8 Comparison of behavior of beam designs.

The comparison is presented graphically in Figure 13-8. It should be evident that significant savings in materials can be obtained by using fixed-end beams whenever possible.

13-4 SUPERPOSITION METHOD

Consider first the supported cantilever shown in Figure 13-9. Because of the restraint at A and the simple support at B, the unknown reactions include:

1. The vertical force R_B
2. The vertical force R_A
3. The restraining moment M_A

The conditions assumed for this beam are that the supports at A and B are

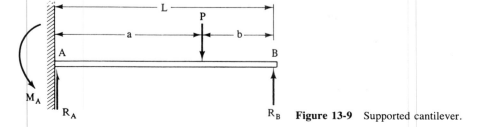

Figure 13-9 Supported cantilever.

absolutely rigid and at the same level, and that the connection at A does not allow any rotation of the beam at that point. Conversely, the support at B will allow rotation and cannot resist moments.

If the support at B is removed, the beam would deflect downward, as shown in Figure 13-10(a), by an amount y_{B1} due to the load P. Now if the load is removed and the reaction force R_B is applied upward at B, the beam would deflect upward by an amount y_{B2}, as shown in Figure 13-10(b). In reality, of course, both forces are applied, and the deflection at B is *zero*. The principal of superposition would then provide the conclusion that

$$y_{B1} + y_{B2} = 0 \tag{13-1}$$

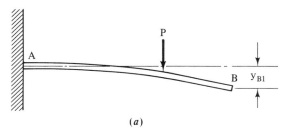

(a)

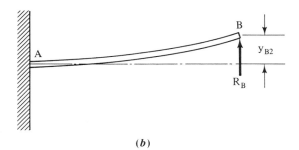

(b)

Figure 13-10 Superposition applied to the supported cantilever.

This equation, along with the normal equations of static equilibrium, will permit the evaluation of all three unknowns, as demonstrated in the example problem which follows. It must be recognized that the principles of static equilibrium are still valid for statically indeterminate beams. However, they are not sufficient to allow a direct solution.

Example Problem 13-2

Determine the support reactions at A and B for the supported cantilever shown in Figure 13-9 if the load P is 2600 N and placed 1.20 m out from A. The total length of the beam is 1.80 m. Then draw the complete shear and bending moment diagrams, and design the beam by specifying a configuration, a material, and the required dimensions of the beam. Use a design factor of 8 based on ultimate strength since the load will be repeated often.
Solution We must first determine the reaction at B using superposition. Equation (13-1) states that, for this situation,

$$y_{B1} + y_{B2} = 0 \tag{13-1}$$

The equation for y_{B1} can be found from the beam deflection formulas in Appendix A-23.

As suggested in Figure 13-10(a), the deflection at the end of a cantilever carrying an intermediate load is required. Then

$$y_{B1} = \frac{-Pa^2}{6EI}(3L - a)$$

where $P = 2600$ N
$a = 1.20$ m
$L = 1.80$ m

The values of E and I are still unknown, but we can express the deflection in terms of EI.

$$y_{B1} = \frac{(-2600 \text{ N})(1.20 \text{ m})^2}{6EI}[3(1.80 \text{ m}) - 1.20 \text{ m})]$$

$$= \frac{-2621 \text{ N} \cdot \text{m}^3}{EI}$$

Now looking at Figure 13-10(b), we need the deflection at the end of the cantilever due to a concentrated load there. Then

$$y_{B2} = \frac{PL^3}{3EI} = \frac{R_B(1.8 \text{ m})^3}{3EI} = \frac{R_B(1.944 \text{ m}^3)}{EI}$$

Putting these values in Equation (13-1) gives

$$\frac{-2621 \text{ N} \cdot \text{m}^3}{EI} + \frac{R_B(1.944 \text{ m}^3)}{EI} = 0$$

The term EI can be canceled out, allowing the solution for R_B.

$$R_B = \frac{2621 \text{ N} \cdot \text{m}^3}{1.944 \text{ m}^3} = 1348 \text{ N}$$

The values of R_A and M_A can now be found using the equations of static equilibrium.

$$\sum F = 0 \qquad \text{(in the vertical direction)}$$

$$R_A + R_B - P = 0$$

$$R_A = P - R_B = 2600 \text{ N} - 1348 \text{ N} = 1252 \text{ N}$$

Summing moments about point A gives

$$0 = M_A - 2600 \text{ N} (1.2 \text{ m}) + 1348 \text{ N} (1.8 \text{ m})$$

$$M_A = 693 \text{ N} \cdot \text{m}$$

The positive sign for the result indicates that the assumed sense of the reaction moment in Figure 13-9 is correct. However, this is a negative moment because it causes the beam to bend concave downward near the support A.

The shear and bending moment diagrams can now be drawn, as shown in Figure 13-11, using conventional techniques. The maximum bending moment occurs at the load where $M = 809$ N $\cdot$ m.

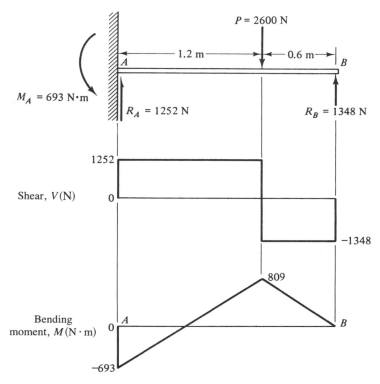

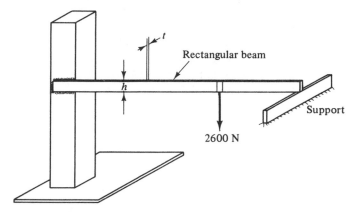

Figure 13-11 Shear and bending moment diagrams for the supported cantilever in Example Problem 13-2.

The beam can now be designed. Let's assume that the actual installation is similar to that sketched in Figure 13-12, with the beam welded at its left end and resting on another beam at the right end. A rectangular bar would work well in this arrangement,

Figure 13-12 Physical implementation of a supported cantilever.

and a ratio of $h = 3t$ will be assumed. A carbon steel such as AISI 1040, hot rolled, provides an ultimate strength of 524 MPa. Its percent elongation—18%—suggests good ductility, which will help it resist the repeated loads. The design should be based on bending stress.

$$\sigma = \frac{M}{S}$$

But let

$$\sigma = \sigma_d = \frac{s_u}{N} = \frac{524 \text{ MPa}}{8} = 65.5 \text{ MPa}$$

Then

$$S = \frac{M}{\sigma_d} = \frac{809 \text{ N} \cdot \text{m}}{65.5 \text{ N/mm}^2} \times \frac{10^3 \text{ mm}}{\text{m}} = 12\ 350 \text{ mm}^3$$

For a rectangular bar,

$$S = \frac{th^2}{6} = \frac{t(3t)^2}{6} = \frac{9t^3}{6} = 1.5t^3$$

Then

$$1.5t^3 = 12\ 350 \text{ mm}^3$$

$$t = 20.2 \text{ mm}$$

Let's use the preferred size of 22 mm for t. Then

$$h = 3t = 3(22 \text{ mm}) = 66 \text{ mm}$$

The final design can be summarized as a rectangular steel bar of AISI 1040, hot rolled, 22 mm thick and 66 mm high, welded to a rigid support at its left end and resting on a simple support at its right end. The maximum stress in the bar would be less than 65.5 MPa, providing a design factor of at least 8 based on ultimate strength.

The superposition method can be applied to any supported cantilever beam analysis for which the equation for the deflection due to the applied load can be found. Either the beam deflection formulas like those in the Appendix, the moment-area method, or the mathematical analysis method developed in Chapter 12 can be used.

Continuous beams can also be analyzed using superposition. Consider the beam on three supports shown in Figure 13-13. The three unknown support reactions make the beam statically indeterminate. The "extra" reaction R_C can be found using the

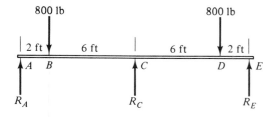

Figure 13-13 Continuous beam.

Statically Indeterminate Beams   Chap. 13

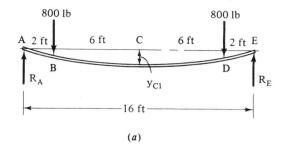

(a)

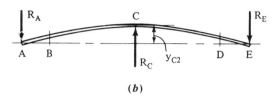

(b)

Figure 13-14 Superposition applied to a continuous beam.

technique suggested in Figure 13-14. Removing the support at C would cause the deflection y_{C1} downward due to the two 800-lb loads. Case c of Appendix A-22 can be used to find y_{C1}. Then if the loads are imagined to be removed and the reaction R_C is replaced, the upward deflection y_{C2} would result. The formulas of case a in Appendix A-22 can be used.

Here again, of course, the actual deflection at C is zero because of the unyielding support. Therefore,

$$y_{C1} + y_{C2} = 0$$

From this relationship, the value of R_C can be computed. The remaining reactions R_A and R_E can then be found in the conventional manner, allowing the creation of the shear and bending moment diagrams.

13-5 CONTINUOUS BEAMS—THEOREM OF THREE MOMENTS

A continuous beam on any number of supports can be analyzed by using the *theorem of three moments*. The theorem actually relates the bending moments at three successive supports to each other and to the loads on the beam. For a beam with only three supports, the theorem allows the direct computation of the moment at the middle support. Known end conditions provide data for computing moments at the ends. Then the principles of statics can be used to find reactions.

For beams on more than three supports, the theorem is applied successively to sets of three adjacent supports (two spans), yielding a set of equations which can be solved simultaneously for the unknown moments.

The theorem of three moments can be used for any combination of loads. Special forms of the theorem have been developed for uniformly distributed loads and concentrated loads. These forms will be used in this chapter.

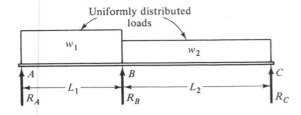

Uniformly distributed loads

w_1

w_2

A B C

L_1 L_2

R_A R_B R_C

Figure 13-15 Uniformly distributed loads on a continuous beam of two spans.

Uniformly distributed loads on adjacent spans. Figure 13-15 shows the arrangement of loads and the definition of terms applicable to Equation (13-2).

$$M_A L_1 + 2M_B(L_1 + L_2) + M_C L_2 = \frac{-w_1 L_1^3}{4} - \frac{w_2 L_2^3}{4} \qquad (13\text{-}2)$$

The values of w_1 and w_2 are expressed in units of force per unit length such as N/m, lb/ft, etc. The bending moments at the supports A, B, and C are M_A, M_B, and M_C. If M_A and M_C at the ends of the beam are known, M_B can be found from Equation (13-2) directly. Example problems will demonstrate the application of this equation.

The special case in which two equal spans carry equal uniform loads allows the simplification of Equation (13-2). If $L_1 = L_2 = L$ and $w_1 = w_2 = w$, then

$$M_A + 4M_B + M_C = \frac{-wL^2}{2} \qquad (13\text{-}3)$$

Concentrated loads on adjacent spans. If adjacent spans carry only one concentrated load each, as shown in Figure 13-16, then Equation (13-4) applies.

$$M_A L_1 + 2M_B(L_1 + L_2) + M_C L_2 = \frac{-P_1 a}{L_1}(L_1^2 - a^2) - \frac{P_2 b}{L_2}(L_2^2 - b^2) \qquad (13\text{-}4)$$

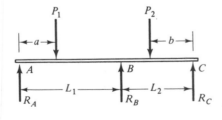

P_1 P_2

a b

A B C

L_1 L_2

R_A R_B R_C

Figure 13-16 Continuous beam of two spans with one concentrated load on each span.

Combinations of uniformly distributed loads and several concentrated loads. This is a somewhat general case, allowing each span to carry a uniformly distributed load and any number of concentrated loads, as suggested in Figure 13-17. The general equation for such a loading is a combination of Equations (13-2) and (13-4), given as Equation (13-5).

$$M_A L_1 + 2M_B(L_1 + L_2) + M_C L_2$$

$$= -\sum \left[\frac{P_i a_i}{L_1}(L_1^2 - a_i^2) \right]_1 - \sum \left[\frac{P_i b_i}{L_2}(L_2^2 - b_i^2) \right]_2 - \frac{w_1 L_1^3}{4} - \frac{w_2 L_2^3}{4} \qquad (13\text{-}5)$$

Statically Indeterminate Beams Chap. 13

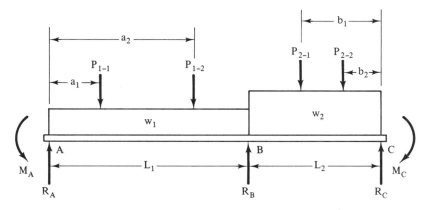

Figure 13-17 General notation for terms in Equation 13-5.

The term in the bracket with the subscript 1 is to be evaluated for *each* concentrated load on span 1 and then summed together. Similarly, the term having the subscript 2 is repeatedly applied for all loads on span 2. Notice that the distances a_i are measured from the reaction at A for each load on span 1, and the distances b_i are measured from the reaction at C for each load on span 2. The moments at the ends A and C can be due to concentrated moments applied there or to loads on overhangs beyond the supports. Any of the terms in Equation (13-5) may be left out of a problem solution if there is no appropriate load or moment existing at a particular section for which the equation is being written. Furthermore, other concentrated loads could be included in addition to those actually shown in Figure 13-17.

Example Problem 13-3

The loading composed of a combination of distributed loads and concentrated loads, shown in Figure 13-18 is to be analyzed to determine the reactions at the three supports and the complete shear and bending moment diagrams. The 17-m beam is to be used as a floor beam in an industrial building.

Solution Because of the combined loading, Equation (13-5) must be used. Remember, the subscript 1 refers to span 1 between A and B, and subscript 2 refers to span 2 between B and C. From the loading, we can say

$$M_C = 0$$

$$M_A = -12 \text{ kN } (2 \text{ m}) - 60 \text{ kN } (1 \text{ m}) = -84 \text{ kN} \cdot \text{m}$$

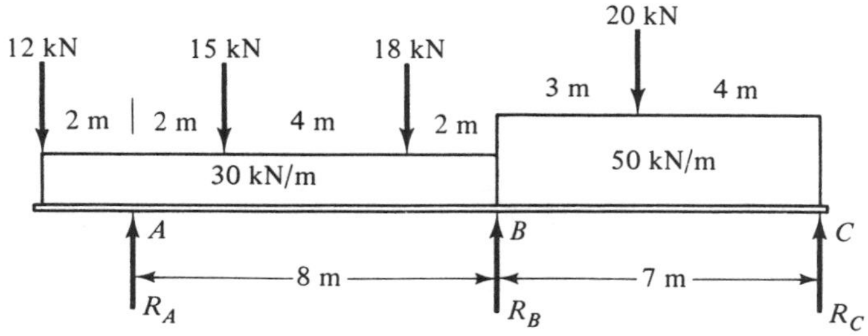

Figure 13-18 Beam for Example Problem 13-3.

Each remaining term in Equation (13-5) will be evaluated.

$$M_A L_1 = -84 \text{ kN} \cdot \text{m}(8 \text{ m}) = -672 \text{ kN} \cdot \text{m}^2$$

$$2M_B(L_1 + L_2) = 2M_B(8 + 7) = M_B(30 \text{ m})$$

$$-\sum \left[\frac{P_i a_i}{L_1} (L_1^2 - a_i^2) \right]_1 = -\frac{15(2)}{8} (8^2 - 2^2) - \frac{18(6)}{8} (8^2 - 6^2)$$

$$= -603 \text{ kN} \cdot \text{m}^2$$

$$-\sum \left[\frac{P_i b_i}{L_2} (L_2^2 - b_i^2) \right]_2 = -\frac{20(4)}{7} (7^2 - 4^2) = -377 \text{ kN} \cdot \text{m}^2$$

$$-\frac{w_1 L_1^3}{4} = -\frac{30(8)^3}{4} = -3840 \text{ kN} \cdot \text{m}^2$$

$$-\frac{w_2 L_2^3}{4} = -\frac{50(7)^3}{4} = -4288 \text{ kN} \cdot \text{m}^2$$

Now putting these values into Equation (13-5) gives

$$-672 \text{ kN} \cdot \text{m}^2 + M_B(30 \text{ m}) + 0 = (-603 - 377 - 3840 - 4288) \text{ kN} \cdot \text{m}^2$$

Solving for M_B yields

$$M_B = 281 \text{ kN} \cdot \text{m}$$

Having the three moments at the supports allows the computation of the reactions using the procedure shown in earlier example problems. From this we get

$$R_A = 183 \text{ kN}$$

$$R_B = 389 \text{ kN}$$

$$R_C = 143 \text{ kN}$$

The maximum positive bending moment is 204 kN · m at a point 2.86 m from C. The maximum negative moment is -281 kN · m at the support B (see Figure 13-19).

PROBLEMS

For Problems 13-1 and 13-2, use the superposition method to determine the reactions at all supports and draw the complete shear and bending moment diagrams. Indicate the maximum shear and bending moment for each beam.

13-1. Use Figure 13-20.
13-2. Use Figure 13-21.

For Problems 13-3 and 13-4, use the theorem of three moments to determine the reactions at all supports and draw the complete shear and bending moment diagrams. Indicate the maximum shear and bending moment for each beam.

13-3. Use Figure 13-20.
13-4. Use Figure 13-22.

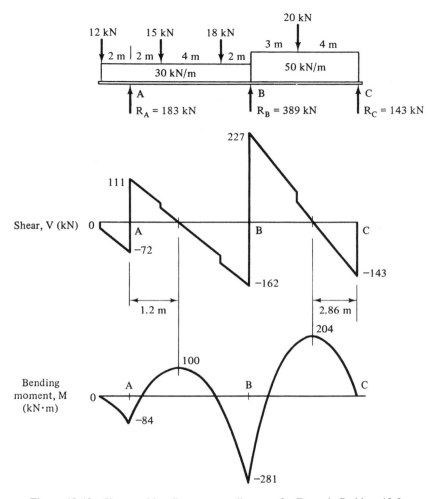

Figure 13-19 Shear and bending moment diagrams for Example Problem 13-3.

13-5. Compare the behavior of the four beams shown in Figure 13-23 with regard to shearing force, bending moment, and maximum deflection. In each case, the beam is designed to support a uniformly distributed load across a given span. Complete the analysis in a manner similar to that used in Example Problem 13-1.

For Problems 13-6 through 13-15, use the formulas in Appendix A-24 to determine the reactions and draw the complete shear and bending moment diagrams.

13-6. Use Figure 13-20(b).
13-7. Use Figure 13-21(a).
13-8. Use Figure 13-21(b).
13-9. Use Case d with $P = 18$ kN, $L = 2.75$ m, and $a = 1.40$ m.
13-10. Use Case f with $P = 8500$ lb, $a = 75$ in., and $b = 34$ in.

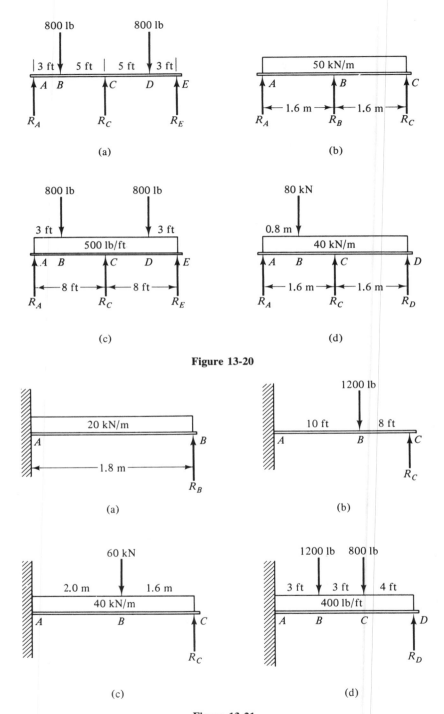

Figure 13-20

Figure 13-21

Statically Indeterminate Beams   Chap. 13

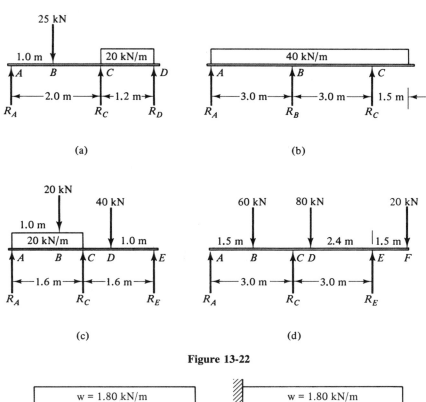

(a)

(b)

(c)

(d)

Figure 13-22

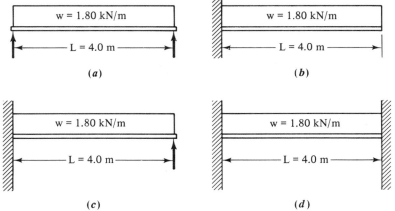

(a)

(b)

(c)

(d)

Figure 13-23 Beams for Problem 13-5. (a) Simply supported beam (b) Cantilever (c) Supported cantilever (d) Fixed end beam.

13-11. Use Case h with $w = 4200$ lb/ft and $L = 16$ ft.

13-12. Use Case i with $w = 50$ kN/m and $L = 3.60$ m.

13-13. Use Case j with $w = 15.0$ lb/in. and $L = 36$ in.

13-14. Use Case e with $P = 140$ lb and $L = 54$ in.

13-15. Use Case b with $P = 250$ N, $a = 15$ mm, and $b = 40$ mm.

13-16. Specify a suitable design for a wooden beam, simply supported at its ends, to carry a uniformly distributed load of 120 lb/ft for a span of 24 ft. The beam must be safe for both bending and shear stresses when made from No. 2 grade southern pine. Then compute the maximum deflection for the beam you have designed.

13-17. Repeat Problem 13-16 except place an additional support at the middle of the beam, 12 ft from either end.

13-18. Repeat Problem 13-16 except use four supports 8 ft apart.

13-19. Compare the behavior of the three beams designed in Problems 13-16, 13-17, and 13-18 with regard to shearing force, bending moment, and maximum deflection. Complete the analysis in the manner used in Example Problem 13-1.

13-20. A plank from a wooden deck carries a load like that shown in Case j in Appendix A-24 with $w = 100$ lb/ft and $L = 24$ in. The plank is a standard 2 × 6 having a thickness of 1.50 in. and a width of 5.50 in. with the long dimension horizontal. Would the plank be safe if it is made from No. 2 grade southern pine?

13-21. Two designs for a diving board are proposed as shown in Figure 13-24. Compare the designs with regard to shearing force, bending moment, and deflection. Express the deflection in terms of the beam stiffness, EI. Note that cases from Appendix A-22 and A-24 can be used.

13-22. For each of the proposed diving board designs described in Problem 13-21 and shown in Figure 13-24, complete the design, specifying the material, the cross section, and the final dimensions. The board is to be 24 in. wide.

13-23. Figure 13-25 shows a roof beam over a loading dock of a factory building. Compute the reactions at the supports and draw the complete shear and bending moment diagrams.

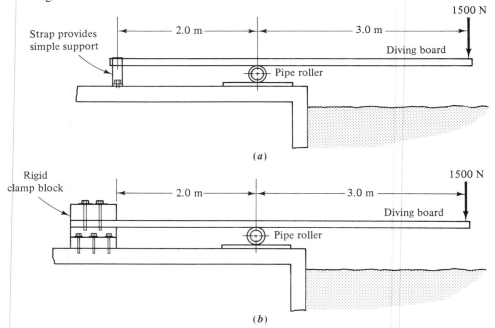

Figure 13-24 Proposals for diving boards for Problems 13-21 and 13-22. (a) Simple supports with overhang (b) Supported cantilever with overhang.

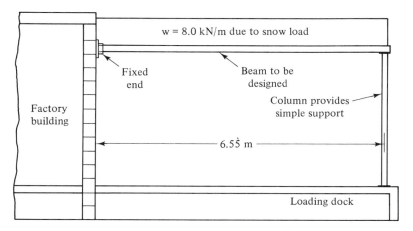

Figure 13-25 Roof beam for a loading dock for Problems 13-23 and 13-24.

13-24. For the roof beam described in Problem 13-23, complete its design, specifying the material, the cross section, and the final dimensions.

COMPUTER PROGRAMMING ASSIGNMENTS

1. Write a program to solve for the critical shearing forces and bending moments for any of the statically indeterminate beam types in Appendix A-24.

ENHANCEMENTS TO ASSIGNMENT 1

    (a) Use the graphics mode to plot the complete shearing force and bending moment diagrams for the beams.

    (b) Compute the required section modulus for the beam cross section to limit the stress due to bending a specified amount.

    (c) Include a table of section properties for steel wide-flange beams and search for suitable beam sizes to carry the load.

    (d) Assuming that the beam cross section will be a solid circular section, compute the required diameter.

    (e) Assuming that the beam cross section will be a rectangle with a given ratio of height to thickness, compute the required dimensions.

    (f) Assuming that the beam cross section will be a rectangle with a given height *or* thickness, compute the other required dimension.

    (g) Assuming that the beam is to be made from wood, in a rectangular shape, compute the required area of the beam cross section to limit the shearing stress to a specified value. Use the special shear formula for rectangular beams from Chapter 9.

    (h) Add the computation of the deflection at specified points on the beam using the formulas of Appendix A-24.

2. Combine two or more of the beam formulas from Appendix A-22 in a program to use the superposition method to solve for the reactions, shearing forces, and bending moments in a statically indeterminate beam using the method outlined in Section 13-4.

3. Write a program to aid in the solution of Equation (13-5), the theorem of three moments applied to a continuous beam of two spans with combinations of uniformly distributed loads and several concentrated loads. Notice that this equation reduces to Equation (13-2), (13-3), or (13-4) when certain terms are zero.

14

Columns

14-1 OBJECTIVES OF THIS CHAPTER

A column is a relatively long member loaded in compression. The analysis of columns is different from what has been studied before because the mode of failure is different. In Chapter 3, when members loaded in compression was discussed, it was assumed that the member failed by yielding of the material. A stress greater than the yield strength of the material would have to be applied. This is true for short members.

A long, slender column fails by *buckling*, a common name given to *elastic instability*. Instead of crushing or tearing the material, the column deflects drastically at a certain critical load and then collapses suddenly. Any thin member can be used to illustrate the buckling phenomenon. Try it with a wood or plastic ruler, a thin rod or strip of metal, or a drinking straw. By gradually increasing the force, applied directly downward, the critical load is reached when the column begins to bend. Normally, the load can be removed without causing permanent damage since yielding does not occur. Thus a column fails by buckling at a stress lower than the yield strength of the material in the column. The objective of column analysis methods is to predict the load or stress level at which a column would become unstable and buckle.

After completing this chapter, you should be able to:

1. Define *column*.
2. Differentiate between a column and a short compression member.
3. Describe the phenomenon of *buckling,* also called *elastic instability.*
4. Define *radius of gyration* for the cross section of a column and be able to compute its magnitude.
5. Understand that a column is expected to buckle about the axis for which the radius of gyration is the minimum.
6. Define *end-fixity factor, K.*
7. Specify the appropriate value of the end-fixity factor, *K*, depending on the manner of supporting the ends of a column.
8. Define *effective length, L_e.*
9. Define *slenderness ratio* and compute its value.
10. Define *transition slenderness ratio,* also called the *column constant, C_c,* and compute its value.
11. Use the values for the slenderness ratio and the column constant to determine when a column is *long* or *short.*
12. Use the *Euler formula* for computing the critical buckling load for long columns.
13. Use the *J. B. Johnson formula* for computing the critical buckling load for short columns.
14. Apply a design factor to the critical buckling load to determine the *allowable load* on a column.
15. Recognize efficient shapes for column cross sections.
16. Design columns to safely carry given axial compression loads.
17. Apply the specifications of the American Institute of Steel Construction (AISC) code to the analysis of columns.
18. Apply the specifications of the Aluminum Association to the analysis of columns.

14-2 SLENDERNESS RATIO

A column has been described as a relatively long, slender member loaded in compression. This description is stated in relative terms and is not very useful for analysis.

The measure of the slenderness of a column must take into account the length, the cross-sectional shape and dimensions of the column, and the manner of attaching the ends of the column to the structures that supply loads and reactions to the column. The commonly used measure of slenderness is the *slenderness ratio,* defined as

$$ \text{SR} = \frac{KL}{r} = \frac{L_e}{r} \tag{14-1} $$

where L = *actual length* of the column between points of support or lateral restraint

L_e = *effective length*, taking into effect the manner of attaching the ends (note that $L_e = KL$)

K = *end-fixity factor*

r = *smallest radius of gyration* of the cross section of the column

Each of these terms is discussed below.

Actual length, L. For a simple column having the load applied at one end and the reaction provided at the other, the actual length is, obviously, the length between its ends. But for components of structures loaded in compression where a means of restraining the member laterally to prevent buckling is provided, the actual length is taken between points of restraint. Each part is then considered to be a separate column.

End-fixity factor, K. The end-fixity factor is a measure of the degree to which each end of the column is restrained against rotation. Three classic types of end connections are typically considered: the pinned end, the fixed end, and the free end. Figure 14-1 shows these end types in several combinations with the corresponding values of K. Note that two values of K are given. One is the theoretical value and the other is the one typically used in practical situations, recognizing that it is difficult to achieve the truly fixed end as discussed below.

Pinned ends of columns are essentially unrestrained against rotation. When a column with two pinned ends buckles, it assumes the shape of a smooth curve between its ends as shown in Figure 14-1(a). This is the basic case of column buckling and the value of $K = 1.0$ is applied to columns having two pinned ends. An ideal implementation of the pinned end is the frictionless spherical ball joint that would permit

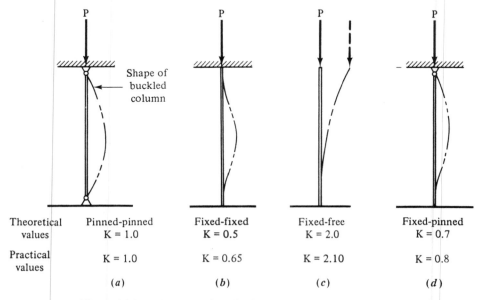

Figure 14-1 Values of K for effective length, $L_e = KL$, for different end connections.

the column to rotate in any direction about any axis. For a cylindrical pin joint, free rotation is permitted about the centerline of the pin, but some restraint is provided in the plane perpendicular to the centerline. Care must be exercised in applying end fixity factors to cylindrical pins for this reason. It is assumed that the pinned end is guided in some way so that the line of action of the axial load remains unchanged.

Fixed ends provide theoretically perfect restraint against rotation of the column at its ends. As the column tends to buckle, the deflected shape of the axis of the column must approach the fixed end with a zero slope as illustrated in Figure 14-1(b). The buckled shape bows out in the middle but exhibits two points of inflection to reverse the direction of curvature near the ends. The theoretical value of the end-fixity factor is $K = 0.5$, indicating that the column acts as if it were only one-half as long as it really is. Columns with fixed ends are much stiffer than pinned-end columns and can therefore take higher loads before buckling. It should be understood that it is very difficult to provide perfectly fixed ends for a column. It requires that the connection to the column is rigid and stiff and that the structure to which the loads are transferred is also rigid and stiff. For this reason, the higher value of $K = 0.65$ is recommended for practical use.

A free end for a column is unrestrained against rotation and also against translation. Because it can move in any direction, this is the worst case for column end fixity. The only practical way of using a column with a free end is to have the opposite end fixed, as illustrated in Figure 14-1(c). Such a column is sometimes referred to as the *flagpole* case because the fixed end is similar to the flagpole inserted deeply into a tight-fitting socket while the other end is free to move in any direction. Called the fixed-free end condition, the theoretical value of K is 2.0. A practical value is $K = 2.10$.

The combination of one fixed end and one pinned end is shown in Figure 14-1(d). Notice that the buckled shape approaches the fixed end with a zero slope while the pinned end rotates freely. The theoretical value of $K = 0.7$ applies to such an end-fixity while $K = 0.80$ is recommended for practical use.

Effective length, L_e.   Effective length combines the actual length with the end-fixity factor; $L_e = KL$. In the problems in this book we use the recommended practical values for the end fixity factor as shown in Figure 14-1. In summary, the following relationships will be used to compute the effective length:

1. Pinned-end columns: $L_e = KL = 1.0(L) = L$
2. Fixed-end columns: $L_e = KL = 0.65(L)$
3. Fixed-free columns: $L_e = KL = 2.10(L)$
4. Fixed-pinned columns: $L_e = KL = 0.80(L)$

Radius of gyration, r.   The measure of slenderness of the cross section of the column is its radius of gyration, r, defined as

$$r = \sqrt{\frac{I}{A}} \tag{14-2}$$

where I = moment of inertia of the cross section of the column with respect to one of the principal axes
A = area of the cross section.

Because both I and A are geometrical properties of the cross section, the radius of gyration, r, is also. Formulas for computing r for several common shapes are given in Appendix A-1. Also, r is listed with other properties for some of the standard shapes in the Appendix. For those for which r is not listed, the values of I and A are available and Equation (14-2) can be used to compute r very simply.

Note that the value of the radius of gyration, r, is dependent on the axis about which it is to be computed. In most cases, it is required that you determine the axis for which the *radius of gyration is the smallest,* because that is the axis about which the column would likely buckle. Consider, for example, a column made from a rectangular section whose width is much greater than its thickness as sketched in Figure 14-2. A simple ruler or a meter stick can be used to demonstrate that when loaded in axial compression with little or no restraint at the ends, the column will

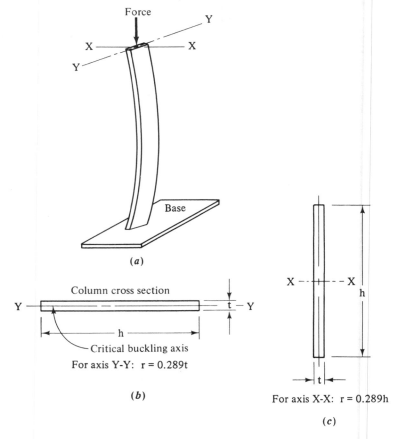

Force

Base

(a)

Column cross section

For axis Y-Y: $r = 0.289t$

Critical buckling axis

(b)

For axis X-X: $r = 0.289h$

(c)

Figure 14-2 Buckling of a thin, rectangular column. (a) General appearance of the buckled column (b) Radius of gyration for Y-Y axis (c) Radius of gyration for X-X axis.

always buckle with respect to the axis through the thinnest dimension. For the rectangle, as shown in Figure 14-2 (b) and (c),

$$r_{min} = r_Y = 0.289t$$

where t is the thickness of the rectangle. Note that

$$r_X = 0.289h$$

where h is the height of the rectangle and that $h > t$. Thus

$$r_X > r_Y$$

and then r_Y is the smallest radius of gyration for the section.

For the wide-flange beams (Appendix A-7) and American Standard beams (Appendix A-8), the minimum value of r is that computed with respect to the Y-Y axis; that is,

$$r_{min} = \sqrt{\frac{I_Y}{A}}$$

Similarly, for rectangular structural tubing (Appendix A-9), the minimum radius of gyration is that with respect to the Y-Y axis. Values for r are listed in the table.

For structural steel angles, called L-shapes, neither the X-X nor the Y-Y axis provides the minimum radius of gyration. As illustrated in Appendix A-5, r_{min} is computed with respect to the Z-Z axis, with the values listed in the table.

For symmetrical sections, the value of r is the same with respect to any principal axis. Such shapes are the solid or hollow circular section and the solid or hollow square section.

Summary of the method for computing the slenderness ratio

1. Determine the actual length of the column, L, between endpoints or between points of lateral restraint.
2. Determine the end-fixity factor from the manner of support of the ends and by reference to Figure 14-1.
3. Compute the effective length, $L_e = KL$.
4. Compute the *smallest* radius of gyration for the cross section of the column.
5. Compute the slenderness ratio from

$$SR = \frac{L_e}{r_{min}}$$

14-3 TRANSITION SLENDERNESS RATIO

When is a column considered long? The answer to this question requires the determination of the *transition slenderness ratio,* or column constant C_c.

$$C_c = \sqrt{\frac{2\pi^2 E}{s_y}} \tag{14-3}$$

If the actual effective slenderness ratio L_e/r is greater than C_c, then the column is long, and the Euler formula, defined in the next section, should be used to analyze the column. If the actual ratio L_e/r is less than C_c, then the column is short. In these cases, either the J. B. Johnson formula, special codes, or the direct compressive stress formula should be used, as discussed in later sections.

Where a given column is being analyzed to determine the load it will carry, the value of C_c and the actual ratio L_e/r should be computed first to determine which method of analysis should be used. Notice that C_c depends on the material properties of yield strength s_y and modulus of elasticity E. In working with steel, E is usually taken to be 30×10^6 psi (207 GPa). Using this value and assuming a range of values for yield strength, we obtain the values for C_c shown in Figure 14-3.

For aluminum, E is approximately 69 GPa (10×10^6 psi). The corresponding values for C_c are shown in Figure 14-4.

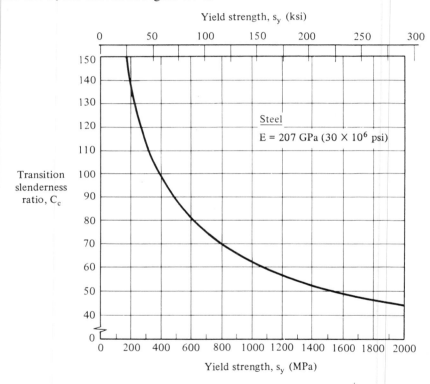

Figure 14-3 Transition slenderness ratio C_c vs. Yield strength for steel.

14-4 THE EULER FORMULA FOR LONG COLUMNS

For long columns having an effective slenderness ratio greater than the transition value C_c, the Euler formula can be used to predict the critical load at which the column would be expected to buckle. The formula is

$$P_{cr} = \frac{\pi^2 EA}{(L_e/r)^2} \tag{14-4}$$

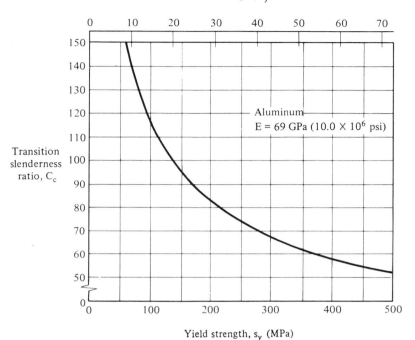

Yield strength, s_y (ksi)

Transition slenderness ratio, C_c

Yield strength, s_y (MPa)

Figure 14-4 Transition slenderness ratio C_c vs. Yield strength for aluminum.

where A is the cross-sectional area of the column. An alternative form can be expressed in terms of the moment of inertia by noting that $r^2 = I/A$. Then the formula becomes

$$P_{cr} = \frac{\pi E I}{L_e^2} \qquad (14\text{-}5)$$

Example Problem 14-1

A round compression member with both ends pinned and made of AISI 1020 cold-drawn steel is to be used in a machine. Its diameter is 25 mm, and its length is 950 mm. What maximum load can the member take before buckling would be expected?

Solution Let's first evaluate C_c and the actual ratio L_e/r for this column.

$$C_c = \sqrt{\frac{2\pi^2 E}{s_y}} = \sqrt{\frac{2\pi^2 (207 \times 10^9 \text{ N/m}^2)}{352 \times 10^6 \text{ N/m}^2}} = 108$$

For a round bar,

$$r = \frac{D}{4} = \frac{25 \text{ mm}}{4} = 6.25 \text{ mm}$$

For a pinned-end column, $K = 1.0$ and $L_e = L$. Then

$$\frac{L_e}{r} = \frac{950 \text{ mm}}{6.25 \text{ mm}} = 152$$

Since L_e/r is greater than C_c, Euler's formula applies.

$$P_{cr} = \frac{\pi^2 EA}{(L_e/r)^2}$$

The area is

$$A = \frac{\pi D^2}{4} = \frac{\pi (25 \text{ mm})^2}{4} = 491 \text{ mm}^2$$

Then

$$P_{cr} = \frac{\pi^2 (207 \times 10^9 \text{ N/m}^2)(491 \text{ mm}^2)}{(152)^2} \times \frac{1 \text{ m}^2}{(10^3 \text{ mm})^2} = 43.4 \text{ kN}$$

14-5 THE J. B. JOHNSON FORMULA FOR SHORT COLUMNS

If a column has an actual effective slenderness ratio L_e/r less than the transition value C_c, the Euler formula predicts an unreasonably high critical load. One formula recommended for machine design applications in the range of L_e/r less than C_c is the J. B. Johnson formula.

$$P_{cr} = As_y \left[1 - \frac{s_y (L_e/r)^2}{4\pi^2 E} \right] \tag{14-6}$$

This is one form of a set of equations called parabolic formulas, and it agrees well with the performance of steel columns in typical machinery.

The Johnson formula gives the same result as the Euler formula for the critical load at the transition slenderness ratio C_c. Then for very short columns the critical load approaches that which would be predicted from the direct compressive stress equation, $\sigma = P/A$. Therefore, it could be said that the Johnson formula applies best to columns of intermediate length.

Example Problem 14-2

Determine the critical load on a steel column having a square cross section 12 mm on a side with a length of 300 mm. The column is to be made of AISI 1040, hot rolled. It will be rigidly welded to a firm support at one end and connected by a pin joint at the other.

Solution The values of the actual ratio L_e/r for the column and C_c are needed to determine which method of analysis should be used. For the fixed-pinned ends, $K = 0.8$, and

$$r = \frac{b}{\sqrt{12}} = \frac{12 \text{ mm}}{\sqrt{12}} = 3.46 \text{ mm}$$

$$\frac{L_e}{r} = \frac{KL}{r} = \frac{(0.8)(300 \text{ mm})}{3.46 \text{ mm}} = 69.4$$

From Figure 14-3, and using $s_y = 290$ MPa, the value of C_c is about 117. Since L_e/r is less than C_c, the Johnson formula, Equation (14-6), should be used.

$$A = b^2 = (12 \text{ mm})^2 = 144 \text{ mm}^2$$

$$P_{cr} = (144 \text{ mm}^2)\left(\frac{290 \text{ N}}{\text{mm}^2}\right)\left[1 - \frac{(290 \times 10^6 \text{ N/m}^2)(69.4)^2}{4\pi^2(207 \times 10^9 \text{ N/m}^2)}\right] = 34.6 \text{ kN}$$

14-6 DESIGN FACTORS FOR COLUMNS AND ALLOWABLE LOAD

Because the mode of failure for a column is buckling rather than yielding or ultimate failure of the material, the methods used before for computing design stress do not apply to columns.

Instead, an *allowable load* is computed by dividing the critical buckling load from either the Euler formula [Equation (14-4)] or the Johnson formula [Equation (14-6)] by a design factor, N. That is,

$$P_a = \frac{P_{cr}}{N} \tag{14-7}$$

where P_a = allowable, safe load
P_{cr} = critical buckling load
N = design factor

The selection of the design factor is the responsibility of the designer unless the project comes under a code. Factors to be considered in the selection of a design factor are similar to those used for design factors applied to stresses. A common factor used in mechanical design is $N = 3.0$, chosen because of the uncertainty of material properties, end fixity, straightness of the column, or the possibility that the load will be applied with some eccentricity rather than along the axis of the column. Larger factors are sometimes used for critical situations and for very long columns.

In building construction, where the design is governed by the specifications of the American Institute of Steel Construction, AISC, a factor of 1.92 is recommended.

Example Problem 14-3

Compute the allowable load for the columns analyzed in Example Problems 14-1 and 14-2 to obtain a design factor of 3.

Solution From Example Problem 14-1, the critical buckling load was found to be $P_{cr} = 43.4$ kN. Then the allowable load is

$$P_a = \frac{P_{cr}}{N} = \frac{43.4 \text{ kN}}{3} = 14.5 \text{ kN}$$

From Example Problem 14-2, the critical buckling load was found to be $P_{cr} = 34.6$ kN. Then the allowable load is

$$P_a = \frac{P_{cr}}{N} = \frac{34.6 \text{ kN}}{3} = 11.5 \text{ kN}$$

14-7 SUMMARY—METHOD OF ANALYZING COLUMNS

The purpose of this section is to summarize the concepts presented in Sections 14-2 through 14-6 into a procedure that can be used to analyze columns. It can be applied to a straight column having a uniform cross section throughout its length, and for which the compressive load is applied in line with the centroidal axis of the column.

To start, it is assumed that the following factors are known:

1. The actual length, L
2. The manner of connecting the column to its supports
3. The shape of the cross section of the column and its dimensions
4. The material from which the column is made

Then the procedure is:

1. Determine the end-fixity factor, K, by comparing the manner of connection of the column to its supports with the information in Figure 14-1.
2. Compute the effective length, $L_e = KL$.
3. Compute the minimum value of the radius of gyration of the cross section from $r_{min} = \sqrt{I_{min}/A}$; or determine r_{min} from tables of data.
4. Compute the maximum slenderness ratio from

$$SR_{max} = \frac{L_e}{r_{min}}$$

5. Using the modulus of elasticity, E, and the yield strength, s_y, for the material, compute the column constant,

$$C_c = \sqrt{\frac{2\pi^2 E}{s_y}}$$

6. Compare the value of SR with C_c.
   a. If $SR > C_c$, the column is long. Use the Euler formula to compute the critical buckling load,

$$P_{cr} = \frac{\pi^2 EA}{(SR)^2}$$

   b. If $SR < C_c$, the column is short. Use the Johnson formula to compute the critical buckling load,

$$P_{cr} = As_y \left[1 - \frac{s_y (SR)^2}{4\pi^2 E} \right]$$

7. Specify the design factor, N.
8. Compute the allowable load, P_a,

$$P_a = \frac{P_{cr}}{N}$$

14-8 EFFICIENT SHAPES FOR COLUMN CROSS SECTIONS

When designing a column to carry a specified load, the designer has the responsibility for selecting the general shape of the column cross section and then to determine the required dimensions. The following principles may aid in the initial selection of the cross-section shape.

An efficient shape is one that uses a small amount of material to perform a given function. For columns, the goal is to maximize the radius of gyration in order to reduce the slenderness ratio. Note also that because $r = \sqrt{I/A}$, maximizing the moment of inertia has the same effect.

When discussing moment of inertia in Chapters 7 and 8, it was noted that it is desirable to put as much of the area of the cross section as far away from the centroid as possible. For beams, discussed in Chapter 8, there was usually only one important axis, the axis about which the bending occurred. For columns, buckling can, in general, occur in any direction. Therefore, it is desirable to have uniform properties with respect to any axis. The hollow circular section, commonly called a pipe, then makes a very efficient shape for a column. Closely approximating that is the hollow square tube. Fabricated sections made from standard structural sections can also be used as shown in Figure 14-5.

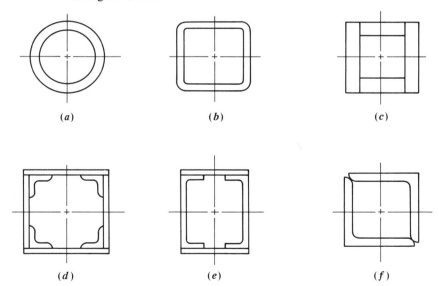

(a) (b) (c)

(d) (e) (f)

Figure 14-5 Examples of efficient column shapes. (a) Hollow circular section, pipe (b) Hollow square tube (c) Box section made from wood beams (d) Equal-leg angles with plates (e) Aluminum channels with plates (f) Two equal-leg angles.

Building columns are often made from special wide-flange shapes called *column sections*. They have relatively wide, thick flanges as compared with the shapes typically selected for beams. This makes the moment of inertia with respect to the *Y-Y* axis more nearly equal to that for the *X-X* axis. The result is that the radii of gyration for the two axes are more nearly equal also. Figure 14-6 shows a comparison of two

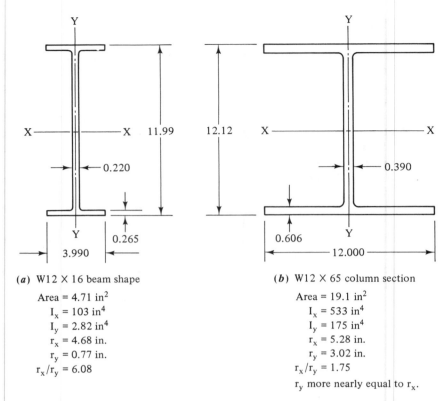

Figure 14-6 Comparison of a wide flange beam shape with a column section.

(*a*) W12 × 16 beam shape

Area = 4.71 in²
I_x = 103 in⁴
I_y = 2.82 in⁴
r_x = 4.68 in.
r_y = 0.77 in.
r_x/r_y = 6.08

(*b*) W12 × 65 column section

Area = 19.1 in²
I_x = 533 in⁴
I_y = 175 in⁴
r_x = 5.28 in.
r_y = 3.02 in.
r_x/r_y = 1.75
r_y more nearly equal to r_x.

12-in. wide-flange shapes: one a column section and one a typical beam shape. Note that the smaller radius of gyration should be used in computing the slenderness ratio.

14-9 SPECIFICATIONS OF THE AISC

Columns are essential elements of most structures. The design and analysis of steel columns in construction applications are governed by the specifications of the AISC, the American Institute of Steel Construction. The specification defines an allowable unit load or stress for columns which is the allowable total load P_a divided by the area of the column cross section. The design formulas are expressed in terms of the transition slenderness ratio C_c, already defined in Equation (14-3), the yield strength of the column material, and the effective slenderness ratio L_e/r. For $L_e/r < C_c$,

$$\frac{P_a}{A} = \left[1 - \frac{(L_e/r)^2}{2C_c^2}\right] \frac{s_y}{FS} \tag{14-8}$$

where P_a = allowable or design load

$$C_c = \sqrt{\frac{2\pi^2 E}{s_y}}$$ (Use $E = 29 \times 10^6$ psi (200 GPa) for structural steel)

FS = factor of safety

Equation (14-8) was developed by the Column Research Council and is identical to the Johnson formula. The factor of safety FS is a function of the ratio of the effective slenderness ratio to C_c in order to include the effect of accidental crookedness, a small eccentricity of the load, residual stresses, and any uncertainties in the evaluation of the effective length factor K. The equation for FS is

$$FS = \frac{5}{3} + \frac{3(L_e/r)}{8C_c} - \frac{(L_e/r)^3}{8C_c^3} \tag{14-9}$$

The value of FS varies from 1.67 for the ratio $\dfrac{L_e/r}{C_c} = 0$ to 1.92 for $\dfrac{L_e/r}{C_c} = 1.0$.

For long columns, $L_e/r > C_c$, Euler's equation is used as defined before but with a factor of safety of 1.92.

$$\frac{P_a}{A} = \frac{\pi^2 E}{(L_e/r)^2(1.92)} \tag{14-10}$$

For structural steel with $E = 29 \times 10^6$ psi,

$$\frac{P_a}{A} = \frac{149 \times 10^6}{(L_e/r)^2} \text{ psi} \tag{14-11}$$

In the SI system, using $E = 200$ GPa for structural steel,

$$\frac{P_a}{A} = \frac{1028}{(L_e/r)^2} \text{ GPa} \tag{14-12}$$

14-10 SPECIFICATIONS OF THE ALUMINUM ASSOCIATION

The Aluminum Association publication, *Specifications for Aluminum Structures* (2), defines allowable stresses for columns for each of several aluminum alloys and their heat treatments. Three different equations are given for short, intermediate, and long columns defined in relation to slenderness limits. The equations are of the form

$$\frac{P_a}{A} = \frac{s_y}{FS} \qquad \text{(short columns)} \tag{14-13}$$

$$\frac{P_a}{A} = \frac{B_c - D_c(L/r)}{FS} \qquad \text{(intermediate columns)} \tag{14-14}$$

$$\frac{P_a}{A} = \frac{\pi^2 E}{FS(L/r)^2} \qquad \text{(long columns)} \tag{14-15}$$

In all three cases, it is recommended that $FS = 1.95$ for buildings and similar structures. The short column analysis assumes that buckling will not occur and that safety is dependent on the yield strength of the material. Equation (14-15) for long columns is the Euler formula with a factor of safety applied. The intermediate column formula [Equation (14-14)] depends on buckling constants B_c and D_c, which are functions of the yield strength of the aluminum alloy and the modulus of elasticity. The division between intermediate and long columns is similar to the C_c used previously in this chapter.

Following are the specific equations for the alloy 6061-T6 used in building structures in the forms of sheet, plate, extrusions, structural shapes, rod, bar, tube, and pipe. The slenderness ratio L/r should be evaluated using the actual L (pinned ends). Any end restraint is assumed to be allowed for in the factor of safety.

Short columns: $L/r < 9.5$

$$\frac{P_a}{A} = 19 \text{ ksi } (131 \text{ MPa})$$ (14-16)

Intermediate columns: $9.5 < L/r < 66$

$$\frac{P_a}{A} = \left(20.2 - 0.126\frac{L}{r}\right) \text{ ksi}$$ (14-17a)

$$\frac{P_a}{A} = \left(139 - 0.869\frac{L}{r}\right) \text{ MPa}$$ (14-17b)

Long columns: $L/r > 66$

$$\frac{P_a}{A} = \frac{51\ 000}{(L/r)^2} \text{ ksi}$$ (14-18a)

$$\frac{P_a}{A} = \frac{352\ 000}{(L/r)^2} \text{ MPa}$$ (14-18b)

REFERENCES

1. American Institute of Steel Construction, *Manual of Steel Construction,* 8th ed., Chicago, 1980.
2. Aluminum Association, *Specifications for Aluminum Structures,* 4th ed., Washington, D.C., 1982.
3. Johnston, B. G., and F. Lin, *Basic Steel Design,* Prentice-Hall, Englewood Cliffs, N.J., 1974.

PROBLEMS

14-1. Determine the critical load for a pinned-end column made of a circular bar of AISI 1020 hot-rolled steel. The diameter of the bar is 20 mm, and its length is 800 mm.

14-2. Repeat Problem 14-1 with the length of 350 mm.

14-3. Repeat Problem 14-1 with the bar made of 6061-T6 aluminum instead of steel.

14-4. Repeat Problem 14-1 with the column ends fixed instead of pinned.

14-5. Repeat Problem 14-1 with a square steel bar with the same cross-sectional area as the circular bar.

14-6. For a 1-in. schedule 40 steel pipe, used as a column, determine the critical load if it is 2.05 m long. The material is similar to AISI 1020 hot-rolled steel. Compute the critical load for each of the four end conditions described in Figure 14-1.

14-7. A rectangular steel bar has cross-sectional dimensions of 12 mm by 25 mm and is 210 mm long. Assuming that the bar has pinned ends and is made from AISI 1141 OQT 1300 steel, compute the critical load when the bar is subjected to an axial compressive load.

14-8. Compute the allowable load on a column with fixed ends if it is 5.45 m long and made from an S6 × 12.5 beam. The material is ASTM A36 steel. Use the AISC formula.

14-9. A raised platform is 20 ft by 40 ft in area and is being designed for 75 pounds per square foot uniform loading. It is proposed that standard 3-in. schedule 40 steel pipe be used as columns to support the platform 8 ft above the ground with the base fixed and the top free. How many columns would be required if a design factor of 3.0 is desired? Use $s_y = 30\ 000$ psi.

14-10. An aluminum I-beam, I10 × 8.646, is used as a column with two pinned ends. It is 2.80 m long and made of 6061-T6 aluminum. Using Equations (14-16) through (14-18b), compute the allowable load on the column.

14-11. Compute the allowable load for the column described in Problem 14-10 if the length is only 1.40 m.

14-12. A column is a W8 × 15 steel beam, 12.50 ft long, and made of ASTM A36 steel. Its ends are attached in such a way that L_e is approximately $0.80L$. Using the AISC formulas, determine the allowable load on the column.

14-13. A built-up column is made of four angles as shown in Figure 14-7. The angles are held together with lacing bars, which can be neglected in the analysis of geometrical properties. Using the standard Johnson or Euler equations with $L_e = L$ and a design

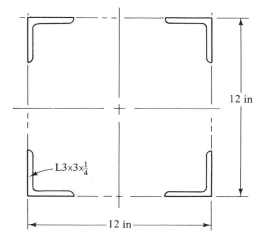

Figure 14-7 Built-up column cross section for Problem 14-13.

factor of 3.0, compute the allowable load on the column if it is 18.4 ft long. The angles are of ASTM A36 steel.

14-14. Compute the allowable load on a built-up column having the cross section shown in Figure 14-8. Use $L_e = L$ and 6061-T6 aluminum. The column is 3.20 m long. Use the Aluminum Association formulas.

14-15. Figure 14-9 shows a beam supported at its ends by pin joints. The inclined bar at the top supports the right end of the beam, but also places an axial compressive force in the beam. Would a standard S6 × 12.5 beam be satisfactory in this application if it carries 1320 kg at its end? The beam is made from ASTM A36 steel.

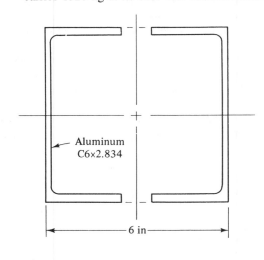

Figure 14-8 Built-up column cross section for Problem 14-14.

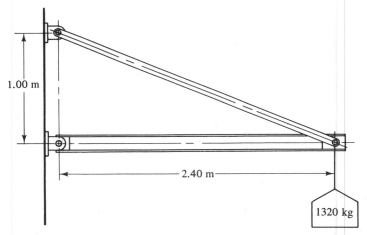

Figure 14-9 Structure for Problem 14-15.

14-16. A link in a mechanism which is 8.40 in. long, has a rectangular cross section $\frac{1}{4}$ in. × $\frac{1}{8}$ in., and is subjected to a compressive load of 50 lb. If the link has pinned ends, is it safe from buckling? Cold-drawn AISI 1040 steel is used in the link.

14-17. A piston rod on a shock absorber is 12 mm in diameter and has a maximum length of 190 mm outside the shock absorber body. The rod is made of AISI 1141 OQT 1300 steel. Consider one end to be pinned and the other to be fixed. What axial compressive load on the rod would be one-third of the critical buckling load?

14-18. A stabilizing rod in an automobile suspension system is a round bar loaded in compression. It is subjected to 1375 lb of axial load and supported at its ends by pin-type connections, 28.5 in. apart. Would a 0.800-in. diameter bar of AISI 1020 hot-rolled steel be satisfactory for this application?

14-19. A structure is being designed to support a large hopper over a plastic extruding machine, as sketched in Figure 14-10. The hopper is to be carried by four columns which share the load equally. The structure is cross-braced by rods. It is proposed that the columns be made from standard 2-in. schedule 40 pipe. They will be fixed at the floor. Because of the cross-bracing, the top of each column is guided so that it behaves as if it was rounded or pinned. The material for the pipe is AISI 1020 steel, hot rolled. The hopper is designed to hold 20 000 lb of plastic powder. Are the proposed columns adequate for this load?

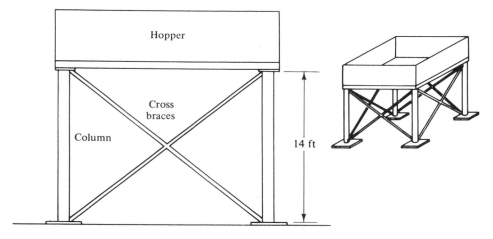

Figure 14-10 Hopper for Problems 14-19 and 14-20.

14-20. Discuss how the column design in Problem 14-19 would be affected if a careless forklift driver runs into the cross braces and breaks them.

14-21. The assembly shown in Figure 14-11 is used to test parts by pulling on them repeatedly with the hydraulic cylinder. A maximum force of 3000 lb can be exerted by the cylinder. The parts of the assembly of concern here are the columns. It is proposed that the two columns be made of $1\frac{1}{4}$-in. square bars of aluminum alloy 6061-T6. The columns are fixed at the bottom and free at the top. Determine the acceptability of the proposal.

14-22. Figure 14-12 shows the proposed design for a hydraulic press used to compact solid waste. A piston at the right is capable of exerting a force of 12 500 lb through the connecting rod to the ram. The rod is straight and centrally loaded. It is made from AISI 1040 OQT 1100 steel. Compute the resulting design factor for this design.

14-23. For the conditions described in Problem 14-22, specify the required diameter of the connecting rod if it is made as a solid circular cross section. Use a design factor of 4.0.

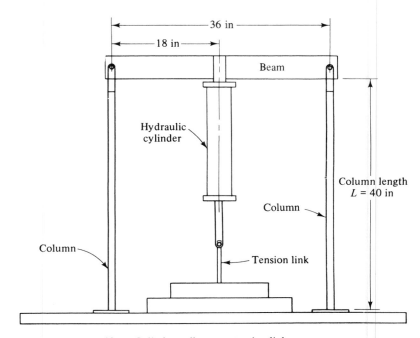

Note: Cylinder pulls up on tension link
and down on beam with a force of 3000 lb.

Figure 14-11 Test fixture for Problem 14-21.

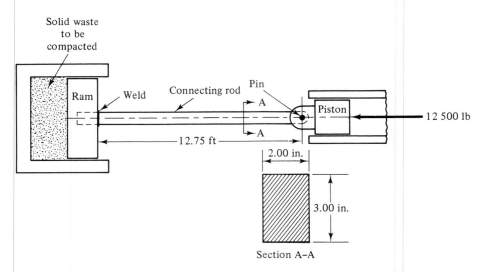

Figure 14-12 Solid waste compactor for Problems 14-22, 14-23, 14-24, and 14-25.

14-24. For the conditions described in Problem 14-22, specify a suitable standard steel pipe for use as the connecting rod. Use a design factor of 4.0. The pipe is to be made from ASTM A501 structural steel.

14-25. For the conditions described in Problem 14-22, specify a suitable standard I-beam for use as the connecting rod. Use a design factor of 4.0. The I-beam is to be made from aluminum alloy 6061-T6. The connection at the piston is as shown in Figure 14-13.

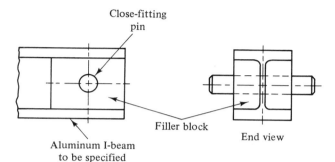

Close-fitting pin

Filler block

End view

Aluminum I-beam to be specified

Figure 14-13 End connection for I-beam for Problem 14-25.

14-26. A hollow square tube, $3 \times 3 \times \frac{1}{4}$, made from ASTM A500 steel, grade B is used as a building column having a length of 16.5 ft. Using $L_e = 0.80L$, compute the allowable load on the column for a design factor of 3.0.

14-27. A hollow rectangular tube, $4 \times 2 \times \frac{1}{4}$, made from ASTM A500 steel, grade B is used as a building column having a length of 16.5 ft. Using $L_e = 0.80L$, compute the allowable load on the column for a design factor of 3.0.

14-28. A column is made by welding two standard steel angles, $3 \times 3 \times \frac{1}{4}$, into the form shown in Figure 14-5(f). The angles are made from ASTM A36 structural steel. If the column has a length of 16.5 ft and $L_e = 0.8L$, compute the allowable load on the column for a design factor of 3.0.

14-29. A rectangular steel bar, made of AISI 1020 hot-rolled steel, is used as a safety brace to hold the ram of a large punch press while dies are installed in the press. The bar has cross-sectional dimensions of 60 mm by 40 mm. Its length is 750 mm and its ends are welded to heavy flat plates which rest on the flat bed of the press and the flat underside of the ram. Specify a safe load that could be applied to the brace.

14-30. It is planned to use an aluminum channel, C4 × 1.738, as a column having a length of 4.25 m. The ends can be considered to be pinned. The aluminum is alloy 6061-T4. Compute the allowable load on the column for a design factor of 4.0.

14-31. In an attempt to improve the load-carrying capacity of the column described in Problem 14-30, alloy 6061-T6 is proposed in place of the 6061-T4 to take advantage of its higher strength. Evaluate the effect of this proposed change on the allowable load.

14-32. Compute the allowable load on the steel W12 × 65 column section shown in Figure 14-6(b) if it is 22.5 ft long, made from ASTM A36 steel, and installed such that $L_e = 0.8L$. Use the ASIC code.

COMPUTER PROGRAMMING ASSIGNMENTS

1. Write a computer program to analyze proposed column designs using the procedure outlined in Section 14-7. Have all the essential design data for material, end fixity, length, and cross-section properties input by the user. Have the program output the critical load, and the allowable load for a given design factor.

ENHANCEMENTS TO ASSIGNMENT 1

(a) Include a table of data for standard schedule 40 steel pipe for use by the program to determine the cross section properties for a specified pipe size.

(b) Design the program to handle columns made from solid circular cross sections and compute the cross-section properties for a given diameter.

(c) Add a table of data for standard structural steel square tubing for use by the program to determine the cross-section properties for a specified size.

(d) Have the program use the specifications of the AISC as stated in Section 14-9 for computing the allowable load and factor of safety for steel columns.

(e) Have the program use the specifications of the Aluminum Association as stated in Section 14-10 for computing the allowable load for columns made from 6061-T6.

2. Write a program to design a column with a solid circular cross section to carry a given load with a given design factor. Note that the program will have to check to see that the correct method of analysis is used, either the Euler formula for long columns or the Johnson formula for short columns, after an initial assumption is made.

3. Write a program to design a column with a solid square cross section to carry a given load with a given design factor.

4. Write a program to select a suitable schedule 40 steel pipe to carry a given load with a given design factor. The program could be designed to search through a table of data for standard pipe sections from the smallest to the largest until a suitable pipe was found. For each trial section, the allowable load could be computed using the Euler or Johnson formula, as required, and compared with the design load.

15

Pressure Vessels

15-1 OBJECTIVES OF THIS CHAPTER

The most common forms of pressure vessels designed to hold liquids or gases under internal pressure are spheres and closed-end cylinders. The internal pressure tends to burst the vessel because there are tensile stresses developed in its walls. The overall objective of this chapter is to describe the manner in which these stresses are developed and to present formulas that can be used to compute the magnitude of the stresses.

After completing this chapter, you should be able to:

1. Determine whether a pressure vessel should be classified as *thin-walled* or *thick-walled*.
2. Draw the free-body diagram for a part of a sphere when subjected to internal pressure to identify the force that must be resisted by the wall of the sphere.
3. Describe *hoop stress* as it is applied to spheres carrying an internal pressure.
4. State the formula for computing the hoop stress developed in the wall of a thin-walled sphere due to internal pressure.
5. Use the hoop stress formula to compute the maximum stress in the wall of a thin-walled sphere.

6. Determine the required wall thickness of the sphere to resist a given internal pressure.

7. Draw the free-body diagram for a part of a cylinder when subjected to internal pressure to identify the force that must be resisted by the wall of the sphere.

8. Describe *hoop stress* as it is applied to cylinders carrying an internal pressure.

9. State the formula for computing the hoop stress developed in the wall of a thin-walled cylinder due to internal pressure.

10. Use the hoop stress formula to compute the maximum stress in the wall of a thin-walled cylinder.

11. Determine the required wall thickness of the cylinder to resist a given internal pressure safely.

12. Describe *longitudinal stress* as it is applied to cylinders carrying internal pressure.

13. State the formula for computing the longitudinal stress in the wall of a thin-walled cylinder due to an internal pressure.

14. Use the longitudinal stress formula to compute the stress in the wall of a thin-walled cylinder that acts in the direction parallel to the axis of the cylinder.

15. Determine the required wall thickness of a thin-walled cylinder to resist a given internal pressure safely.

16. Identify the hoop stress, longitudinal stress, and radial stress developed in the wall of a thick-walled sphere or cylinder due to internal pressure.

17. Apply the formulas for computing the maximum values of the hoop stress, longitudinal stress, and radial stress in the wall of a thick-walled sphere or cylinder.

18. Apply the formulas for computing the magnitudes of the hoop stress, longitudinal stress, and radial stress at any radius within the wall of a thick-walled cylinder or sphere.

15-2 DISTINCTION BETWEEN THIN-WALLED AND THICK-WALLED PRESSURE VESSELS

In general, the magnitude of the stress in the wall of a pressure vessel varies as a function of position within the wall. A precise analysis should enable the computation of the stress at any point. The formulas for making such a computation will be shown in a later section.

However, when the wall thickness of the pressure vessel is small, the assumption that the stress is uniform throughout the wall results in very little error. Also, this assumption permits the development of relatively simple formulas for stress.

The criterion for determining when a pressure vessel can be considered thin-walled is as follows:

If the ratio of the mean radius of the vessel to its wall thickness is 10 or greater, the stress is very nearly uniform and it can be assumed that all the material of the wall shares equally to resist the applied forces. Such pressure vessels are called thin-walled vessels.

The mean radius is defined as the average of the outside radius and the inside radius. That is,

$$R_m = \frac{R_o + R_i}{2} \tag{15-1}$$

Then a pressure vessel is considered to be thin if

$$\frac{R_m}{t} \geq 10 \tag{15-2}$$

where t is the wall thickness of the vessel. We can also define the *mean diameter* as

$$D_m = \frac{D_o + D_i}{2} \tag{15-3}$$

Because the diameter is twice the radius, the criterion for a vessel to be considered thin-walled is

$$\frac{D_m}{t} \geq 20 \tag{15-4}$$

Obviously, if the vessel does not satisfy the criteria listed in Equations (15-2) and (15-4), it is considered to be thick-walled.

In addition to Equations (15-1) and (15-3) for the mean radius and mean diameter, the following forms can be useful.

$$R_i = R_o - t$$

$$R_m = R_0 - \frac{t}{2}$$

$$R_m = R_i + \frac{t}{2}$$

$$D_i = D_o - 2t$$

$$D_m = D_o - t$$

$$D_m = D_i + t$$

The next two sections are devoted to the analysis of *thin-walled* spheres and cylinders. Then Section 15-5 is concerned with *thick-walled* spheres and cylinders.

15-3 THIN-WALLED SPHERES

In analyzing a spherical pressure vessel, the objective is to determine the stress in the wall of the vessel to ensure safety. Because of the symmetry of a sphere, a convenient free body for use in the analysis is one-half of the sphere, as shown in Figure 15-1.

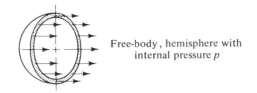

Free-body, hemisphere with
internal pressure p

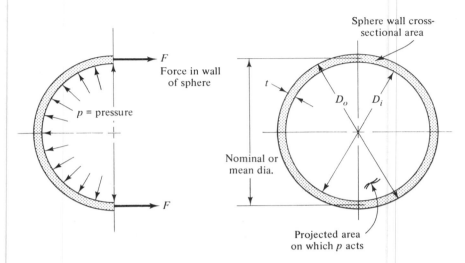

Figure 15-1   Free body diagram for a sphere carrying internal pressure.

The internal pressure of the liquid or gas contained in the sphere acts perpendicular to the walls, uniformly over all the interior surface. Because the sphere was cut through a diameter, the forces in the walls all act horizontally. Therefore, only the horizontal component of the forces due to the fluid pressure needs to be considered in determining the magnitude of the force in the walls. If a pressure P acts on an area A, the force exerted on the area is

$$F = pA \qquad (15\text{-}5)$$

Taking all of the force acting on the entire inside of the sphere and finding the horizontal component, we find the resultant force in the horizontal direction to be

$$F_R = pA_p \qquad (15\text{-}6)$$

where A_p is the *projected area* of the sphere on the plane through the diameter. Therefore,

$$A_p = \frac{\pi D^2}{4} \qquad (15\text{-}7)$$

Because of the equilibrium of the horizontal forces on the free body, the forces in the walls must also equal F_R, as computed in Equation (15-6). These tensile forces acting on the cross-sectional area of the walls of the sphere cause tensile stresses to be developed. That is,

Pressure Vessels   Chap. 15

$$\sigma = \frac{F_R}{A_w} \qquad (15\text{-}8)$$

where A_w is the area of the annular ring cut to create the free body, as shown in Figure 15-1. The actual area is

$$A_w = \frac{\pi}{4}(D_o^2 - D_i^2) \qquad (15\text{-}9)$$

However, for thin-walled spheres with a wall thickness t, less than about $\frac{1}{10}$ of the radius of the sphere, the wall area can be closely approximated as

$$A_w = \pi D t \qquad (15\text{-}10)$$

This is the area of a rectangular strip having a thickness t and a length equal to the circumference of the sphere, πD.

Equations (15-6) and (15-8) can be combined to yield an equation for stress,

$$\sigma = \frac{F_R}{A_w} = \frac{p A_p}{A_w} \qquad (15\text{-}11)$$

Expressing A_p and A_w in terms of D and t from Equations (15-7) and (15-10) gives

$$\sigma = \frac{p(\pi D^2/4)}{\pi D t} = \frac{pD}{4t} \qquad (15\text{-}12)$$

This is the expression for the stress in the wall of a thin-walled sphere subjected to internal pressure.

Example Problem 15-1

A spherical container is being designed to carry nitrogen gas at a pressure of 3500 kPa. The required volume of the container dictates that it be 300 mm in diameter. Compute the required thickness of the sphere if it is to be made of AISI 5160 OQT 1300 steel. Use a design factor of 4 based on yield strength to determine the design stress. Assume that the sphere will be thin-walled and then check that assumption after computing the thickness.

Solution   Equation (15-12) can be used by solving for t.

$$\sigma = \frac{pD}{4t}$$

Letting $\sigma = \sigma_d$, the required thickness is

$$t = \frac{pD}{4\sigma_d} \qquad (15\text{-}13)$$

But using $s_y = 710$ MPa,

$$\sigma_d = \frac{s_y}{4} = \frac{710 \text{ MPa}}{4} = 177.5 \text{ MPa}$$

Then

$$t = \frac{(3500 \times 10^3 \text{ Pa})(300 \text{ mm})}{(4)(177.5 \times 10^6 \text{ Pa})} = 1.48 \text{ mm}$$

Let's let the thickness be $t = 1.50$ mm. Then, using $D_i = 300$ mm, the outside diameter is

$$D_o = D_i + 2t = 300 + 2(1.5) = 303 \text{ mm}$$

Then the mean diameter is

$$D_m = \frac{D_o + D_i}{2} = 301.5 \text{ mm}$$

Now we can compute the ratio of mean diameter to the thickness.

$$\frac{D_m}{t} = \frac{301.5 \text{ mm}}{1.50 \text{ mm}} = 201$$

Because this value is far greater than 20, the sphere is indeed thin-walled.

15-4 THIN-WALLED CYLINDERS

Cylinders are frequently used for pressure vessels, for example, as storage tanks, hydraulic and pneumatic actuators, and for piping of fluids under pressure. The stresses in the walls of cylinders are similar to those found for spheres, although the maximum value is greater.

Two separate analyses are shown here. In one case, the tendency for the internal pressure to pull the cylinder apart in a direction parallel to its axis is found. This is called *longitudinal stress*. Next, a ring around the cylinder is analyzed to determine the stress tending to pull the ring apart. This is called *hoop stress*, or *tangential stress*.

Longitudinal stress. Figure 15-2 shows a part of a cylinder, which is subjected to an internal pressure, cut perpendicular to its axis to create a free body. Assuming that the end of the cylinder is closed, the pressure acting on the circular area of the end would produce a resultant force of

$$F_R = pA = p\left(\frac{\pi D^2}{4}\right) \tag{15-14}$$

This force must be resisted by the force in the walls of the cylinder, which, in turn, creates a tensile stress in the walls. The stress is

$$\sigma = \frac{F_R}{A_w} \tag{15-15}$$

Assuming that the walls are thin, as we did for spheres,

$$A_w = \pi D t$$

where t is the wall thickness.

Now combining Equations (15-14) and (15-15),

$$\sigma = \frac{F_R}{A_w} = \frac{p(\pi D^2/4)}{\pi D t} = \frac{pD}{4t} \tag{15-16}$$

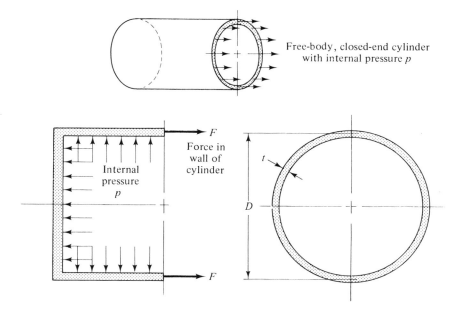

Figure 15-2 Free body diagram of a cylinder carrying internal pressure showing longitudinal stress.

This is the stress in the wall of the cylinder in a direction parallel to the axis, called the longitudinal stress. Notice that it is of the same magnitude as that found for the wall of a sphere.

Hoop stress. The presence of the tangential or hoop stress can be visualized by isolating a ring from the cylinder, as shown in Figure 15-3. The internal pressure pushes outward evenly all around the ring. The ring must develop a tensile stress in a direction tangential to the circumference of the ring to resist the tendency of the pressure to burst the ring. The magnitude of the stress can be determined by using half of the ring as a free body, as shown in Figure 15-3(b).

The resultant of the forces due to the internal pressure must be determined in the horizontal direction and balanced with the forces in the walls of the ring. Using the same reasoning as we did for the analysis of the sphere, we find that the resultant force is the product of the pressure and the *projected area* of the ring. For a ring with a diameter D and a length L,

$$F_R = pA_p = p(DL) \tag{15-17}$$

The tensile stress in the wall of the cylinder is equal to the resisting force divided by the cross-sectional area of the wall. Again assuming that the wall is thin, the wall area is

$$A_w = 2tL \tag{15-18}$$

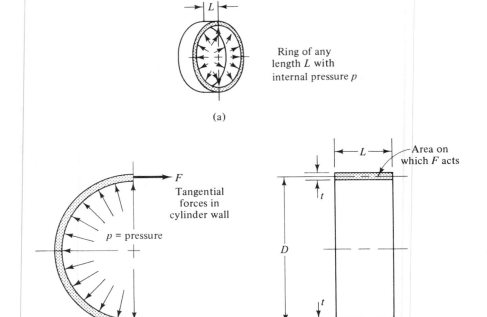

Ring of any
length L with
internal pressure p

(a)

F
Tangential
forces in
cylinder wall

p = pressure

F

(b)

Area on
which F acts

L

t

D

t

Figure 15-3   Free body diagram of a cylinder carrying internal pressure showing hoop stress.

Then the stress is

$$\sigma = \frac{F_R}{A_w} = \frac{F_R}{2tL} \tag{15-19}$$

Combining Equations (15-17) and (15-19) gives

$$\sigma = \frac{F_R}{A_w} = \frac{pDL}{2tL} = \frac{pD}{2t} \tag{15-20}$$

This is the equation for the hoop stress in a thin cylinder subjected to internal pressure. Notice that the magnitude of the hoop stress is *twice* that of the longitudinal stress. Also, the hoop stress is twice that of the stress in a spherical container of the same diameter carrying the same pressure.

Example Problem 15-2

A cylindrical tank holding oxygen at 2000 kPa pressure has a diameter of 450 mm and a wall thickness of 10 mm. Compute the hoop stress and the longitudinal stress in the wall of the cylinder.

Solution   First check to see if the cylinder is thin-walled. The mean diameter can be computed from

$$D_m = D_o - t = 450 \text{ mm} - 10 \text{ mm} = 440 \text{ mm}$$

Then

$$\frac{D_m}{t} = \frac{440 \text{ mm}}{10 \text{ mm}} = 44.0 > 20$$

Therefore, the cylinder is thin-walled. The hoop stress can be found from Equation (15-20).

$$\sigma = \frac{pD}{2t} = \frac{(2000 \times 10^3 \text{ Pa})(450 \text{ mm})}{2(10 \text{ mm})} = 45.0 \text{ MPa}$$

The longitudinal stress, from Equation (15-12), is

$$\sigma = \frac{pD}{4t} = 22.5 \text{ MPa}$$

Example Problem 15-3

Determine the pressure required to burst a standard 8-in. schedule 40 steel pipe if the ultimate tensile strength of the steel is 40 000 psi.

Solution   The dimensions of the pipe are found in the Appendix to be

$$\text{outside diameter} = 8.625 \text{ in.} = D_o$$

$$\text{inside diameter} = 7.981 \text{ in.} = D_i$$

$$\text{wall thickness} = 0.322 \text{ in.} = t$$

We should first check to determine if the pipe should be called a thin-walled cylinder by computing the ratio of the mean diameter to the wall thickness.

$$D_m = \text{mean diameter} = \frac{D_o + D_i}{2} = 8.303 \text{ in.}$$

$$\frac{D_m}{t} = \frac{8.303 \text{ in.}}{0.322 \text{ in.}} = 25.8$$

Since this ratio is greater than 20, the thin-wall equations can be used. The hoop stress is the maximum stress and should be used to compute the bursting pressure.

$$\sigma = \frac{pD}{2t} \qquad (15\text{-}20)$$

Letting $\sigma = 40\ 000$ psi and using the mean diameter gives

$$p = \frac{2t\sigma}{D} = \frac{(2)(0.322 \text{ in.})(40\ 000 \text{ lb/in}^2)}{8.303 \text{ in.}} = 3102 \text{ psi}$$

A design factor of 6 or greater is usually applied to the bursting pressure to get an allowable operating pressure.

15-5 THICK-WALLED CYLINDERS AND SPHERES

The formulas in the preceding sections for thin-walled cylinders and spheres were derived under the assumption that the stress is uniform throughout the wall of the container. As stated, if the ratio of the diameter of the container to the wall thickness is greater than 20, this assumption is reasonably correct. Conversely, if the ratio is less than 20, the walls are considered to be thick, and a different analysis technique is required.

The detailed derivation of the thick-wall formulas will not be given here because of their complexity. But the application of the formulas will be shown.

For a thick-walled cylinder, Figure 15-4 shows the notation to be used. The geometry is characterized by the inner radius a, the outer radius b, and any radial position between a and b, called r. The *longitudinal stress* is called σ_1; the *hoop stress* is σ_2. These have the same meaning as they did for thin-walled vessels, except now they will have varying magnitudes at different positions in the wall. In addition to hoop and longitudinal stresses, a *radial stress* σ_3 is created in a thick-walled vessel. As the name implies, the radial stress acts along a radius of the cylinder or sphere. It is a compressive stress and varies from a magnitude of zero at the outer surface to a maximum at the inner surface, where it is equal to the internal pressure. Table 15-1 shows a summary of the formulas needed to compute the three streses in the walls of thick-walled cylinders and spheres subjected to internal pressure. The terms *longitudinal stress* and *hoop stress* do not apply to spheres. Instead, we refer to the *tangential stress,* which is the same in all directions around the sphere. Then

$$\text{tangential stress} = \sigma_1 = \sigma_2$$

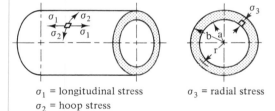

σ_1 = longitudinal stress
σ_2 = hoop stress

σ_3 = radial stress

Figure 15-4 Notation for stresses in thick-walled cylinders.

Example Problem 15-4

Compute the magnitude of the maximum stresses σ_1, σ_2, and σ_3 in a cylinder carrying helium at a pressure of 10 000 psi. The outside diameter is 8.00 in. and the inside diameter if 6.40 in.

Solution   Check first to see if this is a thick-walled cylinder.

$$t = \frac{D_o - D_i}{2} = \frac{8.00 - 6.40}{2} = 0.80 \text{ in.}$$

$$D_m = D_o - t = 8.00 - 0.80 = 7.20 \text{ in.}$$

$$\frac{D_m}{t} = \frac{7.20}{0.80} = 9.00 < 20$$

Therefore, it is a thick-walled cylinder. Referring to Table 15-1, with $a = 3.20$ in. and $b = 4.00$ in., we obtain

$$\sigma_1 = \frac{pa^2}{b^2 - a^2} = \frac{(10\ 000 \text{ psi})(3.20 \text{ in.})^2}{(4.00^2 - 3.20^2) \text{ in}^2} = 17\ 780 \text{ psi}$$

TABLE 15-1 STRESSES IN THICK-WALLED CYLINDERS AND SPHERES*

	Stress at position r	Maximum stress
	Thick-walled cylinder	
Longitudinal	$\sigma_1 = \dfrac{pa^2}{b^2 - a^2}$	$\sigma_1 = \dfrac{pa^2}{b^2 - a^2}$ (uniform throughout wall)
Hoop (tangential)	$\sigma_2 = \dfrac{pa^2(b^2 + r^2)}{r^2(b^2 - a^2)}$	$\sigma_2 = \dfrac{p(b^2 + a^2)}{b^2 - a^2}$ (at inner surface)
Radial	$\sigma_3 = \dfrac{-pa^2(b^2 - r^2)}{r^2(b^2 - a^2)}$	$\sigma_3 = -p$ (at inner surface)
	Thick-walled sphere	
Tangential	$\sigma_1 = \sigma_2 = \dfrac{pa^3(b^3 + 2r^3)}{2r^3(b^3 - a^3)}$	$\sigma_1 = \sigma_2 = \dfrac{p(b^3 + 2a^3)}{2(b^3 - a^3)}$ (at inner surface)
Radial	$\sigma_3 = \dfrac{-pa^3(b^3 - r^3)}{r^3(b^3 - a^3)}$	$\sigma_3 = -p$ (at inner surface)

* Symbols used here are: a = inner radius; b = outer radius; r = any radius between a and b; p = internal pressure, uniform in all directions. Stresses are tensile when positive, compressive when negative.

$$\sigma_2 = \frac{p(b^2 + a^2)}{b^2 - a^2} = \frac{(10\ 000\ \text{psi})(4.00^2 + 3.20^2)\ \text{in}^2}{(4.00^2 - 3.20^2)\ \text{in}^2} = 45\ 560\ \text{psi}$$

$$\sigma_3 = -p = -10\ 000\ \text{psi}$$

All three stresses are a maximum at the inner surface of the cylinder.

Example Problem 15-5

Compute the maximum stresses for a sphere having the same internal and outside diameters as did the cylinder in Example Problem 15-4 for the same internal pressure, 10 000 psi.

Solution   The same check applies for wall thickness as done in Example Problem 15-4. Therefore, this is a thick-walled sphere. From Table 15-1,

$$\sigma_1 = \sigma_2 = \frac{p(b^3 + 2a^3)}{2(b^3 - a^3)} = \frac{(10\ 000\ \text{psi})[4.00^3 + 2(3.20)^3]\ \text{in}^3}{2(4.00^3 - 3.20^3)\ \text{in}^3}$$

$$= 20\ 740\ \text{psi}$$

$$\sigma_3 = -p = -10\ 000\ \text{psi}$$

Each of these stresses is a maximum at the inner surface.

Example problem 15-6

A cylindrical vessel has an outside diameter of 400 mm and an inside diameter of 300 mm. For an internal pressure of 20.1 MPa, compute the hoop stress σ_2 at the inner and outer surfaces and at points within the wall at intervals of 10 mm. Plot a graph of σ_2 versus the radial position in the wall.

Solution   Check first to see if this is a thick-walled cylinder.

$$t = \frac{D_o - D_i}{2} = \frac{400 - 300}{2} = 50 \text{ mm}$$

$$D_m = D_o - t = 400 - 50 = 350 \text{ mm}$$

$$\frac{D_m}{t} = \frac{350}{50} = 7.00 < 20$$

Therefore, it is a thick-walled cylinder. At the inner surface, the maximum value of σ_2 occurs.

$$\sigma_2 = \frac{p(b^2 + a^2)}{b^2 - a^2} = \frac{(20.1 \text{ MPa})(200^2 + 150^2)}{200^2 - 150^2} = 71.8 \text{ MPa}$$

At the outer surface, $r = b = 200$ mm, and

$$\sigma_2 = \frac{pa^2(b^2 + r^2)}{r^2(b^2 - a^2)} = \frac{2pa^2}{b^2 - a^2} = \frac{(2)(20.1 \text{ MPa})(150^2)}{200^2 - 150^2} = 51.7 \text{ MPa}$$

At $r = 160$ mm,

$$\sigma_2 = \frac{(20.1 \text{ MPa})(150^2)(200^2 + 160^2)}{160^2(200^2 - 150^2)} = 66.2 \text{ MPa}$$

Using similar calculations, the stresses σ_2 for other radial positions are:

r (mm)	σ_2 (MPa)
170	61.6
180	57.7
190	54.5

Figure 15-5 shows the graph of σ_2 versus position in the wall. The graph illustrates clearly that the assumption of uniform stress in the wall of a thick-walled cylinder would *not* be valid.

PROBLEMS

15-1. Compute the stress in a sphere having an outside diameter of 200 mm and an inside diameter of 184 mm if an internal pressure of 19.2 MPa is applied.

15-2. A large, spherical storage tank for a compressed gas in a chemical plant is 10.5 m in diameter and is made of AISI 1040 hot-rolled steel plate, 12 mm thick. What internal

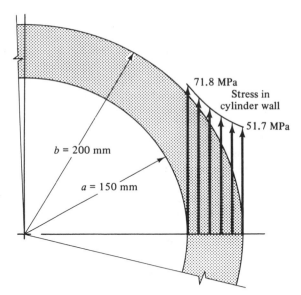

71.8 MPa
Stress in
cylinder wall
51.7 MPa

$b = 200$ mm

$a = 150$ mm

Figure 15-5 Variation of tangential stress in the wall of a thick-walled cylinder.

pressure could the tank withstand if a design factor of 3.0 based on yield strength is desired?

15-3. Titanium 6A1-4V is to be used to make a spherical tank having an outside diameter of 1200 mm. The working pressure in the tank is to be 4.20 MPa. Determine the required thickness of the tank wall if a design factor of 2.5 based on yield strength is desired.

15-4. If the tank of Problem 15-3 was made of aluminum 2014-T6 sheet instead of titanium, compute the required wall thickness. Which design would weigh less?

15-5. Compute the hoop stress in the walls of a 10-in. schedule 40 steel pipe if it carries water at 150 psi.

15-6. A pneumatic cylinder has a bore of 80 mm and a wall thickness of 3.5 mm. Compute the hoop stress in the cylinder wall if an internal pressure of 2.85 MPa is applied.

15-7. A cylinder for carrying acetylene has a diameter of 300 mm and will hold the acetylene at 1.7 MPa. If a design factor of 3 is desired, compute the required wall thickness for the tank. Use AISI 1040 cold-drawn steel.

15-8. The companion oxygen cylinder for the acetylene discussed in the preceding problem carries oxygen at 15.2 MPa. Its diameter is 250 mm. Compute the required wall thickness using the same design criteria.

15-9. A propane tank for a recreation vehicle is made of AISI 1040 hot-rolled steel, 2.20 mm thick. The tank diameter is 450 mm. Determine what design factor would result based on yield strength if propane at 750 kPa is put into the tank.

15-10. The supply tank for propane at the distributor is a cylinder having a diameter of 1800 mm. If it is desired to have a design factor of 3 based on yield strength using AISI 1040 hot-rolled steel, compute the required thickness of the tank walls when the internal pressure is 750 kPa.

15-11. Oxygen on a spacecraft is carried at a pressure of 70.0 MPa in order to minimize the volume required. The spherical vessel has an outside diameter of 250 mm and a wall thickness of 18 mm. Compute the maximum tangential and radial stresses in the sphere.

15-12. Compute the maximum longitudinal, hoop, and radial stresses in the wall of a standard $\frac{1}{2}$-in. schedule 40 steel pipe when carrying an internal pressure of 1.72 MPa (250 psi).

15-13. The barrel of a large field-artillery piece has a bore of 220 mm and an outside diameter of 300mm. Compute the magnitude of the hoop stress in the barrel at points 10 mm apart from the inside to the outside surfaces. The internal pressure is 50 MPa.

15-14. A $1\frac{1}{2}$-in. schedule 40 steel pipe has a mean radius less than 10 times the wall thickness and thus should be classified as a thick-walled cylinder. Compute what maximum stresses would result from both the thin-wall and the thick-wall formulas due to an internal pressure of 10.0 MPa.

15-15. A cylinder has an outside diameter of 50 mm and an inside diameter of 30 mm. Compute the maximum tangential stress in the wall of the cylinder due to an internal pressure of 7.0 MPa.

15-16. For the cylinder of problem 15-15, compute the tangential stress in the wall at increments of 2.0 mm from the inside to the outside. Then plot the results for stress versus radius.

15-17. For the cylinder of Problem 15-15, compute the radial stress in the wall at increments of 2.0 mm from the inside to the outside. Then plot the results for stress versus radius.

15-18. For the cylinder of Problem 15-15, compute the tangential stress that would have been predicted if thin-walled theory were used instead of thick-walled theory. Compare the result with the stress found in Problem 15-15.

15-19. A sphere is made from stainless steel, AISI 501 OQT 1000. Its outside diameter is 500 mm and the wall thickness is 40 mm. Compute the maximum pressure that could be placed in the sphere if the maximum stress is to be one-third of the yield strength of the steel.

15-20. A sphere has an outside diameter of 500 mm and an inside diameter of 420 mm. Compute the tangential stress in the wall at increments of 5.0 mm from the inside to the outside. Then plot the results. Use a pressure of 100 MPa.

15-21. A sphere has an outside diameter of 500 mm and an inside diameter of 420 mm. Compute the radial stress in the wall at increments of 5.0 mm from the inside to the outside. Then plot the results. Use a pressure of 100 MPa.

15-22. To visualize the importance of using the thick-walled formulas for computing stresses in the walls of a cylinder, compute the maximum predicted tangential stress in the wall of a cylinder from both the thin-walled and thick-walled formulas for the following conditions. The outside diameter for all designs is to be 400mm. The wall thickness is to vary from 5.0 mm to 85.0 mm in 10.0 mm increments. Use a pressure of 10.0 MPa. Then compute the ratio of D_m/t and plot percent difference between the stress from the thick-walled and thin-walled theory versus that ratio. Note the increase in the percent difference as the value of D_m/t decreases, that is, as t increases.

15-23. A sphere has an outside diameter of 400 mm and an inside diameter of 325 mm. Compute the variation of the tangential stress from the inside to the outside in increments of 7.5 mm. Use a pressure of 10.0 MPa.

15-24. A sphere has an outside diameter of 400 mm and an inside diameter of 325 mm. Compute the variation of the radial stress from the inside to the outside in increments of 7.5 mm. Use a pressure of 10.0 MPa.

15-25. Appendix A-12 lists the dimensions of American National Standard schedule 40 steel pipe. Which of these pipe sizes should be classified as thick-walled and which can be considered thin-walled?

COMPUTER PROGRAMMING ASSIGNMENTS

1. Write a program to compute the tangential stress in the wall of a thin-walled sphere. Include the computation of the mean diameter and the ratio of mean diameter to thickness to verify that it is thin-walled.

2. Write a program to compute the tangential stress in the wall of a thin-walled cylinder. Include the computation of the mean diameter and the ratio of mean diameter to thickness to verify that it is thin-walled.

3. Write a program to compute the longitudinal stress in the wall of a thin-walled cylinder. Include the computation of the mean diameter and the ratio of mean diameter to thickness to verify that it is thin-walled.

4. Combine the programs of Assignments 2 and 3.

5. Combine the programs of Assignments 1, 2, and 3, and let the user specify whether the vessel is a cylinder or a sphere.

6. Rewrite the programs of Assignments 1, 2, and 5 so that the objective is to compute the required wall thickness for the pressure vessel to produce a given maximum stress for a given internal pressure.

7. Write a program to compute the maximum longitudinal, hoop, and radial stress in the wall of a thick-walled cylinder using the formulas from Table 15-1.

8. Write a program to compute the tangential stress at any radius within the wall of a thick-walled cylinder using the formulas from Table 15-1.

9. Write a program to compute the radial stress at any radius within the wall of a thick-walled cylinder using the formulas from Table 15-1.

10. Write a program to compute the tangential stress at any radius within the wall of a thick-walled sphere using the formulas from Table 15-1.

11. Write a program to compute the radial stress at any radius within the wall of a thick-walled sphere using the formulas from Table 15-1.

12. Combine the programs of Assignments 8 through 11.

13. Write a program to compute the tangential stress distribution within the wall of a thick-walled cylinder using the formulas from Table 15-1. Start at the inside radius and specify a number of increments between the inside and the outside.

14. Write a program to compute the radial stress distribution within the wall of a thick-walled cylinder using the formulas from Table 15-1. Start at the inside radius and specify a number of increments between the inside and the outside.

15. Write a program to compute the tangential stress distribution within the wall of a thick-walled sphere using the formulas from Table 15-1. Start at the inside radius and specify a number of increments between the inside and the outside.

16. Write a program to compute the radial stress distribution within the wall of a thick-walled sphere using the formulas from Table 15-1. Start at the inside radius and specify a number of increments between the inside and the outside.

17. Write a program to perform the computations of the type called for in problem 15-22.

18. Write a program to perform the computations of the type called for in Problem 15-22 except do it for a sphere.

19. Write a program to compute the maximum tangential stress in any standard schedule 40 pipe for a given internal pressure. Include a table of data for the dimensions of the pipe sizes listed in Appendix A-12. Include a check to see if the pipe is thick-walled or thin-walled.

16

Connections

16-1 OBJECTIVES OF THIS CHAPTER

Load carrying members that make up structures and machines must act *together* to perform their desired functions. After completing the design or analysis of the primary members, it is necessary to specify suitable connections between them. As their name implies, connections provide the linkage between members.

The primary objective of this chapter is to provide data and methods of analysis for the safe design of riveted joints, bolted joints, and welded joints. After completing this chapter, you should be able to:

1. Describe the typical geometry of riveted and bolted joints.
2. Identify the probable modes of failure for a joint.
3. Recognize typical styles of rivets.
4. Identify when a fastener is in single shear or double shear.
5. Analyze a riveted or bolted joint for shear force capacity.
6. Analyze a riveted or bolted joint for tensile force capacity.
7. Analyze a riveted or bolted joint for bearing capacity.
8. Use the allowable stresses for steel structural connections as published by the American Institute of Steel Construction (AISC).

9. Describe the difference between a friction-type connection and a bearing-type connection and complete the appropriate analysis.

10. Use the allowable stresses for aluminum structural connections as published by the Aluminum Association.

11. Analyze both symmetrically loaded joints and eccentrically loaded joints.

12. Analyze welded joints with concentric loads.

16-2 TYPES OF CONNECTIONS

Structures and mechanical devices rely on the connections between load-carrying elements to maintain the integrity of the assemblies. The connections provide the path by which loads are transferred from one element to another.

Three common types of connections are riveting, welding, and bolting. Figure 16-1 shows a bulk storage hopper supported by rectangular straps from a tee beam. During fabrication of the hopper, the support tabs were welded to the outside of the side walls. The tabs contain a pattern of holes, allowing the straps to be bolted on at the assembly site. Prior to installation of the tee beam, the straps were riveted to the web.

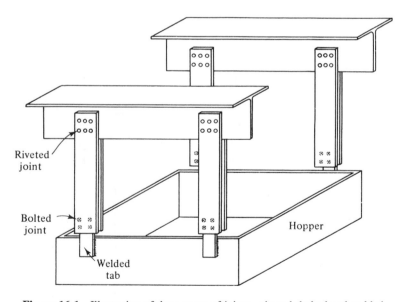

Figure 16-1 Illustration of three types of joints—riveted, bolted and welded.

The load due to the weight of the hopper and the materials in it must be transferred from the hopper walls into the tabs through the welds. Then the four bolts transfer the load to the straps, which act as tension members. Finally, the six rivets transfer the load into the tee.

16-3 MODES OF FAILURE

For riveted and bolted connections, four possible modes of failure exist related to four different types of stresses present in the vicinity of the joint. Figure 16-2 illustrates these kinds of stresses for a simple lap joint in which two flat plates are joined together by two rivets. The joint is prepared by drilling or punching matching holes in each plate. Then the rivets, originally having one of the shapes shown in Figure 16-3, are inserted into the holes, and the heads are upset, gripping the two plates together. In a good riveted joint, the body of the rivet is also upset somewhat, causing the rivet to completely fill the hole. This makes a tight joint which does not allow relative motion of the members joined.

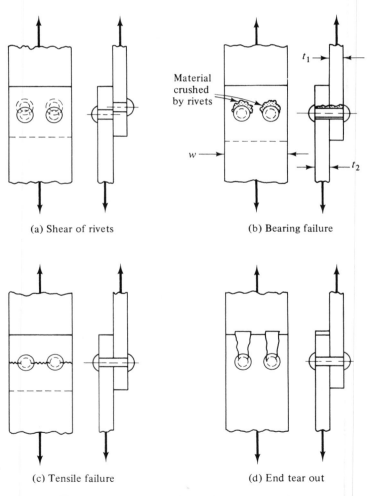

(a) Shear of rivets

(b) Bearing failure

(c) Tensile failure

(d) End tear out

Figure 16-2 Types of failure of riveted connections.

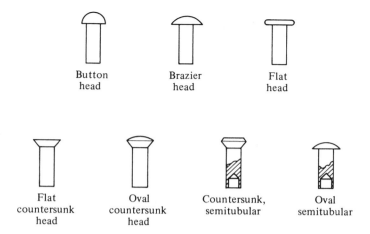

Figure 16-3 Examples of rivet styles.

As the joint is loaded with a tensile force, a shearing force is transmitted across the cross section of the rivets between the two plates. Thus *shear failure* is a mode of joint failure. The body of the rivet must bear against the material of the plates being joined, with the possibility of *failure in bearing*. This would result in a crushing of the material, normally in the plates. *Tensile failure* of the plates being joined must be investigated because the presence of the holes for the rivets causes the cross section of material at the joint to be less than in the main part of the tension member. The fourth possible failure mode is *end tear out,* in which the rivet causes the material between the edge of the plate and the hole to tear out.

Properly designed riveted and bolted joints should have an edge distance from the center of the rivet or bolt to the edge of the plate being joined of at least two times the diameter of the bolt or rivet. The edge distance is measured in the direction toward which the bearing pressure is directed. If this recommendation is heeded, then end tear out should not occur. This will be assumed in example problems for this chapter. Thus shear, bearing, and tensile failure modes only will be considered in evaluating joint strength.

Welded joints fail by shear in the weld material or by fracture of the base metal of the parts joined by the welds. A properly designed and fabricated welded joint will always fail in the base metal. Thus the design of welded connections has the objective of determining the required size of weld and the length of weld in the joint.

16-4 RIVETED CONNECTIONS

In riveted connections, it is assumed that the joined plates are *not* clamped together tightly enough to cause frictional forces between the plates to transmit loads. Therefore, the rivets do bear on the holes, and bearing failure must be investigated. Both shear failure and tensile failure could also occur. The method of analysis of these three modes of failure is outlined below.

Shear failure. The rivet body is assumed to be in direct shear when a tensile load is applied to a joint, provided that the line of action of the load passes through the centroid of the pattern of rivets. It is also assumed that the total applied load is shared equally among all the rivets. The capacity of a joint with regard to shear of the rivets is

$$F_s = \tau_a A_s \tag{16-1}$$

where F_s = capacity of the joint in shear
τ_a = allowable shear stress in rivets
A_s = area in shear

The area in shear is dependent on the number of cross sections of rivets available to resist shear. Calling this number N_s,

$$A_s = \frac{N_s \pi D^2}{4} \tag{16-2}$$

where D is the rivet diameter. In some cases, particularly for rivets driven hot, the body expands to fill the hole, and a larger area is available for resisting shear. However, the increase is small, and only the nominal diameter will be used here.

To determine N_s, it must be observed whether *single shear* or *double shear* exists in the joint. Figure 16-2 shows an example of single shear. Only one cross section of each rivet resists the applied load. Then N_s is equal to the number of rivets in the joint. The straps used to support the bin in Figure 16-1 place the rivets and bolts in double shear. Two cross sections of each rivet resist the applied load. Then N_s is twice the number of rivets in the joint.

Bearing failure. When a cylindrical rivet bears against the wall of a hole in the plate, a nonuniform pressure exists between them. As a simplification of the actual stress distribution, it is assumed that the area in bearing, A_b, is the rectangular area found by multiplying the plate thickness by the diameter of the rivet. This can be considered to be the *projected area* of the rivet hole. Then the bearing capacity of a joint is

$$F_b = \sigma_{ba} A_b \tag{16-3}$$

where F_b = capacity of the joint in bearing
σ_{ba} = allowable bearing stress
A_b = bearing area = $N_b DT$ (16-4)
N_b = number of bearing surfaces

Tensile failure. A direct tensile force applied through the centroid of the rivet pattern would produce a tensile stress. Then the capacity of the joint in tension would be

$$F_t = \sigma_{ta} A_t \tag{16-5}$$

where F_t = capacity of the joint in tension
σ_{ta} = allowable stress in tension
A_t = net tensile area

The evaluation of A_t requires the subtraction of the diameter of all the holes from the width of the plates being joined. Then

$$A_t = (w - ND_H)t \qquad (16\text{-}6)$$

where w = width of plate
$\quad D_H$ = hole diameter (in structures use $D_H = D + \frac{1}{16}$ in. or $D + 2$ mm)
$\quad N$ = number of holes at the section of interest
$\quad t$ = thickness of plate

16-5 ALLOWABLE STRESSES

For members not covered by codes and specifications, the allowable stresses can be determined using the design factors presented in Appendix A-20. For the design of steel building structures, the specifications of the American Institute of Steel Construction (AISC) are usually used. For aluminum structures, the Aluminum Association has published its *Specifications for Aluminum Structures* (1). Table 16-1 shows allowable stresses for steel structures. Table 16-2 summarizes the allowable stresses for aluminum.

TABLE 16-1 ALLOWABLE STRESSES FOR STEEL STRUCTURAL CONNECTIONS*

	Allowable shear stress			Allowable tensile stress	
Rivets ASTM A502	ksi	MPa		ksi	MPa
Grade 1	17.5	121		23	159
Grade 2	22	152		29	200

	Allowable shear stress[†]			Allowable tensile stress	
Bolts	ksi	MPa		ksi	MPa
ASTM A325	17.5	121		44	303
ASTM A490	22	152		54	372

	Allowable bearing stress[‡§]		Allowable tensile stress[‡]
Connected Members			
All alloys	$1.50s_u$		$0.6s_y$

*AISC specifications.

[†] For friction-type connection. For bearing-type connection with no threads in the shear plane, use 30 ksi (207 MPa) for A325, and 40 ksi (276 MPa) for A490.

[‡] See Appendix A-15 for structural steels.

[§] Bearing stress not considered in friction-type bolted joint.

TABLE 16-2 ALLOWABLE STRESSES FOR ALUMINUM STRUCTURAL CONNECTIONS

Rivets			
Alloy and temper		Allowable shear stress	
Before Driving	After Driving*	ksi	MPa
1100-H14	1100-F	4	27
2017-T4	2017-T3	14.5	100
6053-T61	6053-T61	8.5	58
6061-T6	6061-T6	11	76

Bolts				
	Allowable shear stress[†]		Allowable tensile stress[†]	
Alloy and temper	ksi	MPa	ksi	MPa
2024-T4	16	110	26	179
6061-T6	12	83	18	124
7075-T73	17	117	28	193

Connected Members			
		Allowable bearing stress	
Alloy and temper	Allowable tensile stress	ksi	MPa
All alloys	$0.6s_y$ (see Appendix A-17 for s_y)		
1100-H12		11.0	76
2014-T6		49	338
3003-H12		11.5	79
6061-T6		34	234
6063-T6		24	165

Source: Aluminum Association, *Specifications for Aluminum Structures,* Washington, D.C., 1982.

* All cold-driven.

[†] Stresses are based on the area corresponding to the nominal diameter of the bolt unless the threads are in the shear plane. Then the shear area is based on the root diameter.

16-6 BOLTED CONNECTIONS

The analysis of bolted connections is the same as for riveted connections if the bolt is allowed to bear on the hole, a bearing-type connection. This would occur in joints where the clamping force provided by the bolts is small. However, most bolted connections are made with high strength bolts, such as A325 and A490, and tightened to a high tension level. The resulting large clamping forces make a friction-type joint in which the friction forces between the two mating surfaces transmit much of the joint load. The bolts are still designed for shear, using the strengths listed in Table 16-1. But bearing stress is not considered for a friction-type joint.

16-7 EXAMPLE PROBLEMS—RIVETED AND BOLTED JOINTS

Example Problem 16-1

For the single lap joint shown in Figure 16-2, determine the allowable load on the joint if the two plates are $\frac{1}{4}$ in. thick by 2 in. wide and joined by two steel rivets, $\frac{1}{4}$ in. diameter, ASTM A502, grade 1. The plates are ASTM A36 structural steel.

Solution   Shear, bearing, and tensile failure will be investigated to determine the capacity of the joint relative to all three modes. The lowest of the three values is then the limiting load of the joint.

Shear Failure

$$F_s = \tau_a A_s \tag{16-1}$$

From Table 16-1, $\tau_a = 17.5$ ksi. For the two rivets in a single shear, there are two shear planes. Then

$$A_s = \frac{N_s \pi D^2}{4} = \frac{2\pi(0.25 \text{ in.})^2}{4} = 0.098 \text{ in}^2 \tag{16-2}$$

The capacity of the joint in shear is

$$F_s = (17\ 500 \text{ lb/in}^2)(0.098 \text{ in}^2) = 1715 \text{ lb}$$

Bearing Failure

$$F_b = \sigma_{ba} A_b \tag{16-3}$$

In bearing, $\sigma_{ba} = 1.50 s_u$. For ASTM A36 steel, $s_u = 58$ ksi. Then

$$\sigma_{ba} = 1.50\ (58\ 000 \text{ psi}) = 87\ 000 \text{ psi}$$

The bearing area is

$$A_b = N_b D t \tag{16-4}$$

The force on either plate is resisted by two bearing surfaces. Then

$$A_b = 2(0.25 \text{ in.})(0.25 \text{ in.}) = 0.125 \text{ in}^2$$

The capacity of the joint in bearing is

$$F_b = (87\ 000 \text{ lb/in}^2)(0.125 \text{ in}^2) = 10\ 875 \text{ lb}$$

Tensile Failure.
The plates would fail in tension across a section through the rivet holes, as indicated in Figure 16-2(c).

$$F_t = \sigma_{ta} A_t$$

From Table 16-1,

$$\sigma_{ta} = 0.6 s_y = 0.6(36\ 000 \text{ psi}) = 21\ 600 \text{ psi}$$

The net tensile area, assuming that $D_H = D + \frac{1}{16}$ in., is

$$A_t = (w - N D_H)t = [2.0 \text{ in.} - 2(0.25 + 0.063) \text{ in.}](0.25 \text{ in.})$$
$$= 0.344 \text{ in}^2$$

The capacity of the joint in tension is

$$F_t = (21\ 600\ \text{lb/in}^2)(0.344\ \text{in}^2) = 7425\ \text{lb}$$

Since shear failure would occur at a load of 1715 lb, that is the capacity of the joint.

Example Problem 16-2

Determine the allowable load for a joint of the same dimensions as used in Example Problem 16-1, except use two $\frac{3}{8}$-in.-diameter ASTM A490 bolts in a bearing-type connection with no threads in the shear plane.

Solution

Shear Failure

$$F_s = \tau_a A_s$$

$$\tau_a = 40\ 000\ \text{psi} \qquad \text{(Table 16-1)}$$

$$A_s = \frac{2\pi(0.375\ \text{in.})^2}{4} = 0.221\ \text{in}^2$$

Then

$$F_s = (40\ 000\ \text{lb/in}^2)(0.221\ \text{in}^2) = 8840\ \text{lb}$$

Bearing Failure

$$F_b = \sigma_{ba} A_b$$

$$\sigma_{ba} = 1.50(58\ 000\ \text{psi}) = 87\ 000\ \text{psi}$$

$$A_b = N_b D t = (2)(0.375\ \text{in.})(0.25\ \text{in.}) = 0.188\ \text{in}^2$$

Then

$$F_b = (87\ 000\ \text{lb/in}^2)(0.188\ \text{in}^2) = 16\ 356\ \text{lb}$$

Tensile Failure

$$F_t = \sigma_{ta} A_t$$

$$\sigma_{ta} = 0.6(36\ 000\ \text{psi}) = 21\ 600\ \text{psi}$$

$$A_t = [2.0\ \text{in.} - 2(0.375 + 0.063)\ \text{in.}]0.25\ \text{in.} = 0.281\ \text{in}^2$$

Then

$$F_t = (21\ 600\ \text{lb/in}^2)(0.281\ \text{in}^2) = 6070\ \text{lb}$$

Now the capacity in tension is the lowest, so the capacity of the joint is 6070 lb.

16-8 ECCENTRICALLY LOADED RIVETED AND BOLTED JOINTS

Previously considered joints were restricted to cases in which the line of action of the load on the joint passed through the centroid of the pattern of rivets or bolts. In such cases, the applied load is divided equally among all the fasteners. When the load does not pass through the centroid of the fastener pattern, it is called an *eccentrically loaded joint*, and a nonuniform distribution of forces occurs in the fasteners.

In eccentrically loaded joints, the effect of the moment or couple on the fastener must be considered. Figure 16-4 shows a bracket attached to the side of a column and used to support an electric motor. The net downward force exerted by the weight of the motor and the belt tension acts at a distance a from the center of the column flange. Then the total force system acting on the bolts of the bracket consists of the direct shearing force P plus the moment $P \times a$. Each of these components can be considered separately and then added together by using the principle of superposition.

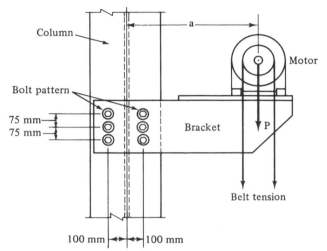

Figure 16-4 Eccentric load on a bolted joint.

Figure 16-5(a) shows that for the direct shearing force P, each bolt is assumed to carry an equal share of the load, just as in concentrically loaded joints. But in part (b) of the figure, because of the moment, each bolt is subjected to a force acting perpendicular to a radial line from the centroid of the bolt pattern. It is assumed that the magnitude of the force in a bolt due to the moment load is proportional to its distance r from the centroid. This magnitude is

$$R_i = \frac{Mr_i}{\sum r^2} \tag{16-7}$$

where R_i = shearing force in bolt i due to the moment M
  r_i = radial distance to bolt i from the centroid of the bolt pattern
  $\sum r^2$ = sum of the radial distances to *all* bolts in the pattern squared

If it is more convenient to work with horizontal and vertical components of forces, they can be found from

$$R_{ix} = \frac{My_i}{\sum r^2} = \frac{My_i}{\sum (x^2 + y^2)} \tag{16-8}$$

$$R_{iy} = \frac{Mx_i}{\sum r^2} = \frac{Mx_i}{\sum (x^2 + y^2)} \tag{16-9}$$

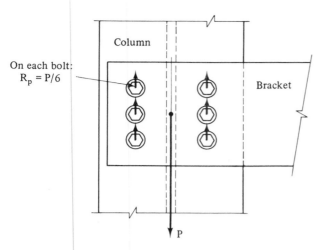

On each bolt:
$R_p = P/6$

Column

Bracket

P

(a) Forces resisting P, the shearing force.

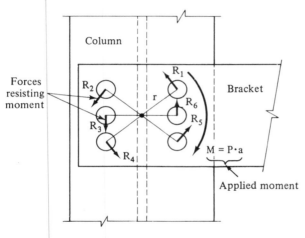

Column

Forces resisting moment

R_2

R_1

r

R_6

R_3

R_5

R_4

Bracket

$M = P \cdot a$

Applied moment

(b) Forces resisting the moment.

Figure 16-5 Loads on bolts that are eccentrically loaded.

where
y_i = vertical distance to bolt i from centroid
x_i = horizontal distance to bolt i from centroid
$\Sigma(x^2 + y^2)$ = sum of horizontal and vertical distances squared for all bolts in the pattern

Finally, all horizontal forces are summed and all vertical forces are summed for any particular bolt. Then the resultant of the horizontal and vertical forces is determined.

Example Problem 16-3

In Figure 16-4 the net downward force P is 26.4 kN on each side plate of the bracket. The distance a is 0.75 m. Determine the required size of ASTM A325 bolts to secure the bracket.

Solution The force to be resisted by the most highly stressed bolt must be determined. Then the required area and diameter will be computed.

One component of the force on each bolt is the share of the 26.4 kN shearing force. Let's call this force R_p.

$$R_p = \frac{P}{6} = \frac{26.4 \text{ kN}}{6} = 4.4 \text{ kN}$$

This is an upward reaction force produced by each bolt, as shown in Figure 16-5.

Now, to find the horizontal and vertical components of the reaction forces due to the eccentric moment, we need the value of

$$\sum (x^2 + y^2) = 6(100 \text{ mm})^2 + 4(75 \text{ mm})^2 = 82\,500 \text{ mm}^2$$

The moment on the joint is

$$P \times a = 26.4 \text{ kN} (0.75 \text{ m}) = 19.8 \text{ kN} \cdot \text{m}$$

Starting first with bolt 1 at the upper right (see Figure 16-6),

$$R_{1x} = \frac{My_1}{\sum (x^2 + y^2)} = \frac{19.8 \text{ kN} \cdot \text{m}(75 \text{ mm})}{82\,500 \text{ mm}^2} \times \frac{10^3 \text{ mm}}{\text{m}}$$

$$= 18.0 \text{ kN} \quad \longleftarrow \quad \text{(acts toward the left)}$$

$$R_{1y} = \frac{Mx_1}{\sum (x^2 + y^2)} = \frac{(19.8 \text{ kN} \cdot \text{m})(100 \text{ mm})}{82\,500 \text{ mm}^2} \times \frac{10^3 \text{ mm}}{\text{m}}$$

$$= 24.0 \text{ kN} \quad \uparrow \quad \text{(acts upward)}$$

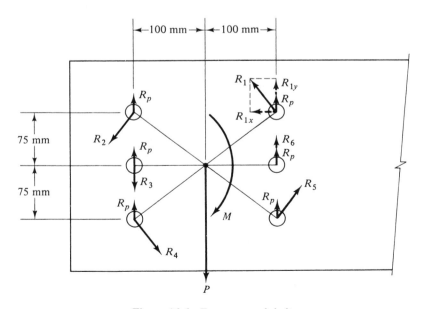

Figure 16-6 Forces on each bolt.

Now the resultant of these forces can be found. In the vertical direction, R_p and R_{1y} both act upward.

$$R_p + R_{1y} = 4.4 \text{ kN} + 24.0 \text{ kN} = 28.4 \text{ kN}$$

Only R_{1x} acts in the horizontal direction. Calling the resultant force on bolt 1, R_{R1},

$$R_{R1} = \sqrt{28.4^2 + 18.0^2} = 33.6 \text{ kN}$$

Investigating the other five bolts in a similar manner would show that bolt 1 is the most highly stressed. Then its diameter will be determined to limit the shear stress to 121 MPa (17.5 ksi) for the ASTM A325 bolts.

$$\tau = \frac{R_{R1}}{A}$$

$$A = \frac{R_{R1}}{\tau_a} = \frac{33.6 \text{ kN}}{121 \text{ N/mm}^2} = 280 \text{ mm}^2$$

$$D = \sqrt{\frac{4A}{\pi}} = \sqrt{\frac{4(280) \text{ mm}^2}{\pi}} = 18.9 \text{ mm}$$

The nearest preferred metric size is 20 mm. If conventional inch units are to be specified,

$$D = 18.9 \text{ mm} \times \frac{1 \text{ in.}}{25.4 \text{ mm}} = 0.74 \text{ in.}$$

Use $D = \frac{3}{4}$ in. $= 0.750$ in.

16-9 WELDED JOINTS WITH CONCENTRIC LOADS

Welding is a joining processs in which heat is applied to cause two pieces of metal to become metallurgically bonded. The heat may be applied by a gas flame, an electric arc, or by a combination of electric resistance heating and pressure.

Types of welds include groove, fillet, and spot welds (as shown in Figure 16-7), and others. Groove and fillet welds are most frequently used in structural connections since they are readily adaptable to the shapes and plates which make up the structures. Spot welds are used for joining relatively light gauge steel sheets and cold-formed shapes.

The variables involved in designing welded joints are the shape and size of the weld, the choice of filler metal, the length of weld, and the position of the weld relative to the applied load.

Fillet welds are assumed to have a slope of 45 deg between the two surfaces joined, as shown in Figure 16-7(b). The size of the weld is denoted as the height of one side of the triangular-shaped fillet. Typical sizes range from $\frac{1}{8}$ in. to $\frac{1}{2}$ in. in steps of $\frac{1}{16}$ in. The stress developed in fillet welds is assumed to be *shear stress* regardless of the direction of application of the load. The maximum shear stress would occur at the throat of the fillet (see Figure 16-7), where the thickness is 0.707 times the nominal size of the weld. Then the shearing stress in the weld due to a load P is

$$\tau = \frac{P}{Lt} \qquad (16\text{-}10)$$

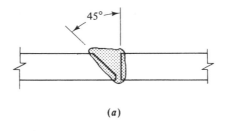

(*a*)

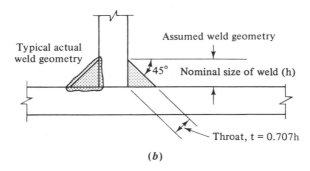

(*b*)

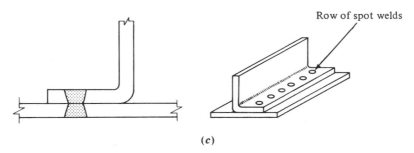

(*c*)

Figure 16-7 Types of welds. (a) Butt weld, single bevel groove (b) Fillet weld (c) Spot weld.

where L is the length of weld and t is the thickness at the throat. Equation (16-10) is used *only* for concentrically loaded members. This requires that the line of action of the force on the welds passes through the centroid of the weld pattern. Eccentricity of the load produces a moment, in addition to the direct shearing force, which must be resisted by the weld metal. References 2 to 5 at the end of this chapter contain pertinent information with regard to eccentrically loaded welded joints.

In electric arc welding, used mostly for structural connections, a filler rod is normally used to add metal to the welded zone. As the two parts to be joined are heated to a molten state, filler metal is added, which combines with the base metal. On cooling, the resulting weld metal is normally stronger than the original base metal. Therefore, a properly designed and made welded joint should fail in the base metal rather than in the weld. In structural welding, the electrodes are given a code beginning with an E and followed by two or three digits, such as E60, E80, or E100. The

TABLE 16-3 PROPERTIES OF WELDING ELECTRODES

Electrode type	Minimum tensile strength		Allowable shear stress		Typical metals joined
	ksi	MPa	ksi	MPa	
E60	60	414	18	124	A36, A500
E70	70	483	21	145	A242, A441
E80	80	552	24	165	A572, Grade 65
E90	90	621	27	186	—
E100	100	690	30	207	—
E110	110	758	33	228	A514

number denotes the ultimate tensile strength in ksi of the weld metal in the rod. Thus an E80 rod would have a tensile strength of 80 000 psi. Other digits may be added to the code number to denote special properties. Complete specifications can be found in the standards of ASTM A233 and A316. The allowable shear stress for fillet welds using electrodes is 0.3 times the tensile strength of the electrode according to AISC. Table 16-3 lists some common electrodes and their allowable stresses.

Aluminum products are welded using either the inert gas-shielded arc process or the resistance welding process. For the inert gas-shielded arc process, filler alloys are specified by the Aluminum Association for joining particular base metal alloys, as indicated in Table 16-4. The allowable shear stress for such welds is also listed. It should be noted that the heat of welding lowers the properties of most aluminum alloys within 1.0 in. of the weld, and allowance for this must be made in the design of welded assemblies.

TABLE 16-4 ALLOWABLE SHEAR STRESSES IN FILLET WELDS IN ALUMINUM BUILDING-TYPE STRUCTURES

Base metal	Filler alloy							
	1100		4043		5356		5556	
	ksi	MPa	ksi	MPa	ksi	MPa	ksi	MPa
1100	3.2	22	4.8	33	—	—	—	—
3003	3.2	22	5.0	34	—	—	—	—
6061	—	—	5.0	34	7.0	48	8.5	59
6063	—	—	5.0	34	6.5	45	6.5	45

Example Problem 16-4

A lap joint is made by placing two $\frac{3}{8}$-in. fillet welds across the full width of two $\frac{1}{2}$-in. ASTM A36 steel plates, as shown in Figure 16-8. The shielded metal-arc method is used, using an E60 electrode. Compute the allowable load P which can be applied to the joint.

Solution The load is assumed to be equally distributed on all parts of the weld, so that Equation (16-10) can be used with $L = 8.0$ in.

$$\tau = \frac{P}{Lt}$$

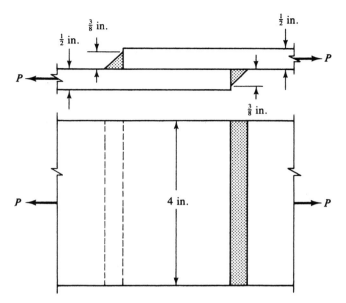

Figure 16-8 Welded lap joint.

Let τ equal the allowable stress of 18 ksi, listed in Table 16-3. The thickness t is

$$t = 0.707 \ (\tfrac{3}{8} \text{ in.}) = 0.265 \text{ in.}$$

Now we can solve for P.

$$P = \tau_a L t = (18\ 000 \text{ lb/in}^2)(8.0 \text{ in.})(0.265 \text{ in.}) = 38\ 200 \text{ lb}$$

REFERENCES

1. Aluminum Association, *Specifications for Aluminum Structures*, 4th ed., Washington, D.C., 1982.

2. American Institute of Steel Construction, *Manual of Steel Construction*, 8th ed., Chicago, 1980.

3. Blodgett, O. W. *Design of Weldments*, James F. Lincoln Arc Welding Foundation, Cleveland, Ohio, 1963.

4. Johnston, B. G., and F. Lin, *Basic Steel Design,* Prentice-Hall, Englewood Cliffs, N.J., 1974.

5. Mott, R. L., *Machine Elements in Mechanical Design*, Charles E. Merrill, Columbus, Ohio, 1985.

PROBLEMS

16-1. Determine the allowable loads on the joints shown in Figure 16-9. The fasteners are all steel rivets, ASTM A502, grade 1. The plates are all ASTM A36 steel.

16-2. Determine the allowable loads on the joints shown in Figure 16-10. The fasteners are all steel rivets, ASTM A502, grade 2. The plates are all ASTM A242 steel.

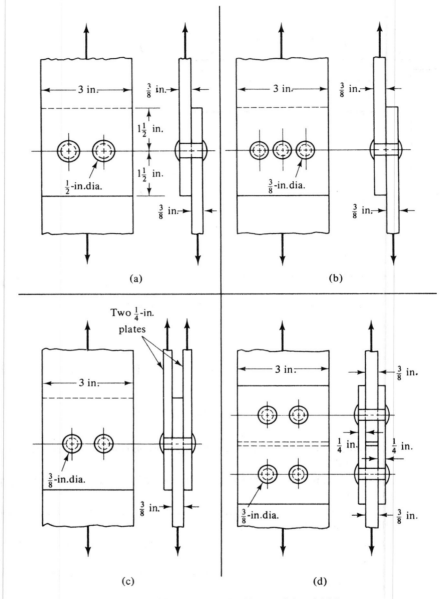

Figure 16-9 Joints for Problems 16-1 and 16-3.

16-3. Determine the allowable loads on the joints shown in Figure 16-9 if the fasteners are all ASTM A325 steel bolts providing a bearing-type connection and the plates are all ASTM A441 steel.

16-4. Determine the allowable loads on the joints shown in Figure 16-10. The fasteners are all ASTM A490 steel bolts providing a friction-type joint. The plates are all ASTM A514 steel.

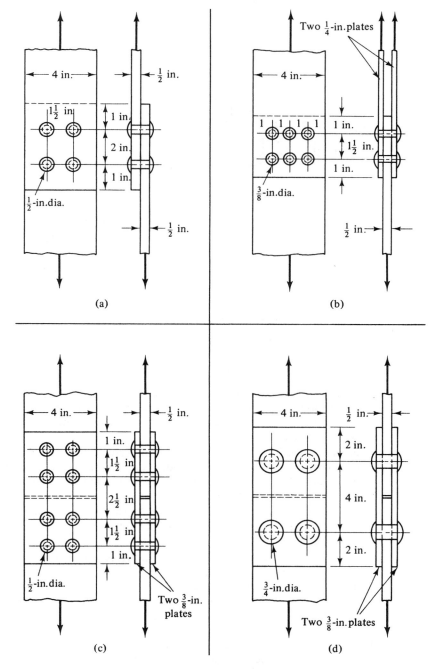

Figure 16-10 Joints for Problems 16-2 and 16-4.

16-5. Determine the required diameter of the bolts used to attach the cantilever beam to the column as shown in Figure 16-11. Use ASTM A325 steel bolts.

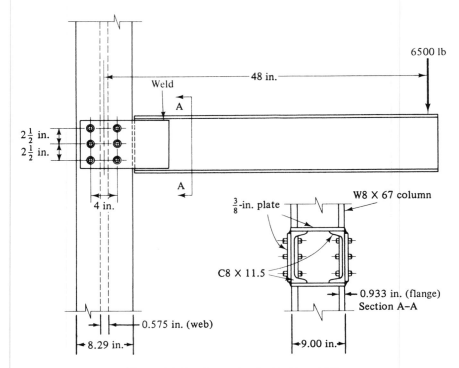

Figure 16-11 Connection for Problem 16-5.

16-6. Design the connection of the channel to the column for the hanger shown in Figure 16-12. Both members are ASTM A36 steel. Specify the type of fastener (rivet, bolt), the pattern and spacing, the number of fasteners, and the material for the fasteners. Use AISC specifications.

16-7. For the connection shown in Figure 16-9(a), asssume that, instead of the two rivets, the two plates were welded across the ends of the 3-in.-wide plates using $\frac{5}{16}$-in. welds. The plates are ASTM A36 steel and the electric arc welding technique is used with E60 electrodes. Determine the allowable load on the connection.

16-8. Determine the allowable load on the joint shown in Figure 16-10(c) if $\frac{1}{4}$-in. welds using E70 electrodes were placed along both ends of both cover plates. The plates are ASTM A242 steel.

16-9. Design the joint at the top of the straps in Figure 16-1 if the total load in the hopper is 54.4 megagrams (Mg). The beam is a WT12 × 34 made of ASTM A36 steel and has a web thickness of 10.6 mm. The clear vertical height of the web is about 250 mm. Use steel rivets and specify the pattern, number of rivets, rivet diameter, rivet material, strap material, and strap dimensions. Specify dimensions, using the preferred metric sizes shown in the Appendix.

16-10. Design the joint at the bottom of the straps in Figure 16-1 if the total load in the hopper is 54.4 Mg. Use steel bolts and a bearing-type connection. Specify the pattern, number of bolts, bolt diameter, bolt material, strap material, and strap dimensions. You may want to coordinate the strap design with the results of Problem 16-9. The design of the tab in Problem 16-11 is also affected by the design of the bolted joint.

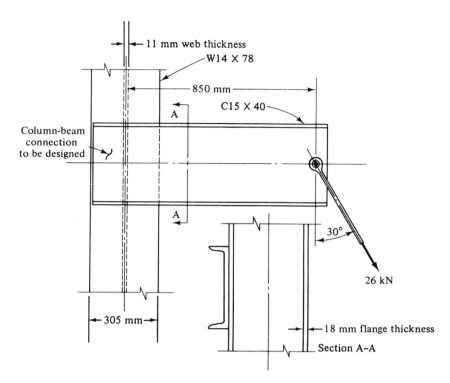

11 mm web thickness

W14 × 78

850 mm

C15 × 40

A

Column-beam
connection
to be designed

A

30°

26 kN

305 mm

18 mm flange thickness

Section A–A

Figure 16-12 Connection for Problem 16-6.

16-11. Design the tab to be welded to the hopper for connection to the support straps as shown in Figure 16-1. The hopper load is 54.4 Mg. The material from which the hopper is made is ASTM A36 steel. Specify the width and thickness of the tab and the design of the welded joint. You may want to coordinate the tab design with the bolted connection called for in Problem 16-10.

Appendix

A-1   PROPERTIES OF AREAS*

Circle

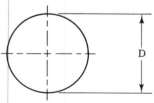

$$A = \frac{\pi D^2}{4}$$

$$I = \frac{\pi D^4}{64}$$

$$S = \frac{\pi D^3}{32}$$

$$r = \frac{D}{4}$$

$$J = \frac{\pi D^4}{32}$$

$$Z_p = \frac{\pi D^3}{16}$$

Hollow circle (tube)

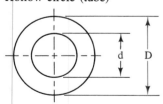

$$A = \frac{\pi(D^2 - d^2)}{4}$$

$$I = \frac{\pi(D^4 - d^4)}{64}$$

$$S = \frac{\pi(D^4 - d^4)}{32D}$$

$$r = \frac{\sqrt{D^2 + d^2}}{4}$$

$$J = \frac{\pi(D^4 - d^4)}{32}$$

$$Z_p = \frac{\pi(D^4 - d^4)}{16D}$$

*Symbols used are:

A = area
I = moment of inertia
S = section modulus

r = radius of gyration = $\sqrt{I/A}$
J = polar moment of inertia
Z_p = polar section modulus

Square

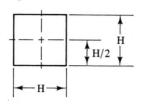

$$A = H^2$$

$$I = \frac{H^4}{12}$$

$$S = \frac{H^3}{6}$$

$$r = \frac{H}{\sqrt{12}}$$

Rectangle

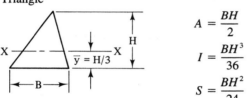

$$A = BH$$

$$I_x = \frac{BH^3}{12}$$

$$S_x = \frac{BH^2}{6}$$

$$r_x = \frac{H}{\sqrt{12}}$$

$$r_y = \frac{B}{\sqrt{12}}$$

Triangle

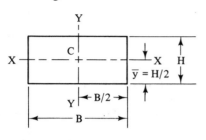

$$A = \frac{BH}{2}$$

$$I = \frac{BH^3}{36}$$

$$S = \frac{BH^2}{24}$$

$$r = \frac{H}{\sqrt{18}}$$

Semicircle

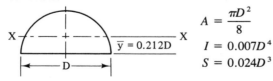

$$A = \frac{\pi D^2}{8}$$

$$I = 0.007D^4$$

$$S = 0.024D^3$$

$$r = 0.132D$$

Regular hexagon

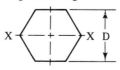

$$A = 0.866D^2$$

$$I = 0.06D^4$$

$$S = 0.12D^3$$

$$r = 0.264D$$

Area under a second-degree curve

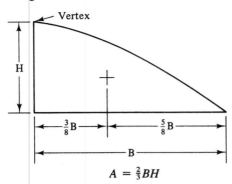

$$A = \tfrac{2}{3}BH$$

Area over a second-degree curve

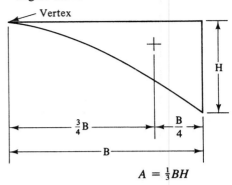

$$A = \tfrac{1}{3}BH$$

Area under a third-degree curve

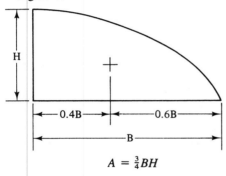

$$A = \tfrac{3}{4}BH$$

Area over a third-degree curve

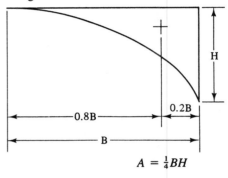

$$A = \tfrac{1}{4}BH$$

A-2 PREFERRED BASIC SIZES

Fractional (in.)				Decimal (in.)			Metric (mm)					
							First	Second	First	Second	First	Second
$\frac{1}{64}$	0.015 625	5	5.000	0.010	2.00	8.50						
$\frac{1}{32}$	0.031 25	$5\frac{1}{4}$	5.250	0.012	2.20	9.00						
$\frac{1}{16}$	0.062 5	$5\frac{1}{2}$	5.500	0.016	2.40	9.50						
$\frac{3}{32}$	0.093 75	$5\frac{3}{4}$	5.750	0.020	2.60	10.00						
$\frac{1}{8}$	0.125 0	6	6.000	0.025	2.80	10.50						
$\frac{5}{32}$	0.156 25	$6\frac{1}{2}$	6.500	0.032	3.00	11.00					100	
$\frac{3}{16}$	0.187 5	7	7.000	0.040	3.20	11.50						110
$\frac{1}{4}$	0.250 0	$7\frac{1}{2}$	7.500	0.05	3.40	12.00					120	
$\frac{5}{16}$	0.312 5	8	8.000	0.06	3.60	12.50						140
$\frac{3}{8}$	0.375 0	$8\frac{1}{2}$	8.500	0.08	3.80	13.00			10		160	
$\frac{7}{16}$	0.437 5	9	9.000	0.10	4.00	13.50				11		180
$\frac{1}{2}$	0.500 0	$9\frac{1}{2}$	9.500	0.12	4.20	14.00			12		200	
$\frac{9}{16}$	0.562 5	10	10.000	0.16	4.40	14.50				14		220
$\frac{5}{8}$	0.625 0	$10\frac{1}{2}$	10.500	0.20	4.60	15.00	1		16		250	
$\frac{11}{16}$	0.687 5	11	11.000	0.24	4.80	15.50		1.1		18		280
$\frac{3}{4}$	0.750 0	$11\frac{1}{2}$	11.500	0.30	5.00	16.00	1.2		20		300	
$\frac{7}{8}$	0.875 0	12	12.000	0.40	5.20	16.50		1.4		22		350
1	1.000	$12\frac{1}{2}$	12.500	0.50	5.40	17.00	1.6		25		400	
$1\frac{1}{4}$	1.250	13	13.000	0.60	5.60	17.50		1.8		28		450
$1\frac{1}{2}$	1.500	$13\frac{1}{2}$	13.500	0.80	5.80	18.00	2		30		500	
$1\frac{3}{4}$	1.750	14	14.000	1.00	6.00	18.50		2.2		35		550
2	2.000	$14\frac{1}{2}$	14.500	1.20	6.50	19.00	2.5		40		600	
$2\frac{1}{4}$	2.250	15	15.000	1.40	7.00	19.50		2.8		45		700
$2\frac{1}{2}$	2.500	$15\frac{1}{2}$	15.500	1.60	7.50	20.00	3		50		800	
$2\frac{3}{4}$	2.750	16	16.000	1.80	8.00			3.5		55		900
3	3.000	$16\frac{1}{2}$	16.500				4		60		1000	
$3\frac{1}{4}$	3.250	17	17.000					4.5		70		
$3\frac{1}{2}$	3.500	$17\frac{1}{2}$	17.500				5		80			
$3\frac{3}{4}$	3.750	18	18.000					5.5		90		
4	4.000	$18\frac{1}{2}$	18.500				6					
$4\frac{1}{4}$	4.250	19	19.000					7				
$4\frac{1}{2}$	4.500	$19\frac{1}{2}$	19.500				8					
$4\frac{3}{4}$	4.750	20	20.000					9				

(a) American Standard thread dimensions, numbered sizes

Size	Basic major diameter, D (in.)	Coarse threads: UNC		Fine threads: UNF	
		Threads per inch, n	Tensile stress area (in²)	Threads per inch, n	Tensile stress area (in²)
0	0.060 0	—	—	80	0.001 80
1	0.073 0	64	0.002 63	72	0.002 78
2	0.086 0	56	0.003 70	64	0.003 94
3	0.099 0	48	0.004 87	56	0.005 23
4	0.112 0	40	0.006 04	48	0.006 61
5	0.125 0	40	0.007 96	44	0.008 30
6	0.138 0	32	0.009 09	40	0.010 15
8	0.164 0	32	0.014 0	36	0.014 74
10	0.190 0	24	0.017 5	32	0.020 0
12	0.216 0	24	0.024 2	28	0.025 8

(b) American Standard thread dimensions, fractional sizes

Size	Basic major diameter, D (in.)	Coarse threads: UNC		Fine threads: UNF	
		Threads per inch, n	Tensile stress area (in²)	Threads per inch, n	Tensile stress area (in²)
$\frac{1}{4}$	0.250 0	20	0.031 8	28	0.036 4
$\frac{5}{16}$	0.312 5	18	0.052 4	24	0.058 0
$\frac{3}{8}$	0.375 0	16	0.077 5	24	0.087 8
$\frac{7}{16}$	0.437 5	14	0.106 3	20	0.118 7
$\frac{1}{2}$	0.500 0	13	0.141 9	20	0.159 9
$\frac{9}{16}$	0.562 5	12	0.182	18	0.203
$\frac{5}{8}$	0.625 0	11	0.226	18	0.256
$\frac{3}{4}$	0.750 0	10	0.334	16	0.373
$\frac{7}{8}$	0.875 0	9	0.462	14	0.509
1	1.000	8	0.606	12	0.663
$1\frac{1}{8}$	1.125	7	0.763	12	0.856
$1\frac{1}{4}$	1.250	7	0.969	12	1.073
$1\frac{3}{8}$	1.375	6	1.155	12	1.315
$1\frac{1}{2}$	1.500	6	1.405	12	1.581
$1\frac{3}{4}$	1.750	5	1.90	—	—
2	2.000	$4\frac{1}{2}$	2.50	—	—

(c) Metric thread dimensions

Basic major diameter, D (mm)	Coarse threads		Fine threads	
	Pitch (mm)	Tensile stress area (mm^2)	Pitch (mm)	Tensile stress area (mm^2)
1	0.25	0.460	—	—
1.6	0.35	1.27	0.20	1.57
2	0.4	2.07	0.25	2.45
2.5	0.45	3.39	0.35	3.70
3	0.5	5.03	0.35	5.61
4	0.7	8.78	0.5	9.79
5	0.8	14.2	0.5	16.1
6	1	20.1	0.75	22.0
8	1.25	36.6	1	39.2
10	1.5	58.0	1.25	61.2
12	1.75	84.3	1.25	92.1
16	2	157	1.5	167
20	2.5	245	1.5	272
24	3	353	2	384
30	3.5	561	2	621
36	4	817	3	865
42	4.5	1 121	—	—
48	5	1 473	—	—

A-4 PROPERTIES OF STANDARD WOOD BEAMS

Nominal size	Actual size in.	Actual size mm	Area of section in²	Area of section mm²×10⁻³	Moment of inertia in⁴	Moment of inertia mm⁴×10⁻⁶	Section modulus in³	Section modulus mm³×10⁻³
2 × 4	1.5 × 3.5	38 × 89	5.25	3.39	5.36	2.23	3.06	50.1
2 × 6	1.5 × 5.5	38 × 140	8.25	5.32	20.8	8.66	7.56	124
2 × 8	1.5 × 7.25	38 × 184	10.87	7.01	47.6	19.8	13.14	215
2 × 10	1.5 × 9.25	38 × 235	13.87	8.95	98.9	41.2	21.4	351
2 × 12	1.5 × 11.25	38 × 286	16.87	10.88	178	74.1	31.6	518
4 × 4	3.5 × 3.5	89 × 89	12.25	7.90	12.51	5.21	7.15	117
4 × 6	3.5 × 5.5	89 × 140	19.25	12.42	48.5	20.2	17.65	289
4 × 8	3.5 × 7.25	89 × 184	25.4	16.39	111.1	46.2	30.7	503
4 × 10	3.5 × 9.25	89 × 235	32.4	20.90	231	96.1	49.9	818
4 × 12	3.5 × 11.25	89 × 286	39.4	25.42	415	172	73.9	1211
6 × 6	5.5 × 5.5	140 × 140	30.3	19.55	76.3	31.8	27.7	454
6 × 8	5.5 × 7.5	140 × 191	41.3	26.65	193	80.3	51.6	846
6 × 10	5.5 × 9.5	140 × 241	52.3	33.74	393	164	82.7	1355
6 × 12	5.5 × 11.5	140 × 292	63.3	40.84	697	290	121	1983
8 × 8	7.5 × 7.5	191 × 191	56.3	36.32	264	110	70.3	1152
8 × 10	7.5 × 9.5	191 × 241	71.3	46.00	536	223	113	1852
8 × 12	7.5 × 11.5	191 × 292	86.3	55.68	951	396	165	2704
10 × 10	9.5 × 9.5	241 × 241	90.3	58.26	679	283	143	2343
10 × 12	9.5 × 11.5	241 × 292	109.3	70.52	1204	501	209	3425
12 × 12	11.5 × 11.5	292 × 292	132.3	85.35	1458	607	253	4146

A-5 PROPERTIES OF ANGLES
EQUAL LEGS AND UNEQUAL LEGS
L-Shapes*

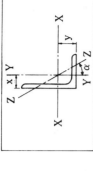

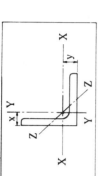

Designation	Area (in²)	Weight per foot (lb)	Axis X-X			Axis Y-Y			Axis Z-Z	
			I (in⁴)	S (in³)	y (in.)	I (in⁴)	S (in³)	x (in.)	r (in.)	α (deg)
L8 × 8 × 1	15.0	51.0	89.0	15.8	2.37	89.0	15.8	2.37	1.56	45.0
L8 × 8 × ½	7.75	26.4	48.6	8.36	2.19	48.6	8.36	2.19	1.59	45.0
L8 × 4 × 1	11.0	37.4	69.6	14.1	3.05	11.6	3.94	1.05	0.846	13.9
L8 × 4 × ½	5.75	19.6	38.5	7.49	2.86	6.74	2.15	0.859	0.865	14.9
L6 × 6 × ¾	8.44	28.7	28.2	6.66	1.78	28.2	6.66	1.78	1.17	45.0
L6 × 6 × ⅜	4.36	14.9	15.4	3.53	1.64	15.4	3.53	1.64	1.19	45.0
L6 × 4 × ¾	6.94	23.6	24.5	6.25	2.08	8.68	2.97	1.08	0.860	23.2
L6 × 4 × ⅜	3.61	12.3	13.5	3.32	1.94	4.90	1.60	0.941	0.877	24.0
L4 × 4 × ½	3.75	12.8	5.56	1.97	1.18	5.56	1.97	1.18	0.782	45.0
L4 × 4 × ¼	1.94	6.6	3.04	1.05	1.09	3.04	1.05	1.09	0.795	45.0
L4 × 3 × ½	3.25	11.1	5.05	1.89	1.33	2.42	1.12	0.827	0.639	28.5
L4 × 3 × ¼	1.69	5.8	2.77	1.00	1.24	1.36	0.599	0.896	0.651	29.2
L3 × 3 × ½	2.75	9.4	2.22	1.07	0.932	2.22	1.07	0.932	0.584	45.0
L3 × 3 × ¼	1.44	4.9	1.24	0.577	0.842	1.24	0.577	0.842	0.592	45.0
L2 × 2 × ⅜	1.36	4.7	0.479	0.351	0.636	0.479	0.351	0.636	0.389	45.0
L2 × 2 × ¼	0.938	3.19	0.348	0.247	0.592	0.348	0.247	0.592	0.391	45.0
L2 × 2 × ⅛	0.484	1.65	0.190	0.131	0.546	0.190	0.131	0.546	0.398	45.0

*Data taken from a variety of sources. Sizes listed represent a small sample of the sizes available.

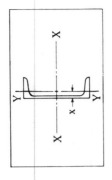

A-6 PROPERTIES OF AMERICAN STANDARD CHANNELS
C-Shapes*

Designation	Area (in.²)	Depth (in.)	Web thickness (in.)	Flange Width (in.)	Flange Thickness (in.)	Axis X-X I (in.⁴)	Axis X-X S (in.³)	Axis Y-Y I (in.⁴)	Axis Y-Y S (in.³)	x (in.)
C15 × 50	14.7	15.00	0.716	3.716	0.650	404	53.8	11.0	3.78	0.798
C15 × 40	11.8	15.00	0.520	3.520	0.650	349	46.5	9.23	3.37	0.777
C12 × 30	8.82	12.00	0.510	3.170	0.501	162	27.0	5.14	2.06	0.674
C12 × 25	7.35	12.00	0.387	3.047	0.501	144	24.1	4.47	1.88	0.674
C10 × 30	8.82	10.00	0.673	3.033	0.436	103	20.7	3.94	1.65	0.649
C10 × 20	5.88	10.00	0.379	2.739	0.436	78.9	15.8	2.81	1.32	0.606
C9 × 20	5.88	9.00	0.448	2.648	0.413	60.9	13.5	2.42	1.17	0.583
C9 × 15	4.41	9.00	0.285	2.485	0.413	51.0	11.3	1.93	1.01	0.586
C8 × 18.75	5.51	8.00	0.487	2.527	0.390	44.0	11.0	1.98	1.01	0.565
C8 × 11.5	3.38	8.00	0.220	2.260	0.390	32.6	8.14	1.32	0.781	0.571
C6 × 13	3.83	6.00	0.437	2.157	0.343	17.4	5.80	1.05	0.642	0.514
C6 × 8.2	2.40	6.00	0.200	1.920	0.343	13.1	4.38	0.693	0.492	0.511
C5 × 9	2.64	5.00	0.325	1.885	0.320	8.90	3.56	0.632	0.450	0.478
C5 × 6.7	1.97	5.00	0.190	1.750	0.320	7.49	3.00	0.479	0.378	0.484
C4 × 7.25	2.13	4.00	0.321	1.721	0.296	4.59	2.29	0.433	0.343	0.459
C4 × 5.4	1.59	4.00	0.184	1.584	0.296	3.85	1.93	0.319	0.283	0.457
C3 × 6	1.76	3.00	0.356	1.596	0.273	2.07	1.38	0.305	0.268	0.455
C3 × 4.1	1.21	3.00	0.170	1.410	0.273	1.66	1.10	0.197	0.202	0.436

*Data taken from a variety of sources. Sizes listed represent a small sample of the sizes available.

A-7 PROPERTIES OF WIDE FLANGE SHAPES
W-Shapes*

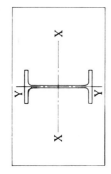

Designation	Area (in²)	Depth (in.)	Web thickness (in.)	Flange Width (in.)	Flange Thickness (in.)	Axis X-X I (in⁴)	Axis X-X S (in³)	Axis Y-Y I (in⁴)	Axis Y-Y S (in³)
W24 × 76	22.4	23.92	0.440	8.990	0.680	2100	176	82.5	18.4
W24 × 68	20.1	23.73	0.415	8.965	0.585	1830	154	70.4	15.7
W21 × 73	21.5	21.24	0.455	8.295	0.740	1600	151	70.6	17.0
W21 × 57	16.7	21.06	0.405	6.555	0.650	1170	111	30.6	9.35
W18 × 55	16.2	18.11	0.390	7.530	0.630	890	98.3	44.9	11.9
W18 × 40	11.8	17.90	0.315	6.015	0.525	612	68.4	19.1	6.35
W14 × 43	12.6	13.66	0.305	7.995	0.530	428	62.7	45.2	11.3
W14 × 26	7.69	13.91	0.255	5.025	0.420	245	35.3	8.91	3.54
W12 × 30	8.79	12.34	0.260	6.520	0.440	238	38.6	20.3	6.24
W12 × 16	4.71	11.99	0.220	3.990	0.265	103	17.1	2.82	1.41
W10 × 15	4.41	9.99	0.230	4.000	0.270	69.8	13.8	2.89	1.45
W10 × 12	3.54	9.87	0.190	3.960	0.210	53.8	10.9	2.18	1.10
W8 × 15	4.44	8.11	0.245	4.015	0.315	48.0	11.8	3.41	1.70
W8 × 10	2.96	7.89	0.170	3.940	0.205	30.8	7.81	2.09	1.06
W6 × 15	4.43	5.99	0.230	5.990	0.260	29.1	9.72	9.32	3.11
W6 × 12	3.55	6.03	0.230	4.000	0.280	22.1	7.31	2.99	1.50
W5 × 19	5.54	5.15	0.270	5.030	0.430	26.2	10.2	9.13	3.63
W5 × 16	4.68	5.01	0.240	5.000	0.360	21.3	8.51	7.51	3.00
W4 × 13	3.83	4.16	0.280	4.060	0.345	11.3	5.46	3.86	1.90

* Data taken from a variety of sources. Sizes listed represent a small sample of the sizes available.

A-8 PROPERTIES OF AMERICAN STANDARD BEAMS
S-Shapes*

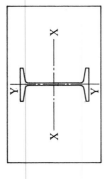

Designation	Area (in²)	Depth (in.)	Web thickness (in.)	Flange Width (in.)	Flange Thickness (in.)	Axis X-X I (in⁴)	Axis X-X S (in³)	Axis Y-Y I (in⁴)	Axis Y-Y S (in³)
S24 × 90	26.5	24.00	0.625	7.125	0.870	2250	187	44.9	12.6
S20 × 96	28.2	20.30	0.800	7.200	0.920	1670	165	50.2	13.9
S20 × 75	22.0	20.00	0.635	6.385	0.795	1280	128	29.8	9.32
S20 × 66	19.4	20.00	0.505	6.255	0.795	1190	119	27.7	8.85
S18 × 70	20.6	18.00	0.711	6.251	0.691	926	103	24.1	7.72
S15 × 50	14.7	15.00	0.550	5.640	0.622	486	64.8	15.7	5.57
S12 × 50	14.7	12.00	0.687	5.477	0.659	305	50.8	15.7	5.74
S12 × 35	10.3	12.00	0.428	5.078	0.544	229	38.2	9.87	3.89
S10 × 35	10.3	10.00	0.594	4.944	0.491	147	29.4	8.36	3.38
S10 × 25.4	7.46	10.00	0.311	4.661	0.491	124	24.7	6.79	2.91
S8 × 23	6.77	8.00	0.441	4.171	0.426	64.9	16.2	4.31	2.07
S8 × 18.4	5.41	8.00	0.271	4.001	0.426	57.6	14.4	3.73	1.86
S7 × 20	5.88	7.00	0.450	3.860	0.392	42.4	12.1	3.17	1.64
S6 × 12.5	3.67	6.00	0.232	3.332	0.359	22.1	7.37	1.82	1.09
S5 × 10	2.94	5.00	0.214	3.004	0.326	12.3	4.92	1.22	0.809
S4 × 7.7	2.26	4.00	0.193	2.663	0.293	6.08	3.04	0.764	0.574
S3 × 5.7	1.67	3.00	0.170	2.330	0.260	2.52	1.68	0.455	0.390

*Data taken from a variety of sources. Sizes listed represent a small sample of the sizes available.

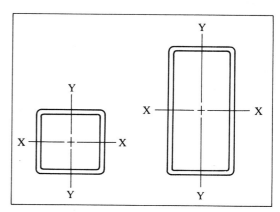

A-9 PROPERTIES OF STRUCTURAL TUBING
SQUARE AND RECTANGULAR*

Size	Area (in²)	Weight per foot (lb)	Axis X-X			Axis Y-Y		
			I (in⁴)	S (in³)	r (in.)	I (in⁴)	S (in³)	r (in.)
$8 \times 8 \times \frac{1}{2}$	14.4	48.9	131	32.9	3.03	131	32.9	3.03
$8 \times 8 \times \frac{1}{4}$	7.59	25.8	75.1	18.8	3.15	75.1	18.8	3.15
$8 \times 4 \times \frac{1}{2}$	10.4	35.2	75.1	18.8	2.69	24.6	12.3	1.54
$8 \times 4 \times \frac{1}{4}$	5.59	19.0	45.1	11.3	2.84	15.3	7.63	1.65
$8 \times 2 \times \frac{1}{4}$	4.59	15.6	30.1	7.52	2.56	3.08	3.08	0.819
$6 \times 6 \times \frac{1}{2}$	10.4	35.2	50.5	16.8	2.21	50.5	16.8	2.21
$6 \times 6 \times \frac{1}{4}$	5.59	19.0	30.3	10.1	2.33	30.3	10.1	2.33
$6 \times 4 \times \frac{1}{4}$	4.59	15.6	22.1	7.36	2.19	11.7	5.87	1.60
$6 \times 2 \times \frac{1}{4}$	3.59	12.2	13.8	4.60	1.96	2.31	2.31	0.802
$4 \times 4 \times \frac{1}{2}$	6.36	21.6	12.3	6.13	1.39	12.3	6.13	1.39
$4 \times 4 \times \frac{1}{4}$	3.59	12.2	8.22	4.11	1.51	8.22	4.11	1.51
$4 \times 2 \times \frac{1}{4}$	2.59	8.81	4.69	2.35	1.35	1.54	1.54	0.770
$3 \times 3 \times \frac{1}{4}$	2.59	8.81	3.16	2.10	1.10	3.16	2.10	1.10
$3 \times 2 \times \frac{1}{4}$	2.09	7.11	2.21	1.47	1.03	1.15	1.15	0.742
$2 \times 2 \times \frac{1}{4}$	1.59	5.41	0.766	0.766	0.694	0.766	0.766	0.694

*Data taken from a variety of sources. Sizes listed represent a small sample of the sizes available.

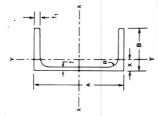

A-10 ALUMINUM ASSOCIATION STANDARD CHANNELS: DIMENSIONS, AREAS, WEIGHTS, AND SECTION PROPERTIES

| Size | | Area (in³) | Weight (lb/ft) | Flange thickness, t_1 (in.) | Web thickness, t (in.) | Fillet radius, R (in.) | Section properties | | | | | | |
| Depth, A (in.) | Width, B (in.) | | | | | | Axis X-X | | | Axis Y-Y | | | |
							I (in⁴)	S (in³)	r (in.)	I (in⁴)	S (in³)	r (in.)	x (in.)
2.00	1.00	0.491	0.577	0.13	0.13	0.10	0.288	0.288	0.766	0.045	0.064	0.303	0.298
2.00	1.25	0.911	1.071	0.26	0.17	0.15	0.546	0.546	0.774	0.139	0.178	0.391	0.471
3.00	1.50	0.965	1.135	0.20	0.13	0.25	1.41	0.94	1.21	0.22	0.22	0.47	0.49
3.00	1.75	1.358	1.597	0.26	0.17	0.25	1.97	1.31	1.20	0.42	0.37	0.55	0.62
4.00	2.00	1.478	1.738	0.23	0.15	0.25	3.91	1.95	1.63	0.60	0.45	0.64	0.65
4.00	2.25	1.982	2.331	0.29	0.19	0.25	5.21	2.60	1.62	1.02	0.69	0.72	0.78
5.00	2.25	1.881	2.212	0.26	0.15	0.30	7.88	3.15	2.05	0.98	0.64	0.72	0.73
5.00	2.75	2.627	3.089	0.32	0.19	0.30	11.14	4.45	2.06	2.05	1.14	0.88	0.95
6.00	2.50	2.410	2.834	0.29	0.17	0.30	14.35	4.78	2.44	1.53	0.90	0.80	0.79
6.00	3.25	3.427	4.030	0.35	0.21	0.30	21.04	7.01	2.48	3.76	1.76	1.05	1.12
7.00	2.75	2.725	3.205	0.29	0.17	0.30	22.09	6.31	2.85	2.10	1.10	0.88	0.84
7.00	3.50	4.009	4.715	0.38	0.21	0.30	33.79	9.65	2.90	5.13	2.23	1.13	1.20
8.00	3.00	3.526	4.147	0.35	0.19	0.30	37.40	9.35	3.26	3.25	1.57	0.96	0.93
8.00	3.75	4.923	5.789	0.41	0.25	0.35	52.69	13.17	3.27	7.13	2.82	1.20	1.22
9.00	3.25	4.237	4.983	0.35	0.23	0.35	54.41	12.09	3.58	4.40	1.89	1.02	0.93
9.00	4.00	5.927	6.970	0.44	0.29	0.35	78.31	17.40	3.63	9.61	3.49	1.27	1.25
10.00	3.50	5.218	6.136	0.41	0.25	0.35	83.22	16.64	3.99	6.33	2.56	1.10	1.02
10.00	4.25	7.109	8.360	0.50	0.31	0.40	116.15	23.23	4.04	13.02	4.47	1.35	1.34
12.00	4.00	7.036	8.274	0.47	0.29	0.40	159.76	26.63	4.77	11.03	3.86	1.25	1.14
12.00	5.00	10.053	11.822	0.62	0.35	0.45	239.69	39.95	4.88	25.74	7.60	1.60	1.61

Source: Aluminum Association, *Aluminum Standards and Data*, 8th ed., Washington, D.C., ©1984, p. 184.

A-11 ALUMINUM ASSOCIATION STANDARD I-BEAMS: DIMENSIONS, AREAS, WEIGHTS, AND SECTION PROPERTIES

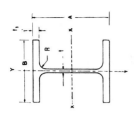

| Size | | Area* | Weight[†] | Flange thickness, t_1 | Web thickness, t | Fillet radius, R | Section properties[‡] | | | | | |
| Depth, A (in.) | Width B (in.) | (in.²) | (lb/ft) | (in.) | (in.) | (in.) | Axis X-X | | | Axis Y-Y | | |
							I (in.⁴)	S (in.³)	r (in.)	I (in.⁴)	S (in.³)	r (in.)
3.00	2.50	1.392	1.637	0.20	0.13	0.25	2.24	1.49	1.27	0.52	0.42	0.61
3.00	2.50	1.726	2.030	0.26	0.15	0.25	2.71	1.81	1.25	0.68	0.54	0.63
4.00	3.00	1.965	2.311	0.23	0.15	0.25	5.62	2.81	1.69	1.04	0.69	0.73
4.00	3.00	2.375	2.793	0.29	0.17	0.25	6.71	3.36	1.68	1.31	0.87	0.74
5.00	3.50	3.146	3.700	0.32	0.19	0.30	13.94	5.58	2.11	2.29	1.31	0.85
6.00	4.00	3.427	4.030	0.29	0.19	0.30	21.99	7.33	2.53	3.10	1.55	0.95
6.00	4.00	3.990	4.692	0.35	0.21	0.30	25.50	8.50	2.53	3.74	1.87	0.97
7.00	4.50	4.932	5.800	0.38	0.23	0.30	42.89	12.25	2.95	5.78	2.57	1.08
8.00	5.00	5.256	6.181	0.35	0.23	0.30	59.69	14.92	3.37	7.30	2.92	1.18
8.00	5.00	5.972	7.023	0.41	0.25	0.30	67.78	16.94	3.37	8.55	3.42	1.20
9.00	5.50	7.110	8.361	0.44	0.27	0.30	102.02	22.67	3.79	12.22	4.44	1.31
10.00	6.00	7.352	8.646	0.41	0.25	0.40	132.09	26.42	4.24	14.78	4.93	1.42
10.00	6.00	8.747	10.286	0.50	0.29	0.40	155.79	31.16	4.22	18.03	6.01	1.44
12.00	7.00	9.925	11.672	0.47	0.29	0.40	255.57	42.60	5.07	26.90	7.69	1.65
12.00	7.00	12.153	14.292	0.62	0.31	0.40	317.33	52.89	5.11	35.48	10.14	1.71

Source: Aluminum Association, *Aluminum Standards and Data*, 8th ed., Washington, D.C., © 1984, p. 184.

* Areas listed are based on nominal dimensions.

[†] Weights per foot are based on nominal dimensions and a density of 0.098 pound per cubic inch which is the density of alloy 6061.

[‡] I = moment of inertia; S = section modulus; r = radius of gyration.

A-12 PROPERTIES OF AMERICAN NATIONAL STANDARD SCHEDULE 40 WELDED AND SEAMLESS WROUGHT STEEL PIPE

Nominal	Diameter (in.) Actual inside	Diameter (in.) Actual outside	Wall thickness (in.)	Cross-sectional area of metal (in²)	Moment of inertia, I (in⁴)	Radius of gyration (in.)	Section modulus, S (in³)	Polar section modulus, Z_p (in³)
$\frac{1}{8}$	0.269	0.405	0.068	0.072	0.00106	0.122	0.00525	0.01050
$\frac{1}{4}$	0.364	0.540	0.088	0.125	0.00331	0.163	0.01227	0.02454
$\frac{3}{8}$	0.493	0.675	0.091	0.167	0.00729	0.209	0.02160	0.04320
$\frac{1}{2}$	0.622	0.840	0.109	0.250	0.01709	0.261	0.04070	0.08140
$\frac{3}{4}$	0.824	1.050	0.113	0.333	0.03704	0.334	0.07055	0.1411
1	1.049	1.315	0.133	0.494	0.08734	0.421	0.1328	0.2656
$1\frac{1}{4}$	1.380	1.660	0.140	0.669	0.1947	0.539	0.2346	0.4692
$1\frac{1}{2}$	1.610	1.900	0.145	0.799	0.3099	0.623	0.3262	0.6524
2	2.067	2.375	0.154	1.075	0.6658	0.787	0.5607	1.121
$2\frac{1}{2}$	2.469	2.875	0.203	1.704	1.530	0.947	1.064	2.128
3	3.068	3.500	0.216	2.228	3.017	1.163	1.724	3.448
$3\frac{1}{2}$	3.548	4.000	0.226	2.680	4.788	1.337	2.394	4.788
4	4.026	4.500	0.237	3.174	7.233	1.510	3.215	6.430
5	5.047	5.563	0.258	4.300	15.16	1.878	5.451	10.90
6	6.065	6.625	0.280	5.581	28.14	2.245	8.496	16.99
8	7.981	8.625	0.322	8.399	72.49	2.938	16.81	33.62
10	10.020	10.750	0.365	11.91	160.7	3.674	29.91	59.82
12	11.938	12.750	0.406	15.74	300.2	4.364	47.09	94.18
16	15.000	16.000	0.500	24.35	732.0	5.484	91.50	183.0
18	16.876	18.000	0.562	30.79	1172	6.168	130.2	260.4

A-13 TYPICAL PROPERTIES OF CARBON AND ALLOY STEELS*

Material AISI No.	Condition†	Ultimate strength, s_u ksi	MPa	Yield strength, s_y ksi	MPa	Percent elongation
1020	Hot-rolled	55	379	30	207	25
1020	Cold-drawn	61	421	51	352	15
1040	Hot-rolled	76	524	42	290	18
1040	Cold-drawn	85	586	71	490	12
1040	Annealed	75	517	51	352	30
1040	WQT 700	127	876	93	641	19
1040	WQT 900	118	814	90	621	22
1040	WQT 1100	113	779	80	552	24
1040	WQT 1300	87	600	63	434	32
1080	Annealed	89	614	54	372	25
1080	OQT 700	189	1303	141	972	12
1080	OQT 900	179	1234	129	889	13
1080	OQT 1100	145	1000	103	710	17
1080	OQT 1300	117	807	70	483	23
1141	Annealed	87	600	51	352	26
1141	OQT 700	193	1331	172	1186	9
1141	OQT 900	146	1007	129	889	15
1141	OQT 1100	116	800	97	669	20
1141	OQT 1300	94	648	68	469	28
4140	Annealed	95	655	60	414	26
4140	OQT 700	231	1593	212	1462	12
4140	OQT 900	187	1289	173	1193	15
4140	OQT 1100	147	1014	131	903	18
4140	OQT 1300	118	814	101	696	23
5160	Annealed	105	724	40	276	17
5160	OQT 700	263	1813	238	1641	9
5160	OQT 900	196	1351	179	1234	12
5160	OQT 1100	149	1027	132	910	17
5160	OQT 1300	115	793	103	710	23

* Other properties approximately the same for all carbon and alloy steels:
    Modulus of elasticity in tension = 30 000 000 psi (207 GPa)
    Modulus of elasticity in shear   = 11 500 000 psi (80 GPa)
    Density = 0.283 lb/in^3 (7680 kg/m^3)

† OQT means oil-quenched and tempered. WQT means water-quenched and tempered.

A-14 TYPICAL PROPERTIES OF STAINLESS STEELS AND NONFERROUS METALS

Material and condition	Ultimate strength, s_u		Yield strength, s_y		Percent elonga-tion	Density		Modulus of elasticity, E	
	ksi	MPa	ksi	MPa		lb/in.³	kg/m³	psi × 10⁻⁶	GPa
Stainless steels									
AISI 301 annealed	110	758	40	276	60	0.290	8030	28	193
AISI 301 cold-worked	185	1280	140	965	9	0.290	8030	28	193
AISI 430 annealed	75	517	40	276	25	0.280	7750	29	200
AISI 430 cold-worked	90	621	80	552	15	0.280	7750	29	200
AISI 501 annealed	70	483	30	207	28	0.280	7750	29	200
AISI 501 OQT 1000	175	1210	135	931	15	0.280	7750	29	200
17-4PH H900	210	1450	185	1280	14	0.281	7780	28.5	197
PH 13-8 Mo H1000	215	1480	205	1410	13	0.279	7720	29.4	203
Copper and its alloys									
C14500 copper, hard	48	331	44	303	10	0.323	8940	17	117
C17200 beryllium copper, hard	175	1210	150	1030	5	0.298	8250	19	131
C22000 bronze, hard	61	421	54	372	5	0.318	8800	17	117
C26000 brass, hard	76	524	63	434	8	0.308	8530	16	110
Magnesium—cast									
ASTM AZ 63A-T6	40	276	19	131	5	0.066	1830	6.5	45
Zinc—cast-ZA 12									
ASTM B669-80	47	324	31	214	3	.218	6030	12	83
Titanium alloy									
Ti-6A1-4V, aged	170	1170	155	1070	8	0.160	4430	16.5	114

A-15 PROPERTIES OF STRUCTURAL STEELS

Material ASTM no. and products	Ultimate strength, s_u*		Yield strength, s_y*		Percent elongation in 2 in.
	ksi	MPa	ksi	MPa	
A36—Carbon steel shapes, plates, and bars	58	400	36	248	21
A242, 441—High-strength low-alloy shapes, plates, and bars					
$\leq \frac{3}{4}$ in. thick	70	483	50	345	21
$\frac{3}{4}$ to $1\frac{1}{2}$ in. thick	67	462	46	317	21
$1\frac{1}{2}$ to 4 in. thick	63	434	42	290	21
A500—Cold-formed structural tubing					
Round, grade A	45	310	33	228	25
Round, grade B	58	400	42	290	23
Round, grade C	62	427	46	317	21
Shaped, grade A	45	310	39	269	25
Shaped, grade B	58	400	46	317	23
Shaped, grade C	62	427	50	345	21
A501—Hot-formed structural tubing, round or shaped	58	400	36	248	23
A514—High-yield strength, quenched and tempered alloy steel plate					
$\frac{3}{4}$ to $2\frac{1}{2}$ in. thick	110	758	100	690	18
$2\frac{1}{2}$ to 6 in. thick	100	690	90	620	16
A572—High-strength low-alloy columbium-vanadium steel shapes, plates, and bars					
Grade 42	60	414	42	290	24
Grade 50	65	448	50	345	21
Grade 60	75	517	60	414	18
Grade 65	80	552	65	448	17

*Minimum values; may range higher.
The American Institute of Steel Construction specifies $E = 29 \times 10^6$ psi (200 GPa) for structural steel.

A-16 TYPICAL PROPERTIES OF CAST IRON*

Material type and grade	s_u† ksi	s_u† MPa	s_{uc}‡ ksi	s_{uc}‡ MPa	s_{us} ksi	s_{us} MPa	s_{yt}‡ ksi	s_{yt}‡ MPa	psi × 10^{-6}	GPa	Percent elongation
Gray iron ASTM A48											
Grade 20	20	138	80	552	32	221	—	—	12.2	84	< 1
Grade 40	40	276	140	965	57	393	—	—	19.4	134	< 0.8
Grade 60	55	379	170	1170	72	496	—	—	21.5	148	< 0.5
Ductile iron ASTM A536											
60-40-18	60	414	—	—	57	393	40	276	24	165	18
80-55- 6	80	552	—	—	73	503	55	379	24	165	6
100-70- 3	100	690	—	—	—	—	70	483	24	165	3
120-90- 2	120	827	180	1240	—	—	90	621	23	159	2
Austempered ductile iron (ADI)											
Grade 1	125	862	—	—	—	—	85	586	24	165	10
Grade 2	150	1034	—	—	—	—	100	690	24	165	7
Grade 3	175	1207	—	—	—	—	120	827	24	165	4
Grade 4	200	1379	—	—	—	—	140	965	24	165	2
Malleable iron ASTM A220											
45008	65	448	240	1650	49	338	45	310	26	170	10
60004	80	552	240	1650	65	448	60	414	27	186	4
80002	95	655	240	1650	75	517	80	552	27	186	2

* The density of cast iron ranges from 0.25 to 0.27 lb/in³ (6920 to 7480 kg/m³).

† Minimum values; may range higher.

‡ Approximate values; may range higher or lower by about 15%.

A-17 TYPICAL PROPERTIES OF ALUMINUM ALLOYS*

Alloy and temper	Ultimate strength, s_u		Yield strength, s_y		Percent elonga- tion	Shear strength, s_{us}	
	ksi	MPa	ksi	MPa		ksi	MPa
1100-H12	16	110	15	103	25	10	69
1100-H18	24	165	22	152	15	13	90
2014-0	27	186	14	97	18	18	124
2014-T4	62	427	42	290	20	38	262
2014-T6	70	483	60	414	13	42	290
3003-0	16	110	6	41	40	11	76
3003-H12	19	131	18	124	20	12	83
3003-H18	29	200	27	186	10	16	110
5154-0	35	241	17	117	27	22	152
5154-H32	39	269	30	207	15	22	152
5154-H38	48	331	39	269	10	28	193
6061-0	18	124	8	55	30	12	83
6061-T4	35	241	21	145	25	24	165
6061-T6	45	310	40	276	17	30	207
7075-0	33	228	15	103	16	22	152
7075-T6	83	572	73	503	11	48	331
Casting alloys (permanent mold castings)							
204.0-T4	48	331	29	200	8	—	—
356.0-T6	33	228	22	152	3	—	—

*Modulus of elasticity E for most aluminum alloys, including 1100, 3003, 6061, and 6063, is 10×10^6 psi (69 GPa). For 2014, $E = 10.6 \times 10^6$ psi (73 GPa). For 5154, $E = 10.2 \times 10^6$ psi (70 GPa). For 7075, $E = 10.4 \times 10^6$ psi (72 GPa). Density of most aluminum alloys is approximately 0.10 lb/in³ (2770 kg/m³).

A-18 TYPICAL PROPERTIES OF WOOD

	Allowable stress												
	Bending		Tension parallel to grain		Horizontal shear		Compression				Modulus of elasticity	
							Perpendicular to grain		Parallel to grain			
Type and grade	psi	MPa	psi	MPa	psi	MPa	psi	MPa	psi	MPa	ksi	GPa
Douglas fir—2 to 4 in. thick, 6 in. and wider												
No. 1	1500	10.3	1000	6.9	95	0.66	385	2.65	1250	8.62	1900	13.1
No. 2	1250	8.6	825	5.7	95	0.66	385	2.65	1050	7.24	1700	11.7
No. 3	725	5.0	475	3.3	95	0.66	385	2.65	675	4.65	1500	10.3
Hemlock—2 to 4 in. thick, 6 in. and wider												
No. 1	1200	8.3	800	5.5	75	0.52	245	1.69	1000	6.90	1500	10.3
No. 2	1000	6.9	650	4.5	75	0.52	245	1.69	850	5.86	1400	9.7
No. 3	575	4.0	375	2.6	75	0.52	245	1.69	550	3.79	1200	8.3
Southern pine—2$\frac{1}{2}$ to 4 in. thick, 6 in. and wider												
No. 1	1200	8.3	800	5.5	80	0.55	270	1.86	825	5.69	1600	11.0
No. 2	850	5.9	550	3.8	70	0.48	230	1.59	600	4.14	1300	9.0
No. 3	575	4.0	375	2.6	70	0.48	230	1.59	425	2.93	1300	9.0

A-19 TYPICAL PROPERTIES OF SELECTED PLASTICS

Type*	Tensile strength		Flexural strength		Tensile modulus		Density (lb/in³)
	ksi	MPa	ksi	MPa	ksi	GPa	
ABS	7	48	11	76	360	2.5	0.038
Acetal copolymer	16	110	24	165	1200	8.3	0.057
Epoxy, molded	10	69	21	145	2000	13.8	0.069
Nylon 6/6	—	—	35	241	1800	12.4	0.053
Phenolic	8	55	27	186	3150	21.7	0.066
Polycarbonate	19	131	23	159	1250	8.6	0.049
Polyester PET	22	152	31	214	1700	11.7	0.059
Polyimide	28	193	56	386	4500	31.0	0.069
Polystyrene	14	97	17	117	1200	8.3	0.047
Polypropylene	8	55	10	69	800	5.5	0.041

* Reinforced with glass or other fibers.

A-20 DESIGN FACTORS

Manner of loading	Design factor, N		Brittle metals
	Ductile metals		
	Yield strength basis	Ultimate strength basis	Ultimate strength basis
Static	2	—	6
Repeated	—	8	10
Impact or shock	—	12	15

A-21-1 AXIALLY LOADED GROOVED ROUND BAR IN TENSION

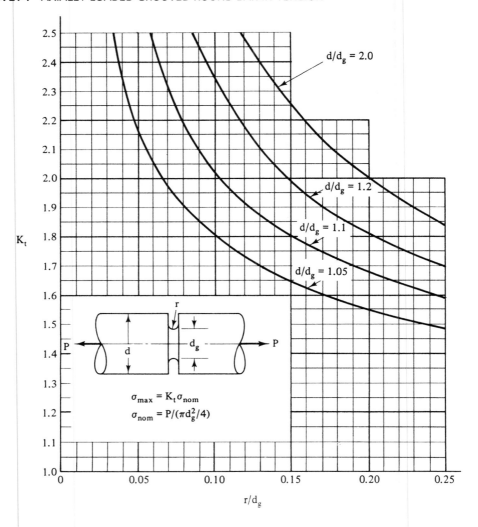

K_t

$d/d_g = 2.0$

$d/d_g = 1.2$

$d/d_g = 1.1$

$d/d_g = 1.05$

$\sigma_{max} = K_t \sigma_{nom}$

$\sigma_{nom} = P/(\pi d_g^2/4)$

r/d_g

A-21-2 AXIALLY LOADED STEPPED ROUND BAR

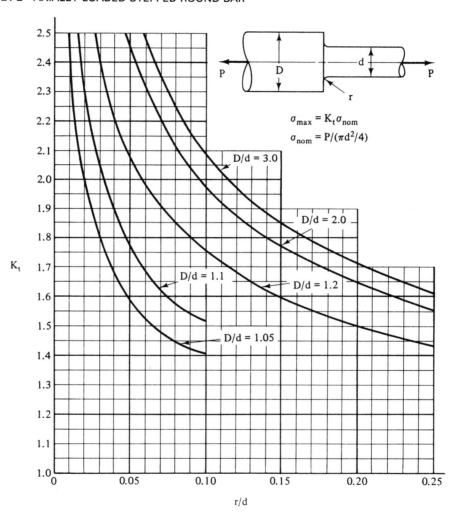

A-21-3 GROOVED ROUND BAR IN TORSION

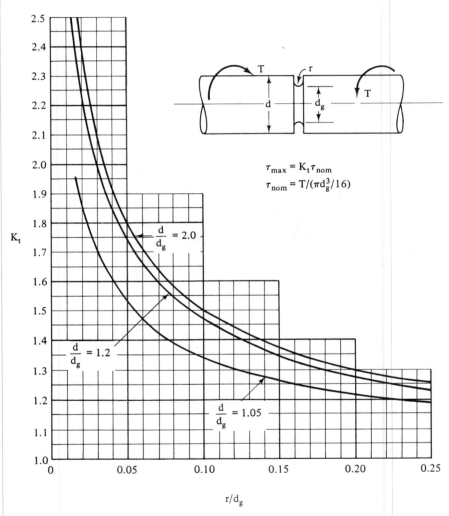

$$\tau_{max} = K_t \tau_{nom}$$
$$\tau_{nom} = T/(\pi d_g^3/16)$$

A-21-4 STEPPED ROUND BAR IN TORSION

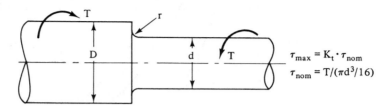

$$\tau_{max} = K_t \cdot \tau_{nom}$$
$$\tau_{nom} = T/(\pi d^3/16)$$

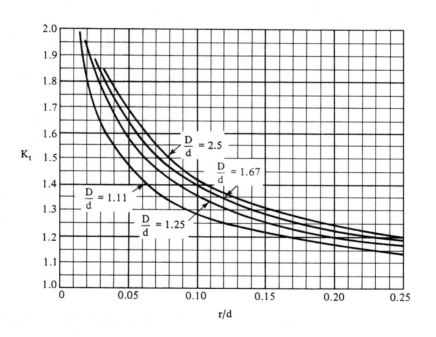

A-21-5 GROOVED ROUND BAR IN BENDING

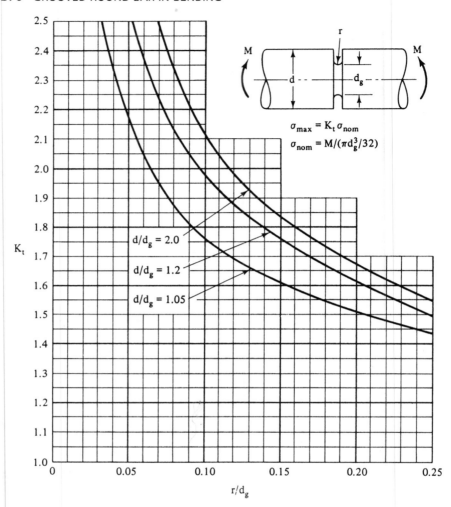

$$\sigma_{max} = K_t \sigma_{nom}$$

$$\sigma_{nom} = M/(\pi d_g^3/32)$$

$d/d_g = 2.0$

$d/d_g = 1.2$

$d/d_g = 1.05$

A-21-6 SHAFTS WITH KEYSEATS—BENDING AND TORSION

Type of key seat	K_t*
Sled-runner	1.6
Profile	2.0

*K_t is to be applied to the stress computed for the full nominal diameter of the shaft where the key seat is located.

A-21-7 STEPPED ROUND BAR IN BENDING

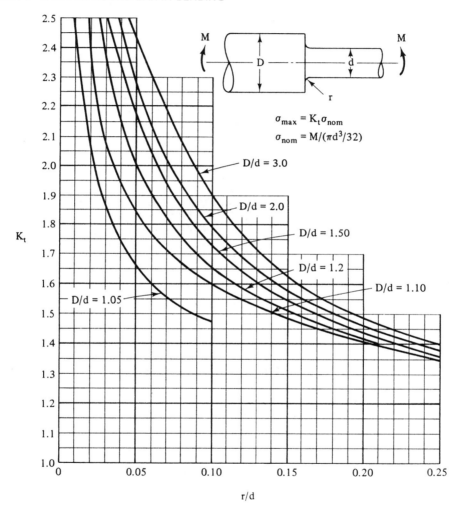

$\sigma_{max} = K_t \sigma_{nom}$

$\sigma_{nom} = M/(\pi d^3/32)$

D/d = 3.0

D/d = 2.0

D/d = 1.50

D/d = 1.2

D/d = 1.10

D/d = 1.05

K_t

r/d

A-21-8 ROUND BAR WITH TRANSVERSE HOLE IN TENSION, BENDING, AND TORSION

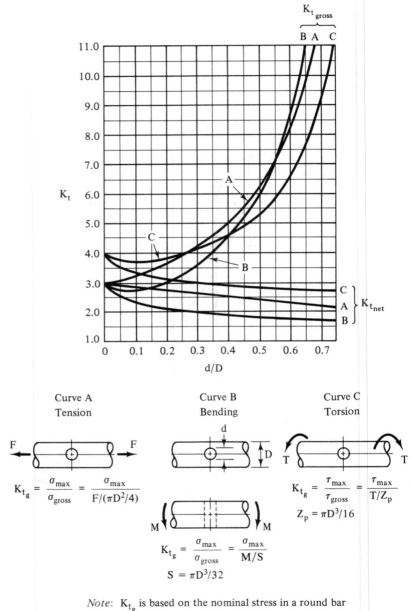

Curve A
Tension

$$K_{t_g} = \frac{\sigma_{max}}{\sigma_{gross}} = \frac{\sigma_{max}}{F/(\pi D^2/4)}$$

Curve B
Bending

$$K_{t_g} = \frac{\sigma_{max}}{\sigma_{gross}} = \frac{\sigma_{max}}{M/S}$$

$$S = \pi D^3/32$$

Curve C
Torsion

$$K_{t_g} = \frac{\tau_{max}}{\tau_{gross}} = \frac{\tau_{max}}{T/Z_p}$$

$$Z_p = \pi D^3/16$$

Note: K_{t_g} is based on the nominal stress in a round bar without a hole (gross section).

$K_{t_{net}}$ is based on the nominal stress for the net section.

A-21-9 FLAT PLATE WITH CENTRAL HOLE IN TENSION AND BENDING

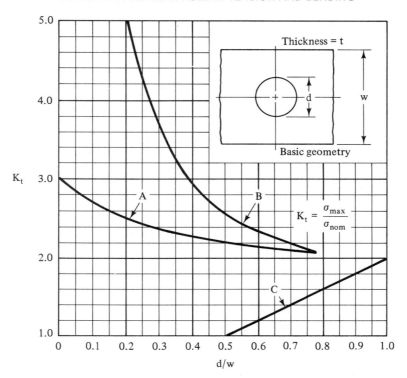

Curve A
Direct tension
on plate

$$\sigma_{nom} = \frac{F}{(w - d)t}$$

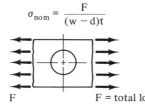

F   F = total load

Curve B
Tension-load
applied through
a pin in the hole

$$\sigma_{nom} = \frac{F}{(w - d)t}$$

Curve C
Bending in
the plane
of the plate

$$\sigma_{nom} = \frac{Mc}{I_{net}} = \frac{6Mw}{(w^3 - d^3)t}$$

Note: $K_t = 1.0$ for $d/w < 0.5$

A-21-10 STEPPED FLAT PLATE IN TENSION AND BENDING

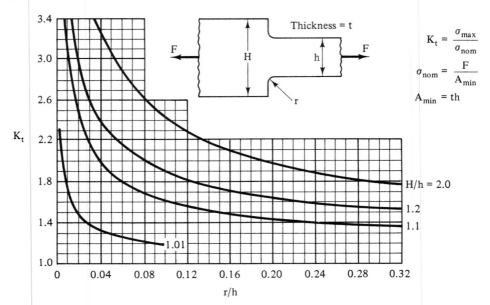

$$K_t = \frac{\sigma_{max}}{\sigma_{nom}}$$

$$\sigma_{nom} = \frac{F}{A_{min}}$$

$$A_{min} = th$$

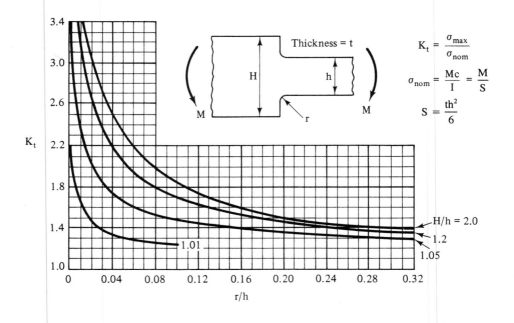

$$K_t = \frac{\sigma_{max}}{\sigma_{nom}}$$

$$\sigma_{nom} = \frac{Mc}{I} = \frac{M}{S}$$

$$S = \frac{th^2}{6}$$

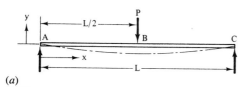

(a)

$$y_B = y_{max} = \frac{-PL^3}{48EI} \text{ at center}$$

Between A and B:

$$y = \frac{-Px}{48EI} (3L^2 - 4x^2)$$

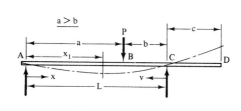

(b)

$a > b$

$$y_{max} = \frac{-Pab(L + b)\sqrt{3a(L + b)}}{27EIL}$$

at $x_1 = \sqrt{a(L + b)/3}$

$$y_B = \frac{-Pa^2b^2}{3EIL} \text{ at load}$$

Between A and B: (the longer segment)

$$y = \frac{-Pbx}{6EIL} (L^2 - b^2 - x^2)$$

Between B and C: (the shorter segment)

$$y = \frac{-Pav}{6EIL} (L^2 - v^2 - a^2)$$

At end of overhang at D:

$$y_D = \frac{Pabc}{6EIL} (L + a)$$

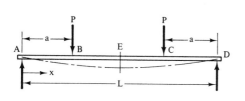

(c)

$$y_E = y_{max} = \frac{-Pa}{24EI} (3L^2 - 4a^2) \text{ at center}$$

$$y_B = y_C = \frac{-Pa^2}{6EI} (3L - 4a) \text{ at loads}$$

Between A and B:

$$y = \frac{-Px}{6EI} (3aL - 3a^2 - x^2)$$

Between B and C:

$$y = \frac{-Pa}{6EI} (3Lx - 3x^2 - a^2)$$

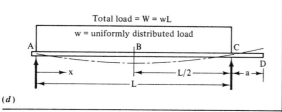

Total load = W = wL

w = uniformly distributed load

(d)

$$y_B = y_{max} = \frac{-5wL^4}{384EI} = \frac{5WL^3}{384EI} \text{ at center}$$

Between A and B:

$$y = \frac{-wx}{24EI} (L^3 - 2Lx^2 + x^3)$$

At D at end:

$$y_D = \frac{wL^3a}{24EI}$$

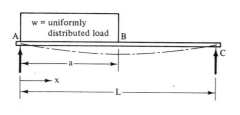

w = uniformly distributed load

(e)

Between A and B:

$$y = \frac{-wx}{24EIL} [a^2(2L - a)^2 - 2ax^2(2L - a) + Lx^3]$$

Between B and C:

$$y = \frac{-wa^2(L - x)}{24EI} (4Lx - 2x^2 - a^2)$$

Appendix

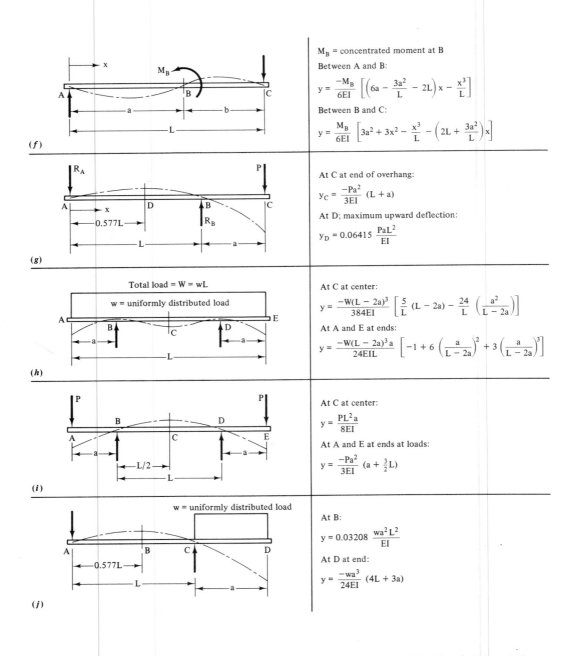

(f)

M_B = concentrated moment at B

Between A and B:

$$y = \frac{-M_B}{6EI}\left[\left(6a - \frac{3a^2}{L} - 2L\right)x - \frac{x^3}{L}\right]$$

Between B and C:

$$y = \frac{M_B}{6EI}\left[3a^2 + 3x^2 - \frac{x^3}{L} - \left(2L + \frac{3a^2}{L}\right)x\right]$$

(g)

At C at end of overhang:

$$y_C = \frac{-Pa^2}{3EI}(L + a)$$

At D; maximum upward deflection:

$$y_D = 0.06415\,\frac{PaL^2}{EI}$$

(h)

Total load = W = wL

w = uniformly distributed load

At C at center:

$$y = \frac{-W(L - 2a)^3}{384EI}\left[\frac{5}{L}(L - 2a) - \frac{24}{L}\left(\frac{a^2}{L - 2a}\right)\right]$$

At A and E at ends:

$$y = \frac{-W(L - 2a)^3 a}{24EIL}\left[-1 + 6\left(\frac{a}{L - 2a}\right)^2 + 3\left(\frac{a}{L - 2a}\right)^3\right]$$

(i)

At C at center:

$$y = \frac{PL^2 a}{8EI}$$

At A and E at ends at loads:

$$y = \frac{-Pa^2}{3EI}\left(a + \frac{3}{2}L\right)$$

(j)

w = uniformly distributed load

At B:

$$y = 0.03208\,\frac{wa^2 L^2}{EI}$$

At D at end:

$$y = \frac{-wa^3}{24EI}(4L + 3a)$$

Source: *Engineering Data for Aluminum Structures* (Washington, D.C.: The Aluminum Association, 1981), pp 63–77.

Appendix

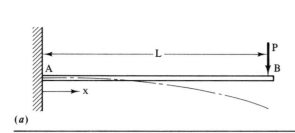

At B at end:

$$y_B = y_{max} = \frac{-PL^3}{3EI}$$

Between A and B:

$$y = \frac{-Px^2}{6EI}(3L - x)$$

(a)

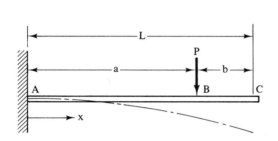

At B at load:

$$y_B = \frac{-Pa^3}{3EI}$$

At C at end:

$$y_C = y_{max} = \frac{-Pa^2}{6EI}(3L - a)$$

Between A and B:

$$y = \frac{-Px^2}{6EI}(3a - x)$$

Between B and C:

$$y = \frac{-Pa^2}{6EI}(3x - a)$$

(b)

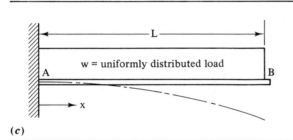

W = total load = wL

At B at end:

$$y_B = y_{max} = \frac{-WL^3}{8EI}$$

Between A and B:

$$y = \frac{-Wx^2}{24EIL}[2L^2 + (2L - x)^2]$$

(c)

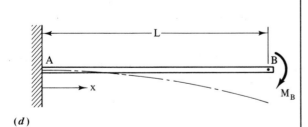

M_B = concentrated moment at end

At B at end:

$$y_B = y_{max} = \frac{-M_B L^2}{2EI}$$

Between A and B:

$$y = \frac{-M_B x^2}{2EI}$$

(d)

Source: *Engineering Data for Aluminum Structures* (Washington, D.C.: The Aluminum Association, 1981), pp 63–77.

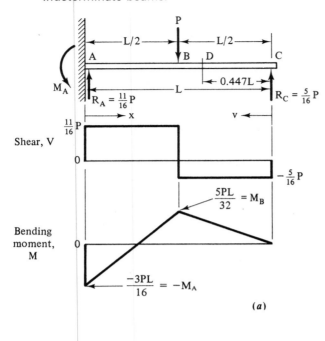

Deflections

At B at load:

$$y_B = \frac{-7}{768} \frac{PL^3}{EI}$$

y_{max} is at v = 0.447L at D:

$$y_D = y_{max} = \frac{-PL^3}{107EI}$$

Between A and B:

$$y = \frac{-Px^2}{96EI} (9L - 11x)$$

Between B and C:

$$y = \frac{-Pv}{96EI} (3L^2 - 5v^2)$$

(a)

Reactions

$$R_A = \frac{Pb}{2L^3} (3L^2 - b^2)$$

$$R_C = \frac{Pa^2}{2L^3} (b + 2L)$$

Moments

$$M_A = \frac{-Pab}{2L^2} (b + L)$$

$$M_B = \frac{Pa^2b}{2L^3} (b + 2L)$$

Deflections

At B at load:

$$y_B = \frac{-Pa^3b^2}{12EIL^3} (3L + b)$$

Between A and B:

$$y = \frac{-Px^2b}{12EIL^3} (3C_1 - C_2x)$$

$$C_1 = aL(L + b); \quad C_2 = (L + a)(L + b) + aL$$

Between B and C:

$$y = \frac{-Pa^2v}{12EIL^3} [3L^2b - v^2(3L - a)]$$

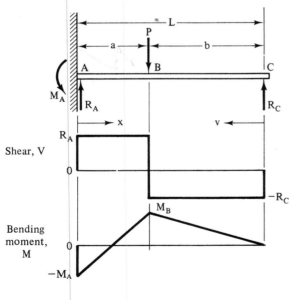

(b)

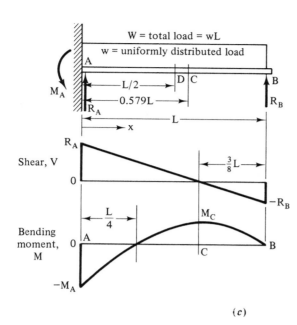

Reactions

$$R_A = \frac{5}{8}W$$

$$R_B = \frac{3}{8}W$$

Moments

$$M_A = -0.125WL$$

$$M_C = 0.0703WL$$

Deflections

At C at x = 0.579L:

$$y_C = y_{max} = \frac{-WL^3}{185EI}$$

At D at center:

$$y_D = \frac{-WL^3}{192EI}$$

Between A and B:

$$y = \frac{-Wx^2(L-x)}{48EIL}(3L-2x)$$

(c)

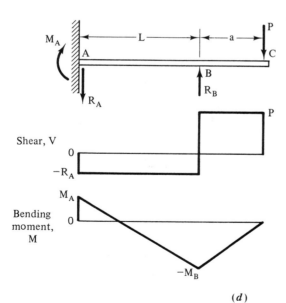

Reactions

$$R_A = \frac{-3Pa}{2L}$$

$$R_B = P\left(1 + \frac{3a}{2L}\right)$$

Moments

$$M_A = \frac{Pa}{2}$$

$$M_B = -Pa$$

Deflection

At C at end:

$$y_C = \frac{-PL^3}{EI}\left(\frac{a^2}{4L^2} + \frac{a^3}{3L^3}\right)$$

(d)

Appendix

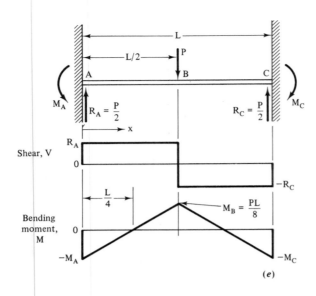

Moments

$$M_A = M_B = M_C = \frac{PL}{8}$$

Deflections

At B at center:

$$y_B = y_{max} = \frac{-PL^3}{192EI}$$

Between A and B:

$$y = \frac{-Px^2}{48EI}(3L - 4x)$$

(e)

Reactions

$$R_A = \frac{Pb^2}{L^3}(3a + b)$$

$$R_C = \frac{Pa^2}{L^3}(3b + a)$$

Moments

$$M_A = \frac{-Pab^2}{L^2}$$

$$M_B = \frac{2Pa^2b^2}{L^3}$$

$$M_C = \frac{-Pa^2b}{L^2}$$

Deflections

At B at load:

$$y_B = \frac{-Pa^3b^3}{3EIL^3}$$

At D at $x_1 = \frac{2aL}{3a + b}$

$$y_D = y_{max} = \frac{-2Pa^3b^2}{3EI(3a + b)^2}$$

Between A and B (longer segment):

$$y = \frac{-Px^2b^2}{6EIL^3}[2a(L - x) + L(a - x)]$$

Between B and C (shorter segment):

$$y = \frac{-Pv^2a^2}{6EIL^3}[2b(L - v) + L(b - v)]$$

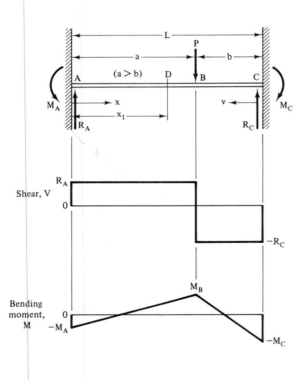

(f)

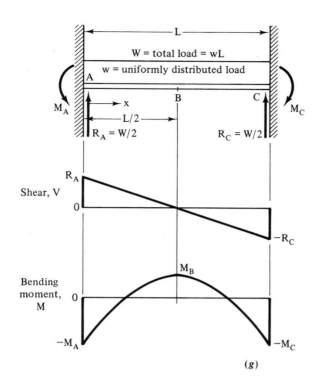

Moments

$$M_A = M_C = \frac{-WL}{12}$$

$$M_B = \frac{WL}{24}$$

Deflections

At B at center:

$$y_B = y_{max} = \frac{-WL^3}{384EI}$$

Between A and C:

$$y = \frac{-wx^2}{24EI}(L - x)^2$$

(g)

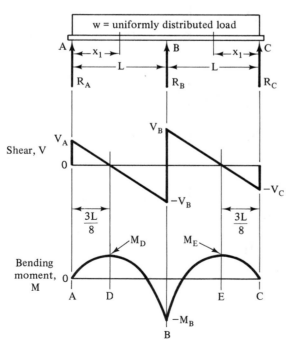

Reactions

$$R_A = R_C = \frac{3wL}{8}$$

$$R_B = 1.25wL$$

Shearing forces

$$V_A = V_C = R_A = R_C = \frac{3wL}{8}$$

$$V_B = \frac{5wL}{8}$$

Moments

$$M_D = M_E = 0.0703wL^2$$

$$M_B = -0.125wL^2$$

Deflections

At $x_1 = 0.4215L$ from A or C:

$$y_{max} = \frac{-wL^4}{185EI}$$

Between A and B:

$$y = \frac{-w}{48EI}(L^3x - 3Lx^3 + 2x^4)$$

(h)

Appendix

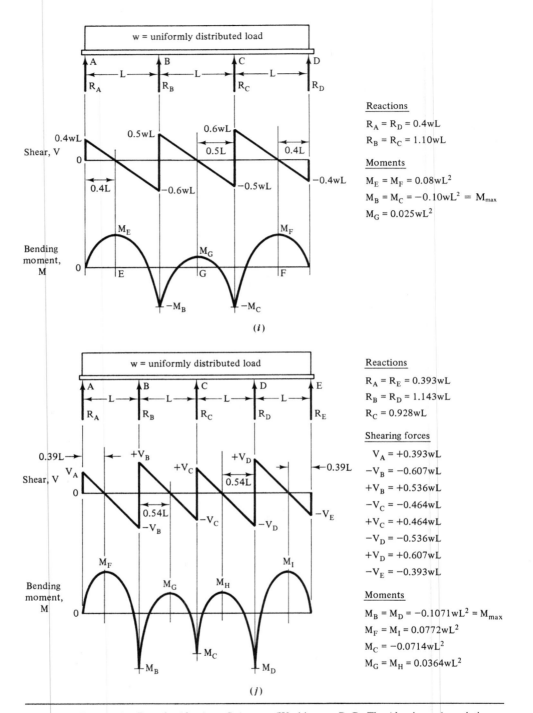

Reactions

$R_A = R_D = 0.4wL$

$R_B = R_C = 1.10wL$

Moments

$M_E = M_F = 0.08wL^2$

$M_B = M_C = -0.10wL^2 = M_{max}$

$M_G = 0.025wL^2$

(i)

Reactions

$R_A = R_E = 0.393wL$

$R_B = R_D = 1.143wL$

$R_C = 0.928wL$

Shearing forces

$V_A = +0.393wL$

$-V_B = -0.607wL$

$+V_B = +0.536wL$

$-V_C = -0.464wL$

$+V_C = +0.464wL$

$-V_D = -0.536wL$

$+V_D = +0.607wL$

$-V_E = -0.393wL$

Moments

$M_B = M_D = -0.1071wL^2 = M_{max}$

$M_F = M_I = 0.0772wL^2$

$M_C = -0.0714wL^2$

$M_G = M_H = 0.0364wL^2$

(j)

Source: *Engineering Data for Aluminum Structures* (Washington, D.C.: The Aluminum Association, 1981), pp 63–77.

Appendix

Quantity	U.S. Customary unit	SI unit	Multiply given value by factor to convert from:	
			U.S. Customary to SI	SI to U.S. Customary
Area	in^2	mm^2	645.16	1.550×10^{-3}
	ft^2	m^2	0.0929	10.76
Loading	lb/in^2	kPa	6.895	0.1450
	lb/ft^2	kPa	0.0479	20.89
Bending moment	$lb \cdot in.$	$N \cdot m$	0.1130	8.851
	$lb \cdot ft$	$N \cdot m$	1.356	0.7376
Density	lb_m/in^3	kg/m^3	2.768×10^4	3.613×10^{-5}
	lb_m/ft^3	kg/m^3	16.02	0.0624
Force	lb	N	4.448	0.2248
	kip	kN	4.448	0.2248
Length	in.	mm	25.4	0.03937
	ft	m	0.3048	3.281
Mass	lb_m	kg	0.454	2.205
Torque	$lb \cdot in.$	$N \cdot m$	0.1130	8.851
Power	hp	kW	0.7457	1.341
Stress or pressure	psi	kPa	6.895	0.1450
	ksi	MPa	6.895	0.1450
	psi	GPa	6.895×10^{-6}	1.450×10^5
	ksi	GPa	6.895×10^{-3}	145.0
Section modulus	in^3	mm^3	1.639×10^4	6.102×10^{-5}
Moment of inertia	in^4	mm^4	4.162×10^5	2.403×10^{-6}

Answers
to Selected
Problems

Chapter 1

1-17. 7.85 kN front
11.77 kN rear
1-19. 54.5 mm
1-23. 1765 lb front
2646 lb rear
1-25. 55.1 lb
25.7 lb/in.
2.14 in.
1-27. 398 slugs
1-29. 8274 kPa
1-31. 96.5 to 524 MPa
1-33. 9097 mm^2
1-35. Area = 324 in^2
Area = 2.09×10^5 mm^2
Vol. = 3888 in^3
Vol. = 6.37×10^7 mm^3
Vol. = 6.37×10^{-2} m^3
1-37. 40.7 MPa
1-39. 5375 psi

1-41. 79.8 MPa
1-43. 803 psi
1-45. σ_{AB} = 107.4 MPa
σ_{BC} = 75.2 MPa
σ_{BD} = 131.1 MPa
1-47. σ_{AB} = 167 MPa Tension
σ_{BC} = 77.8 MPa Tension
σ_{CD} = 122 MPa Tension
1-49. σ_{AB} = 20 471 psi Tension
σ_{BC} = 3 129 psi Tension
1-51. Forces: AD = CD = 10.5 KN
AB = BC = 9.09 kN
Stresses: $\sigma_{AB} = \sigma_{BC}$ = 25.3 MPa
Tension
σ_{BD} = 17.5 MPa Tension
$\sigma_{AD} = \sigma_{CD}$ = 21.0 MPa
Compression
1-53. 50.0 MPa
1-55. 11 791 psi
1-57. 72.9 MPa
1-59. 151 MPa

1-61. 81.1 MPa

1-63. 24.7 MPa

1-65. Pin: $\tau = 50\ 930$ psi
Collar: $\tau = 38\ 800$ psi

1-67. 183 MPa

1-69. 73.9 MPa

1-71. 22.6 MPa

1-73. a) 2187 psi
b) 417 psi

1-75. 18 963 psi

1-77. 28.3 MPa

1-79. 5.39 MPa

Chapter 2

2-15. 1020 HR

2-19. 16.4 lb

2-21. Magnesium

2-29. $s_{ut} = 40$ ksi; $s_{uc} = 140$ Ksi

2-31. Bending: $\sigma_d = 1250$ psi
Tension: $\sigma_d = 825$ psi
Compression: $\sigma_d = 1050$ psi
parallel to grain
Compression: $\sigma_d = 385$ psi
perpendicular to grain
Shear: $\tau_d = 95$ psi

Chapter 3

3-1. Required $s_y = 216$ MPa

3-3. Required $s_u = 86\ 000$ psi

3-5. No. Tensile stress too high.

3-7. $d_{min} = 0.824$ in.

3-9. $d_{min} = 12.4$ mm

3-11. Required $\sigma_d > 803$ psi

3-13. 16.7 kN

3-15. $d_{min} = 0.412$ in.

3-17. For sides B and H: $B_{min} = 22.2$ mm; $H_{min} = 44.4$ mm

3-19. Required $s_y = 360$ MPa

3-21. Required $s_u = 400$ MPa

3-23. 20 260 lb

3-25. 0.577 m

3-27. 52.3 kips

3-29. $\sigma_b = 450$ psi; $\sigma_{bd} = 35\ 100$ psi

3-31. $L_{min} = 0.764$ in. based on shear

3-33. $d_{min} = 0.868$ in. based on shear
For $d = 1.00$ in., $L_{min} = 0.329$ in. for bearing

3-35. $\tau = 151$ MPa; Required $s_u = 1473$ MPa

3-37. 83 kN

3-39. 256 kN

3-41. 18 300 lb

3-43. 54 900 lb

3-45. 119 500 lb

3-47. 119 700 lb

3-49. Pin A: $d_{min} = 15.4$ mm
Pins B and C: $d_{min} = 19.9$ mm

3-51. Lifting force = 2184 lb
$\tau = 6275$ psi if pin is in double shear

3-53. $d_{min} = 0.199$ in.

3-57. 19.15 MPa

Chapter 4

4-1. 0.044 in.

4-3. Force = 2357 lb
$\sigma = 3655$ psi

4-5. a) and b) $d_{min} = 10.63$ mm;
Mass = 0.430 kg
c) $d_{min} = 18.4$ mm;
Mass = 0.465 kg

4-7. a) 0.857 mm
b) 0.488 mm

4-9. Elongation = 0.0040 in.
Compression = 0.00045 in.

4-11. Force = 3214 lb; Unsafe

4-13. $\delta = 0.016$ mm
$\sigma = 27.9$ MPa

4-15. 0.804 mm

4-17. 1.52 mm shorter

4-19. a) $\delta = 0.276$ in.; $\sigma = 37\ 300$ psi
(Close to s_y)
b) $\sigma = 62\ 200$ psi − Greater than s_u. Wire will break.

4-21. Force = 7350 lb
$\delta = 0.060$ in.

4-23. Mass = 132 kg
$\sigma = 183$ MPa

4-25. 0.806 in.

4-27. 186 MPa

4-29. a) 0.459 mm
      b) 213 MPa

4-31. 365 psi

4-33. 234.8°C

4-35. Brass: δ = 6.30 mm
      Stainless steel: δ = 3.51 mm

4-37. 38.7 MPa

4-39. σ = 37 500 psi Compression.
      Bar should fail in compression
      and/or buckle.

4-41. 0.157 in.

4-43. 154 MPa

4-45. σ_c = 13.9 MPa; σ_s = 208 MPa

4-47. 9.60 in.

4-49. d_{min} = 6.20 mm

4-51. σ_s = 427 MPa; σ_a = 49.4 MPa

Chapter 5

5-1. 178 MPa

5-3. 4042 psi

5-5. 83.8 MPa

5-7. τ = 6716 psi; Safe

5-9. τ = 5190 psi; Required
     s_y = 62 300 psi

5-11. τ = 52.8 MPa; θ = 0.030 rad;
      Required s_y = 211 MPa

5-13. D_i = 46.5 mm; D_o = 58.1 mm

5-15. d_{min} = 0.512 in.

5-17. Power = 0.0686 hp;
      τ = 8488 psi; Required
      s_y = 67 900 psi

5-19. D_i = 12.09 in.; D_o = 15.11 in.

5-21. 1.96 N · m

5-23. 0.1509 rad

5-25. 0.267 rad

5-27. 0.0756 rad

5-29. 0.278 rad

5-31. θ_{AB} = 0.0636 rad;
      θ_{AC} = 0.0976 rad

5-33. τ = 9.06 MPa; θ = 0.0046 rad

5-35. 49.0 MPa

5-37. 1370 lb · in.

5-39. 2902 lb · in.

5-41. 0.083 rad

5-43. 0.112 rad

5-45. 0.0667 rad

5-47. 1.82 MPa

5-49. 0.0034 rad

5-51. 0.0042 rad

5-53. 82 750 lb · in.

5-55. 153 600 lb · in.

5-57. τ_{tube}/τ_{pipe} = 1.028
      $\theta_{tube}/\theta_{pipe}$ = 1.186

Chapter 6

NOTE: The following answers refer to Figures 6-36 to 6-55. For reactions, R_1 is the left; R_2 is the right. V and M refer to the maximum absolute values for shear and bending moment respectively. The complete solutions require the construction of the complete shearing force and bending moment diagrams.

6-36. (a) $R_1 = R_2$ = 325 lb
          V = 325 lb
          M = 4550 lb · in.
      (c) R_1 = 11.43 K; R_2 = 4.57 K
          V = 11.43 K
          M = 45.7 K · ft

6-37. (a) R_1 = 575 N; R_2 = 325 N
          V = 575 N
          M = 195 N · m
      (c) $R_1 = R_2$ = 50 kN

          V = 50 kN
          M = 85 kN · m

6-38. (a) R_1 = 1557 lb
          R_2 = 1743 lb
          V = 1557 lb
          M = 6228 lb · in.
      (c) R_1 = 7.5 K
          R_2 = 37.5 K
          V = 20 K
          M = 60 K · ft

6-39. (a) $R_1 = R_2 = 250$ N
  $V = 850$ N
  $M = 362.5$ N·m
  (c) $R_1 = 37.4$ kN (down)
  $R_2 = 38.3$ kN (up)
  $V = 24.9$ kN
  $M = 50$ kN·m
6-40. (a) $R = 120$ lb
  $V = 120$ lb
  $M = 960$ lb·in.
  (c) $R = 24$ K
  $V = 24$ K
  $M = 168$ K·ft
6-41. (a) $R = 1800$ N
  $V = 1800$ N
  $M = 1020$ N·m
  (c) $R = 120$ kN
  $V = 120$ kN
  $V = 120$ kN
  $M = 240$ kN·m
6-42. (a) $R_1 = R_2 = 180$ lb
  $V = 180$ lb
  $M = 810$ lb·in.
  (c) $R_1 = R_2 = 12$ K
  $V = 12$ K
  $M = 84$ K·ft
6-43. (a) $R_1 = 99.2$ N;
  $R_2 = 65.8$ N
  $V = 99.2$ N
  $M = 9.9$ N·m
  (c) $R_1 = R_2 = 40$ kN
  $V = 40$ kN
  $M = 140$ kN·m
6-44. (a) $R_1 = R_2 = 440$ lb
  $V = 240$ lb
  $M = 360$ lb·in.
  (c) $R_1 = 1456$ N;
  $R_2 = 644$ N
  $V = 956$ N
  $M = 125$ N·m
  (e) $R_1 = R_2 = 54$ kN
  $V = 30$ kN
  $M = 24$ kN·m
6-45. (a) $R = 360$ lb
  $V = 360$ lb
  $M = 1620$ lb·in.

(c) $R = 600$ N
  $V = 600$ N
  $M = 200$ N·m
6-46. (a) $R_1 = R_2 = 330$ lb
  $V = 330$ lb
  $M = 4200$ lb·in.
  (c) $R_1 = 36.6$ K; $R_2 = 30.4$ K
  $V = 36.6$ K
  $M = 183.2$ K·ft
6-47. (a) $R_1 = R_2 = 450$ N
  $V = 450$ N
  $M = 172.5$ N·m
  (c) $R_1 = R_2 = 175$ kN
  $V = 175$ kN
  $M = 600$ kN·m
6-48. (a) $R_1 = 636$ lb; $R_2 = 1344$ lb
  $V = 804$ lb
  $M = 2528$ lb·in.
  (c) $R_1 = 4950$ N;
  $R_2 = 3100$ N
  $V = 2950$ N
  $M = 3350$ N·m
6-49. (a) $R = 236$ lb
  $V = 236$ lb
  $M = 1504$ lb·in.
  (c) $R = 1130$ N
  $V = 1130$ N
  $M = 709$ N·m
  (e) $R = 230$ kN
  $V = 230$ kN
  $M = 430$ kN·m
6-50. (a) $R = 1400$ lb
  $V = 1500$ lb
  $M = 99\,000$ lb·in.
  (c) $R = 1250$ N
  $V = 1250$ N
  $M = 1450$ N·m
6-51. (a) $R_1 = 1333$ lb
  $R_2 = 2667$ lb
  $V = 2667$ lb
  $M = 5132$ lb·ft
  (c) $R_1 = R_2 = 75$ N
  $V = 75$ N
  $M = 15$ N·m
6-52. (a) $R_1 = 8.60$ *kN*
  $R_2 = 12.2$ kN

$$V = 12.2 \text{ kN}$$
$$M = 9.30 \text{ kN} \cdot \text{m}$$
(c) $R_1 = R_2 = 5400$ lb
$$V = 5400 \text{ lb}$$
$$M = 19\ 800 \text{ lb} \cdot \text{ft}$$

6-53. (a) $R = 10.08$ kN
$$V = 10.08 \text{ kN}$$
$$M = 8.064 \text{ kN} \cdot \text{m}$$
(c) $R = 7875$ lb
$$V = 7875 \text{ lb}$$
$$M = 21\ 063 \text{ lb} \cdot \text{ft}$$

For Problems 6-54 and 6-55, results are shown for the main horizontal section only.

6-54. (a) $R_1 = R_2 = 282$ N
$$V = 282 \text{ N}$$
$$M = 120 \text{ N} \cdot \text{m}$$
(c) $R_1 = R_2 = 162$ N
$$V = 162 \text{ N}$$
$$M = 42.2 \text{ N} \cdot \text{m}$$

6-55. (a) $R_1 = 165.4$ N
$$R_2 = 18.4 \text{ N}$$
$$V = 165.4 \text{ N}$$
$$M = 16.54 \text{ N} \cdot \text{m}$$
(c) $R_1 = 4.35$ N
$$R_2 = 131.35 \text{ N}$$
$$V = 127 \text{ N}$$
$$M = 6.35 \text{ N} \cdot \text{m}$$

Chapter 7

NOTE: The following answers refer to Figures 7-14 to 7-22. The first number is the distance to the centroid from the bottom of the section. The second number is the moment of inertia with respect to the centroidal axis.

7-14. (a) 0.900 in.; 0.366 in^4
(c) 4.00 in.; 166 in^4
7-15. (a) 152.5 mm; 4.64×10^7 mm^4
(c) 25.33 mm; 3.38×10^5 mm^4
7-16. (a) 12.50 mm; 6.167×10^4 mm^4
(c) 20.00 mm; 5.36×10^4 mm^4
7-17. (a) 0.533 in.; 0.367 in^4
(c) 1.28 in.; 0.826 in^4
7-18. (a) 7.33 in.; 1155 in^4
(c) 7.40 in.; 423.5 in^4

7-19. (a) 3.50 in.; 88.9 in^4
(c) 3.00 in. from center of either pipe; 22.9 in^4
7-20. (a) 2.72 in.; 46.6 in^4
(c) 2.61 in.; 58.2 in^4
7-21. (a) 4.25 in.; 151.4 in.^{4}
(c) 2.25 in.; 107.2 in^4
7-22. (a) 30.0 mm; 2.64×10^5 mm^4
(c) 12.39 mm; 1.696×10^4 mm^4

Chapter 8

8-1.   94.4 MPa
8-3.   (a) 20 620 psi
        (b) 41 240 psi
8-5.   26.9 mm
8-7. $\sigma = 92.0$ MPa; OK
8-9.   Required S = 0.294 in^3;
        $1\frac{1}{2}$-in. schedule 40 pipe
8-11. $\sigma = 504$ psi; OK
8-13.   21 050 psi

8-15. $\sigma_t = 6882$ psi;
        $\sigma_c = 12\ 970$ psi
8-17.   39.7 MPa; OK
8-19.   6880 psi (47.4 MPa)
8-21.   Required S = 1.41×10^5 mm^3
        (a) $d_{min} = 112.8$ mm;
            A = 9998 mm^2
        (b) $b_{min} = 94.6$ mm;
            A = 8945 mm^2

(c) b = 37.5 mm; h = 150 mm;
A = 5625 mm^2

(d) S8 × 18.4; A = 3490 mm^2
(Lightest)

8-23. Required S = 1.35 in^3;
3-in. schedule 40 pipe

8-25. Required S = 2.61 in^3;
4 × 4 × $\frac{1}{4}$ or 6 × 2 × $\frac{1}{4}$

8-27. Required S = 2.83 in^3;
W8 × 10

8-29. Required S = 2.83 in^3;
4-in. schedule 40 pipe

8-31. Required S = 22.1 in^3;
2 × 12 wood beam

8-33. 2 × 10 wood beam

Problems 8-35 to 8-43 are design problems for which multiple solutions are possible.

8-45. σ_d = 4.00 MPa
At A: σ = 3.81 MPa; OK
At B: σ = 5.16 MPa; Unsafe
At C: σ = 4.62 MPa; Unsafe

8-47. 5.31 N/mm
8-49. 979 N
8-51. 1045 lb
8-53. 102 lb
8-55. 7.86 lb/in.
8-57. 4.94 lb/in.
8-59. 822 N
8-61. 625 N
8-63. 21.0 kN
8-65. 9.69 kN/m
8-67. 126 kN

8-69. Required S = 23.1 in^3;
W14 × 26

8-71. Required S = 16.1 in^3;
W12 × 16

8-73. Required S = 63.3 in^3;
W18 × 40

8-75. Required S = 20.2 in^3;
W14 × 26

8-77. Required S = 0.540 in^3;
W8 × 10

8-79. Required S = 23.1 in^3;
S10 × 25.4

8-81. Required S = 16.1 in^3;
S8 × 23

8-83. Required S = 63.3 in^3;
S15 × 50

8-85. Required S = 20.2 in^3;
S10 × 25.4

8-87. Required S = 0.540 in^3;
S3 × 5.7

8-89. Required S = 13.9 in^3;
W12 × 16

8-91. Required S = 9.66 in^3;
W10 × 12

8-93. Required S = 38.0 in^3;
W12 × 30

8-95. Required S = 12.1 in^3;
W10 × 15

8-97. Required S = 0.320 in^3;
W8 × 10

8-99. Required S = 22.73 in^3;
W14 × 26

8-101. For composite beam,
w = 4.18 kips/ft
For S12 × 50 alone,
w = 3.58 kips/ft

8-103. M_{max} = 24.0 kN · m; Required
S = 1.6 × 10^5 mm^3 (9.76 in^3)
W10 × 12

8-105. 3.50 ft
8-107. 398 MPa

8-109. At fulcrum: σ = 8000 psi
At hole B_1 nearest fulcrum:
σ = 5067 psi
At next hole B_2: σ = 3800 psi
At next hole B_3: σ = 2534 psi
At next hole B_4: σ = 1267 psi

8-111. a) Pivot in end hole:
σ_{max} = 8064 psi at hole
nearest to fulcrum.
b) Pivot in second hole:
σ_{max} = 6800 psi at fulcrum.
c) Pivot in third hole:
σ_{max} = 5600 psi at fulcrum.

d) Pivot in fourth hole:
   σ_{max} = 4400 psi at fulcrum.
e) Pivot in fifth hole:
   σ_{max} = 3200 psi at fulcrum.

8-113. 108.8 MPa
8-115. σ = 149.3 MPa; Required
       s_u = 1195 MPa
8-117. 1422 N
8-119. Impossible
8-121. Yes; d_{max} = 37.2 mm
8-123. σ_{max} = 118.3 MPa; Required
       s_u = 946 MPa
8-125. Maximum L_1 = 206.5 mm

Maximum L_2 = 83.4 mm
Maximum L_3 = 24.7 mm

8-127.

x (mm)	σ (MPa)
0	0
40	52.1
80	76.5
120	87.9
160	92.6
200	93.8
240	112.5

8-129. h_1 = 22 mm;
       h_2 = 8 mm − Approximately

8-131.

Thickness (mm)	I (mm⁴)	e′ to flexural center
0.50	71 731	13.91 mm
1.60	219 745	13.14 mm
3.00	389 674	12.16 mm

8-133.

Thickness (in.)	e′ to flexural center (in.)
0.020	0.940
0.063	0.860
0.125	0.781

8-135.

Thickness (mm)	e′ to flexural center (mm)
0.50	30.0
1.60	28.2
3.00	25.5

8-137. e = 0.345 in.

Chapter 9

9-1.   1.125 MPa
9-3.   1724 psi
9-5.   3.05 MPa
9-7.   3180 psi
9-9.   1661 psi
9-11. 69.3 psi
9-13. 7.46 MPa
9-15. 2.79 MPa
9-17. 10.3 MPa
9-19. 2342 psi
9-21. 245 lb
9-23. 788 lb
9-25. 5098 lb
9-27. 661 lb
9-29. 787 lb

9-31. Let y = distance from bottom
      of the I-shape:

y (in.)	τ (psi)
0.0	0.0
0.5	5.65
1.0(−)	10.54
1.0(+)	63.25
1.5	67.39
2.0	70.78
2.5	73.42
3.0	75.30
3.5	76.43
4.0	76.81

9-33. 8698 psi

9-35. From web shear formula:
$\tau = 8041$ psi
Approximately 8% low compared with τ_{max} from 9-33.

9-37. Let y = distance from bottom of the I-shape:

y (in.)	τ (psi)
0.0	0
0.175	155
0.35(−)	303
0.35(+)	6582
1.0	7073
2.0	7638
3.0	7976
4.0	8089

9-39. $\tau = 4352$ psi;
$\tau_d = 14\,400$ psi − OK
$\sigma = 26\,100$ psi;
$\sigma_d = 23760$ psi − Unsafe

9-41. W18 × 55; $\tau = 6300$ psi;
$\tau_d = 14\,400$ psi − OK

9-43. $1\frac{1}{2}$-in. schedule 40 pipe;
$\tau = 2013$ psi;
$\tau_d = 5000$ psi − OK

9-45. 4 × 10 beam

9-47. 1256 lb

9-49. 116.3 MPa

9-51. (a) $\tau = 1125$ psi
(b) $\sigma = 6750$ psi
(c) $s_y = 20\,250$ psi; Any steel

9-53. 35.4 mm; $\tau = 1.59$ MPa;
N = 86.7 for shear

9-55. 24.5 mm; $\tau = 1.28$ MPa;
$\sigma_d = 120$ MPa

9-57. Required S = 0.72 in³; $2\frac{1}{2}$-in. schedule 40 pipe.

9-59. q = 736 lb/in; $\tau = 526$ psi

9-61. 431 lb/ft based on strength of rivets

9-63. 1484 lb based on bending

9-65. Maximum spacing = 4.36 in.

9-67. Maximum spacing = 3.03 in.

Chapter 10

NOTE: The complete solutions for Problems 10-1 to 10-27 require the construction of the complete Mohr's circle and the drawing of the principal stress element and the maximum shear stress element. Listed below are the significant numerical results.

Prob. No.	σ_1	σ_2	ϕ (deg)	τ_{max}	σ_{avg}	ϕ' (deg)
1	315.4 MPa	−115.4 MPa	10.9 cw	215.4 MPa	100.0 MPa	34.1 ccw
3	110.0 MPa	−40.0 MPa	26.6 cw	75.0 MPa	35.0 MPa	18.4 ccw
5	23.5 ksi	−8.5 ksi	19.3 ccw	16.0 ksi	7.5 ksi	64.3 ccw
7	79.7 ksi	−9.7 ksi	31.7 ccw	44.7 ksi	35.0 ksi	76.7 ccw
9	677.6 kPa	−977.6 kPa	77.5 ccw	827.6 kPa	−150.0 kPa	57.5 cw
11	327.0 kPa	−1202.0 kPa	60.9 ccw	764.5 kPa	−437.5 kPa	74.1 cw
13	570.0 psi	−2070.0 psi	71.3 cw	1320.0 psi	−750.0 psi	26.3 cw
15	4180.0 psi	−5180.0 psi	71.6 cw	4680.0 psi	−500.0 psi	26.6 cw
17	360.2 MPa	−100.2 MPa	27.8 ccw	230.2 MPa	130.0 MPa	72.8 ccw
19	23.9 ksi	−1.9 ksi	15.9 ccw	12.9 ksi	11.0 ksi	29.1 ccw
21	4.4 ksi	−32.4 ksi	20.3 cw	18.4 ksi	−14.0 ksi	24.7 ccw
23	321.0 MPa	−61.0 MPa	66.4 ccw	191.0 MPa	130.0 MPa	68.6 cw
25	225.0 MPa	−85.0 MPa	0.0	155.0 MPa	70.0 MPa	45.0 ccw
27	775.0 kPa	−145.0 kPa	0.0	460.0 kPa	315.0 kPa	45.0 ccw

For Problems 10-29 to 10-39, the Mohr's circle from the given data results in both principal stresses having the same sign. For this class of problems, the supplementary circle is drawn using the procedures discussed in Section 10-11 of the text. The results include three principle stresses where $\sigma_1 > \sigma_2 > \sigma_3$. Also, the maximum shear stress is found from the radius of the circle containing σ_1 and σ_3, and is equal to $\frac{1}{2}\sigma_1$ or $\frac{1}{2}\sigma_3$, whichever has the greatest magnitude. Angles of rotation of the resulting elements are not requested.

Prob. No.	σ_1	σ_2	σ_3	τ_{max}
29	328.1 MPa	71.9 MPa	0.0 MPa	164.0 MPa
31	214.5 MPa	75.5 MPa	0.0 MPa	107.2 MPa
33	35.0 ksi	10.0 ksi	0.0 ksi	17.5 ksi
35	55.6 ksi	14.4 ksi	0.0 ksi	27.8 ksi
37	0.0 kPa	−307.9 kPa	−867.1 kPa	433.5 kPa
39	0.0 psi	−295.7 psi	−1804.3 psi	902.1 psi

For Problems 10-41 to 10-49, the Mohr's circles from earlier problems are used to find the stress condition on the element at some specified angle of rotation. The listed results include the two normal stresses and the shear stress on the specified element.

Prob. No.	σ_A	$\sigma_{A'}$	τ_A
41	130.7 MPa	69.3 MPa	213.2 MPa cw
43	−37.9 MPa	197.9 MPa	31.6 MPa ccw
45	3.6 ksi	−21.6 ksi	43.9 ksi cw
47	−2010.3 psi	510.3 psi	392.6 psi cw
49	8363.5 psi	86.5 psi	1421.2 psi cw

10-51. $\tau_{max} = 230.2$ MPa

10-53. $\tau_{max} = 12.9$ ksi

Chapter 11

11-1. −10 510 psi

11-3. $\sigma_N = 9480$ psi;
$\sigma_M = -7530$ psi

11-5. $\sigma_N = 13\ 980$ psi;
$\sigma_M = -15\ 931$ psi

11-7. −64.3 MPa

11-9. 336 N

11-11. $\sigma = 724$ MPa; Required
$s_y = 1448$ MPa;
AISI 4140 OQT 700

11-13. At B: $\sigma = 24\ 183$ psi tension
on top of beam
$\sigma = -18\ 674$ psi compression
on bottom of beam.

11-15. Load $= 6783$ N;
Mass $= 691$ kg

11-17. 51.6 MPa

11-19. $\tau = 9923$ psi; Required
$s_y = 119$ ksi;
AISI 1141 OQT 900

11-21. 51.5 MPa

11-23. L4 × 3 × $\frac{1}{2}$

11-25. 7019 psi

11-27. 8168 psi

11-29. 7548 psi

11-31. 7149 psi

11-33. 61.5 MPa

11-35. (a) 3438 psi
(b) 3438 psi
(c) 344 psi
(d) 2300 psi
(e) 1186 psi

11-37.   Middle of beam   Near supports   15 in. from A

(a)	1875 psi	0 psi	1406 psi
(b)	1875 psi	0 psi	1406 psi
(c)	0 psi	375 psi	188 psi
(d)	1250 psi	208 psi	943 psi
(e)	625 psi	333 psi	498 psi

11-39.   1488 MPa   11-45.   0.720 inch
11-41.   75.3 MPa   11-47.   0.832 inch
11-43. (a) P = 43 167 lb;
             τ_{max} = 8333 psi
        (b) τ_{max} = 8862 psi; N = 2.82

Chapter 12

12-1.   −2.01 mm   12-27. 0.020 in.
12-3.   −0.503 mm   12-29. y = −0.0078 in.
12-5.   −5.40 mm   at x = 8.56 in.
12-7.   At loads: −0.251 in.   12-31. y = −3.79 mm
         At center: −0.385 in.   12-33. D = 69.2 mm
12-9.   −0.424 in.   12-35. I7 × 5.800 aluminum beam
12-11.   −0.271 in.   y = −5.37 mm
12-13.   +0.093-in. deflection at   12-37. D = 109 mm
         x = 69.2 in.   12-39. −0.0078 in.
12-15.   At 500-N load: −16.18 mm   12-41. −3.79 mm
         At 400-N load: −21.63 mm   12-43. −3.445 mm
12-17.   −0.0078 in.   12-45. −5.13 mm
12-19.   −0.869 mm   13-47. −0.0163 in.
12-21.   −3.997 mm   12-49. −0.257 in.
12-23.   64.8 mm   12-51. −71.7 mm
12-25.   Required $I = 1.66 \times 10^8$ mm^4
         W18 × 40 beam

Chapter 13

13-1. (a) $R_A = R_E$ = 371 lb   M = 8.1 kN · m
            R_C = 858 lb   (c) R_A = 127.4 kN
            V = 429 lb R_C = 76.6 kN
            M = 1113 lb · ft   V = 127.4 kN
       (c) $R_A = R_E$ = 1873 lb   M = 103.4 kN · m
            R_C = 5854 lb   13-3. (a) Same as 13-1(a)
            V = 2927 lb   (c) Same as 13-1(c)
            M = 5016 lb · ft   13-4. (a) R_A = 8.90 kN
13-2. (a) R_A = 22.5 kN R_C = 34.1 kN
            R_B = 13.5 kN R_D = 6.00 kN
            V = 22.5 kN   V = 18 kN

$$M = 8.9 \text{ kN} \cdot \text{m}$$
(c) $R_A = 15.8$ kN
$R_C = 68.9$ kN

$R_E = 7.3$ kN
$V = 36.2$ kN
$M = 12.3$ kN $\cdot$ m

For Problems 13-6 to 13-15, the complete solutions required the drawing of the complete shearing force and bending moment diagrams. Given here are only the reaction forces at the supports, the maximum shearing force, and the maximum bending moment.

13-6. $R_1 = 30$ kN; $R_2 = 100$ kN;
$R_3 = 30$ kN; $V = 50$ kN;
$M = -16.0$ kN $\cdot$ m

13-8. $R_1 = 747$ lb; $R_2 = 453$ lb;
$V = 747$ lb;
$M = -3852$ lb $\cdot$ ft

13-10. $R_1 = 1965$ lb; $R_2 = 6535$ lb;
$V = -6535$ lb;
$M = -136\ 800$ lb $\cdot$ in.

13-12. $R_1 = R_4 = 72$ kN;
$R_2 = R_3 = 198$ kN;
$V = 108$ kN;
$M = -64.8$ kN $\cdot$ m

13-14. $R_1 = R_2 = 70$ lb; $V = 70$ lb;
$M = 945$ lb $\cdot$ in.

13-21.

	Design (a)	Design (b)
V	2250 N	3375 N
M	4500 N $\cdot$ m	4500 N $\cdot$ m
y	(22 500)/EI	(20 250)/EI

13-23. $R_1 = 32.75$ kN;
$R_2 = 19.65$ kN;
$V = 32.75$ kN;
$M = -42.9$ kN $\cdot$ m

Chapter 14

14-1. 25.1 kN
14-3. 8.35 kN
14-5. 26.2 kN
14-7. 111 kN
14-9. $P_a = 7318$ lb/column; Use 9 columns
14-11. 499 kN
14-13. 65 300 lb
14-15. Axial force = 31.1 kN;
Critical load = 260 kN;
N = 8.37 OK
14-17. 15.1 kN

14-19. Critical load = 10 914 lb;
Actual load = 5 000 lb;
N = 2.18 − Low
14-21. Critical load = 2 849 lb;
Actual load = 1 500 lb;
N = 1.90 − Low
14-23. 2.68 in.
14-25. Required $I = 5.02$ in^4;
17 × 5.800
14-27. 5869 lb
14-29. 158 kN
14-31. No improvement

Chapter 15

15-1. 115 MPa
15-3. 2.94 mm
15-5. 2134 psi
15-7. 1.56 mm

15-9. N = 3.80
15-11. $\sigma_2 = 212$ MPa;
$\sigma_3 = -70.0$ MPa

Radius (mm)	σ_2 (MPa)
110	166 Inner surface
120	149
130	136
140	125
150	116 Outer surface

15-15. 14.88 MPa

15-17.

Radius (mm)	σ_3 (MPa)
15	−7.00 Inner surface
17	−4.58
19	−2.88
21	−1.64
23	−0.71
25	0.00 Outer surface

15-19. 115.7 MPa

15-21.

Radius (mm)	σ_3 (MPa)
210	−100.0 Inner surface
215	−83.3
220	−68.0
225	−54.1
230	−41.4
235	−29.7
240	−19.0
245	−9.1
250	0.00 Outer surface

15-23.

Radius (mm)	σ_2 (MPa)
162.5	22.35 Inner surface
170	20.99
177.5	19.84
185	18.88
192.5	18.06
200	17.35 Outer surface

15-25. Sizes 1/8 through 4 are thick walled.
Sizes 5 through 18 are thin walled.

Chapter 16

16-1. (a) F_s = 6872 lb (Allowable);
F_b = 32 625 lb;
F_t = 15 179 lb
(c) F_s = 7732 lb (Allowable);
F_b = 24 470 lb;
F_t = 17 204 lb

16-2. (a) F_s = 17 279 lb (Allowable);
F_b = 105 000 lb;
F_t = 43 110 lb
(c) F_s = 34 558 lb (Allowable);
F_b = 105 000 lb;
F_t = 43 110 lb

16-3. (a) F_s = 11 780 lb (Allowable);
F_b = 39 375 lb;
F_t = 21 083 lb
(c) F_s = 13 253 lb (Allowable);
F_b = 29 531 lb;
F_t = 23 895 lb

16-4. (a) F_s = 17 279 lb (Allowable);
F_t = 86 220 lb
(c) F_s = 34 558 lb (Allowable);
F_t = 86 220 lb

16-5. 1.25 in.

16-7. 23 860 lb on welds

INDEX

Unified Numbering System (UNS),
48–49
uses, 41–42
Modulus of elasticity:
in shear, 20, 146–47
in tension, 20, 45–46
Mohr's circle for stress, 323–43
construction, 324–28
examples of use, 330–35
stress on selected planes, 335–38
when both principal stresses have
the same sign, 338–43
Moment of inertia, 216–26
commercially available shapes, 224
complex shapes, 218–26
mathematical definition, 217
simple shapes, 216–17
transfer–of–axis theorem, 220–21

N

Neutral axis, 241–42
Noncircular cross sections, 151–54

P

Percent elongation, 46–47
Pipe dimensions, 516
Plastics, 61–62, 523
Poisson's ratio, 18
values, 19
Polar moment of inertia, 131–33
Polar section modulus, 136–37
Power, 127
Preferred sizes, 21, 505
Pressure vessels, 467–78
distinction between thin–walled and
thick–walled, 468–69
hoop stress, 473
longitudinal stress, 472
radial stress, 476–77
tangential stress, 476–77
thick–walled cylinders and spheres,
475–78

thin–walled cylinders, 472–75
thin–walled spheres, 469–72
Principal stresses, 321–22
Properties of areas, 502–4
Proportional limit, 43–45

R

Radius of gyration, 448–51
Riveted connections, 485–94
Rotational speed, 127–28

S

Screw threads, 506–7
Section modulus, 248
polar, 136–37
Shear center, 253–56
Shear stress:
direct, 9
double shear, 9
single shear, 9
web shear formula, 297–98
Shearing forces in beams, 173–79
Shearing strain, 18
Shearing stresses in beams, 275–301
design shear stress, 298
distribution, 288–91
general shear formula, 280
development, 291–94
use, 283–88
importance, 278–80
maximum, 288
shear flow, 298–301
special shear formulas, 294–98
statical moment, 280–82
visualization, 277
SI unit system, 3
prefixes for, 3
Slenderness ratio, 447–48
Standard shapes, 21, 22
Statically indeterminate beams,
422–40
beam diagrams, 536–40